Brain and Human Body Modeling 2020

Sergey N. Makarov • Gregory M. Noetscher
Aapo Nummenmaa

Editors

Brain and Human Body Modeling 2020

Computational Human Models Presented
at EMBC 2019 and the BRAIN Initiative®
2019 Meeting

 Springer

Editors
Sergey N. Makarov
Electrical and Computer Engineering
Department
Worcester Polytechnic Institute
Worcester, MA, USA

Athinoula A. Martinos Center
for Biomedical Imaging
Massachusetts General Hospital
Charlestown, MA, USA

Aapo Nummenmaa
Athinoula A. Martinos Center
for Biomedical Imaging
Massachusetts General Hospital
Charlestown, MA, USA

Gregory M. Noetscher
Electrical and Computer Engineering
Department
Worcester Polytechnic Institute
Worcester, MA, USA

ISBN 978-3-030-45625-2   ISBN 978-3-030-45623-8   (eBook)
https://doi.org/10.1007/978-3-030-45623-8

This Springer imprint is published by the registered company Springer Nature Switzerland AG
The registered company address is: Gewerbestrasse 11, 6330 Cham, Switzerland

Foreword

Scientists and engineers have long used abstraction to represent complex systems. The advantages of such practice are manyfold, and they have enabled substantial advances in numerous disciplines of knowledge. First year students in electrical engineering are introduced to the concept of resistance without the impacts of heat conduction, wire induction, or component capacitance, factors that would greatly obscure and severely complicate the goal of learning the predominant behavior of this critical circuit element. Students of physics will likely remember learning of a rigid body falling through air and accounting for the motion solely from the force of gravity on the object while entirely neglecting the contributions of air friction. This idealization is helpful when attempting to understand the underlying equations of motion without the need to account for second-order (or higher) effects that do not dramatically change the global system. Computer engineers begin the study of operational amplifiers assuming the open-loop gain and the input resistance of the component are infinite, the amplifier draws zero current, and the bandwidth is constant across all frequencies. These conditions are entirely fictitious but allow the introduction of the topic without overcomplicating the matter and postpone the study of more realistic conditions to a later time. Incidentally, it is worth realizing that similar approaches of simplified – sometimes simplistic – abstractions and models are applied in human biology and medicine. Even the conceptualization of diseases as clear nosological entities is an example of such abstractions. Ultimately clinicians need to consider the many individual differences in phenotypical and pathophysiological characteristics of a given disease. To truly understand, it is not sufficient to learn about the disease that a patient has, but it is essential – as Hippocrates admonished – to learn about the person who has the disease.

All these examples are rather obvious but they illustrate a pattern of simplifying a problem to some most basic components to facilitate the understanding of core concepts. Once this is accomplished, more complex conditions can be considered, likely introducing a greater number of relevant variables and using more involved analysis tools and methods.

One of the greatest talents an engineer or scientist can possess is the ability to know when these approximations break down and when to apply more advanced techniques to achieve the desired solution. Thankfully, as demonstrated by the research contained within this *Volume II of Brain and Human Body Modeling*, there are a number of technical choices available to understand extremely complicated, highly detailed systems.

The present volume offers an important, timely, and needed extension to Volume I of this series on *Brain and Human Body Modeling* which compiled a selection of extended papers presented during the 3rd Annual Invited Session on Computational Human Models, a component of the 40th Annual International Conference of the IEEE Engineering in Medicine and Biology Society (EMBS), given from July 17 to 21, 2018, in Honolulu, HI. This Volume was distributed as an open access text through Springer and may be found here:

https://link.springer.com/book/10.1007%2F978-3-030-21293-3

Given the success of that first Volume, with over 30k downloads from the Springer website, a second collection was requested and is presented herein. This second Volume has been expanded to include extended works that were presented and discussed at two highly respected and relevant conferences:

(1) The 41st Annual International Conference of the IEEE EMBS, took place between July 23 and 27, 2019, in Berlin, Germany. The focus was on "Biomedical engineering ranging from wellness to intensive care." This conference provided an opportunity for researchers from academia and industry to discuss a variety of topics relevant to EMBS and hosted the 4th Annual Invited Session on Computational Human Models. At this session, a bevy of research related to the development of human phantoms was presented, together with a substantial variety of practical applications explored through simulation.

(2) The 5th Annual Brain Research through Advancing Innovative Neurotechnologies (BRAIN)® Initiative Investigators Meeting was held between April 11and 13, 2019, in Washington, DC. This meeting offered an open forum for government and non-government agencies to engage in scientific collaboration and discussion on ongoing and potential research.

When viewed collectively, the work presented at these two conferences represents the state-of-the-art in brain modeling, human phantom development, and the application of these models through simulation to examine cutting edge diagnostic methods and therapeutic techniques. It also offers a view into exciting directions to be explored in the future.

Brain and Human Body Modeling Volume II is divided into seven thematic parts. The first part covers tumor-treating fields (TTFields), a promising treatment for glioblastoma that has received approval from the US Food and Drug Administration. This part describes TTFields in great depth, gives a background on pathogen formulation, and details treatment methods that may be used to combat tumor formation. These treatments range from non-invasive placement of electrodes on the skin surface to more intrusive means, including removing portions of the skull to provide a more direct field connection. Detailed models and efficient simulation techniques are critical to the exploration of these technologies and treatments.

The second part is devoted to non-invasive neurostimulation with specific emphasis on the brain. Topics include recent work on simulation of electric field magnitude and orientation during Transcranial Direct Current Stimulation, computational modeling of brain stimulation that utilizes white matter tractography, and personalization of multi-electrode configurations to be used during Transcranial Electrical Stimulation. Brain modeling plays a pivotal role in estimating field strength, direction, and stimulation in this work with more complex models appearing as more illustrative medical data becomes available.

Part three presents recent research on non-invasive neurostimulation with specific emphasis on the spinal cord and peripheral nervous system. An entire chapter is devoted to the non-invasive electric and magnetic stimulation of the spinal cord, describing potential neuromodulation of spinal sensory and motor pathways and modeling of electric field distributions during stimulation. A second chapter describes application of a highly focal miniature magnetic stimulator for peripheral nervous system applications.

The fourth part covers advancements in modeling with emphasis on neurophysiological signals. The first chapter of this part examines fusion of electromagnetic source modeling with hemodynamic measures to potentially produce a multimodal neuroimaging methodology. The second chapter presents a novel simulation approach that provides tremendous numerical accuracy at unprecedented computational speeds. The final chapter builds on the previous by applying this new simulation approach to the cerebral cortex and examining the potential for multiscale modeling, covering large tissue areas and groups of neurons.

Part five is dedicated to the study of neural circuits at both small and large scales. The first chapter presents necessary definitions applicable to neural circuits with specific discussion on modeling very large and complex systems. The second chapter looks at modeling of selected retinal ganglion cells. The third chapter explores functional connectivity, making substantial use of the Human Connectome Project.

In part six, numerous models that offer valuable resources to explore the effects of high- and radio frequencies are presented. Chapter one discusses potential simplifications to numerical human models that may be employed to facilitate the calculation of Specific Absorption Rate (SAR). The second chapter calculates SAR in a human phantom during a 3T MRI procedure. The third chapter examines unintended stimulation of implanted medical devices while undergoing an MRI. Chapter four provides an overview of electric field and SAR estimation when tissues are in close proximity radiofrequency sources. The final chapter of this part presents a new CAD compatible male model based on the US Library of Medicine's Visible Human Project.

The final, seventh, part examines several topics that are crucial to the construction of brain and human body models. Chapter one provides instruction on the preparation of head models for a new simulation method that combines the advantages of the Boundary Element Method with those of the Fast Multipole Method (FMM).

Chapter two explores the performance of the FMM when applied to human head modeling. The final chapter covers an analytical solution for field response of a conducting object representing a human head in an external electric field.

The exciting work presented here is certainly not all-inclusive and there are always new avenues to explore. The upcoming 42nd Annual International Conference of the IEEE EMBS to be held in Montreal, Canada, from July 20 to 24, 2020, and the 6th Annual BRAIN® Initiative Investigators Meeting planned from June 1 to 3, 2020, in Arlington, VA, will certainly provide more exceptional examples of brain and human body modeling, and perhaps enable further volumes in this valuable collection of publications on *Brain and Human Modelling* as essential abstractions to help us advance the complex realities of the interface between human biology with engineering tools and solutions.

Alvaro Pascual-Leone
Senior Scientist at the Hinda and Arthur
Marcus Institute for Aging Research at
Hebrew SeniorLife, Boston, MA, USA

Professor of Neurology, Harvard
Medical School, Boston, MA, USA

Director, Guttmann Brain Health
Institut, Institut Guttman de
Neurorehabilitación, Universitat
Autónoma de Barcelona, Barcelona,
Spain

The original version of the book was revised due to incorrect chapter order: The correction to the book is available at https://doi.org/10.1007/978-3-030-45623-8_24

Contents

Part I
Tumor Treating Fields

Tumor-Treating Fields at EMBC 2019: A Roadmap to Developing a Framework for TTFields Dosimetry and Treatment Planning

Ze'ev Bomzon, Cornelia Wenger, Martin Proescholdt, and Suyash Mohan

1 Introduction

Tumor-treating fields (TTFields) are electric fields with intensities of 1–5 V/cm in the frequency range of 100–500 kHz known to inhibit the growth of cancerous tumors. TTFields have been approved for the treatment of glioblastoma multiforme (GBM) since 2011 [19–21]. Recently, the therapy was FDA-approved for the treatment of malignant pleural mesothelioma (MPM) [4]. TTFields are delivered noninvasively through two pairs of transducer arrays that are placed on the patient's skin in close proximity to the tumor (see Fig. 1). At any instance, only one pair of arrays is used to create the field, while the second pair is switched off. The pairs of arrays are placed such that the fields created are roughly orthogonal, and switching of the active arrays occurs about once per second. This results in the creation of an alternating electric field, which switches direction periodically. The field is generated by a portable field generator. Treatment is continuous as analysis of clinical data has shown a positive connection between device usage (fraction of time patient is on therapy) and patient outcomes [2].

Preclinical studies have shown that the antimitotic effect of TTFields is frequency- and intensity-dependent. The inhibitory effect on different cell types is observed at cell-specific frequencies [6, 9, 10], and the higher the intensity of the

Z. Bomzon (✉) · C. Wenger
Novocure Ltd., Haifa, Israel
e-mail: zbomzon@novocure.com

M. Proescholdt
Department of Neurosurgery, University Regensburg Hospital Medical Center, Regensburg, Germany

S. Mohan
Department of Radiology, Division of Neuroradiology, Perelman School of Medicine University of Pennsylvania, Philadelphia, PA, USA

© The Author(s) 2021
S. N. Makarov et al. (eds.), *Brain and Human Body Modeling 2020*,
https://doi.org/10.1007/978-3-030-45623-8_1

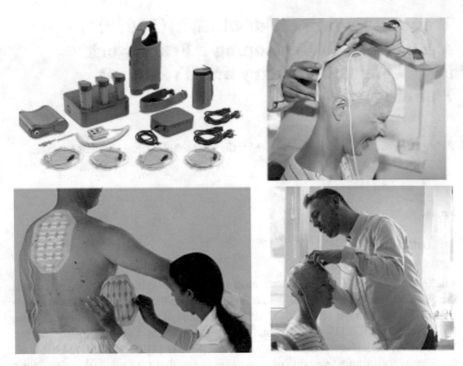

Fig. 1 Top left image shows Optune™ device used to deliver TTFields to the brain. The device comprises a portable battery-powered field generator, connected to four transducer arrays which are placed on the scalp as shown in top and bottom-right images. The image in the bottom-left corner shows placement of transducer arrays for treatment of thoracic tumors with TTFields

field, the stronger the inhibitory effect. As mentioned above, the effect of TTFields is also time-dependent, with higher usage associated with improved patient survival [18]. Posthoc analysis of the EF-14 trials showed overall survival for patients treated with TTFields+temozolomide (TMZ) with usage of 90% or more was 24 months compared to 16.03 months in patients treated with TMZ alone [18]. More recently, a study by Ballo et al. [2] showed that in newly diagnosed GBM patients, survival correlated with TTFields dose delivered to the tumor bed. Dose was defined as power loss density multiplied by usage. These findings suggest that patient outcome could be significantly improved with rigorous treatment planning, in which numerical simulations are used to identify array layouts that optimize delivery of TTFields to the tumor bed. The plan could be adapted periodically as the tumor evolves to maximize the effect of treatment in regions where tumor progression occurs.

Performing such adaptive planning in a practical and meaningful manner requires a rigorous and scientifically proven framework defining TTFields dose and showing how dose distribution influences disease progression in different malignancies (TTFields dosimetry). The adaptive planning also requires a set of principles on how best to perform treatment planning, along with numerical methods and algorithms devised to optimize therapy based on the principles mentioned above. The principles should be derived from our understanding of TTFields dosimetry and how dose distributions influence disease. An effective treatment planning strategy also requires quality assurance and uncertainty analysis to understand how uncertainties in the model, numerical solver, and positions of the array influence the field distribution to create a quality assurance system to ensure that the plan is adhered to within the allowed uncertainties.

At EMBC 2019, several talks discussing key components related to TTFields dosimetry and treatment planning were presented. The purpose of this manuscript is to provide a short overview of this work and discuss how it sets the foundations for the emerging field of TTFields dosimetry and treatment planning.

2 An Outline for TTFields Dosimetry and Treatment Planning

Figure 2 is a flow chart describing the steps required in order to realize an effective scheme for TTFields treatment planning.

The first step in the process is clinical evaluation and contouring. In this step, the planning physician examines imaging data of the patient, identifies regions of active

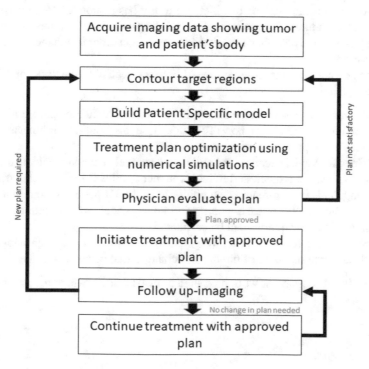

Fig. 2 Flowchart describing the steps involved in treatment planning and how these could be integrated into an effective workflow in the clinic

tumor, and selects the target regions in which TTFields dose should be optimized. The physician may also identify areas to avoid, like areas on the skin over which transducer arrays should not be placed. Next, the imaging data are used to create a patient-specific computational model, which can be used to simulate delivery of TTFields to the patient. In the context of TTFields treatment planning, the model involves the creation of a three-dimensional representation of the patient, in which electric conductivity is assigned to each point. The model, target regions, and avoidance areas are inputs for an optimization algorithm that seeks to find a transducer array layout that optimizes the dose in the target regions while avoiding placement of transducer arrays over the avoidance structures. The optimization algorithm will typically entail iterative use of a numerical solver that simulates delivery of TTFields to the patient for different array layouts. The output from this process will be an optimal array layout for treating the patient, as well as quantitative and visual aids that enable the physician to evaluate the quality of the plan. These aids could include color maps describing the field distribution within the patient's body and dose-volume histogram (DVH) describing the distribution of TTFields dose within the target regions and within other areas of interest. Once a plan that the physician deems satisfactory has been generated, the patient is instructed on how to place the transducer arrays on their body, and treatment commences. Patient follow-ups occur periodically. During these follow-ups, additional imaging of the patient may be acquired as physician assesses for disease progression. The new imaging data may demonstrate regions in which the tumor has responded to therapy and/or regions in which tumor has progressed. Depending on patient's patterns of response and progression, the physician may decide to re-plan in order to enhance treatment to new target regions.

Three key components required to establish an effective framework for TTFields treatment planning are:

TTFields Dosimetry: An understanding of how to define TTFields dose accurately and an understanding on how TTFields dose distributions influence disease progression and patient outcomes.

Patient-Specific Model Creation: An ability to accurately calculate field distributions within patients in a quick and reliable manner is crucial for TTFields treatment planning. This in turn requires an ability to build 3D patient-specific models in which conductivity at each point within the model is well-defined, so that accurate dose distributions can be calculated.

Advanced Imaging for Monitoring Response to Therapy: Imaging technologies that enable accurate mapping of tumors and changes that occur within the tumor.

Below is a discussion of work presented at EMBC 2019, which touches on these three topics.

3 TTFields Dosimetry

As mentioned above, preclinical research has shown that the effect of TTFields on cancer cells depends on the frequency of the field, its intensity, and the duration of exposure to the fields. The EF-14 trial compared patient outcome in patients treated with chemoradiation+TTFields with outcome in patients treated with chemoradiation alone. Post hoc analysis of this trial has shown that patient outcome positively correlates with device usage (% of time on active treatment) [2].

More recently, Ballo et al. [2] published a study in which they defined TTFields dose as power loss density multiplied by usage and showed that within the EF-14 trial population, patients who received higher doses to the tumor bed exhibited overall improvement. Bomzon et al. presented a summary of this work at EMBC 2019 [3].

A total of $N = 340$ patients who received TTFields as part of the EF-14 trial were included in the study. Realistic head models of the patients were derived from T1-contrast-enhanced images captured at baseline using a previously described method [24]. The transducer array layout on each patient was obtained from EF-14 records, and average usage and average electrical current delivered to the patient during the first 6 months of treatment were derived from log files of the TTFields devices used by patients. Finite element simulations of TTFields delivery to the patients were performed using Sim4Life (ZMT Zurich, Switzerland). The average field intensity, power loss density, and dose density within a tumor bed comprising the gross tumor volume and the 3-mm-wide peritumoral boundary zone were calculated. The values of average field intensity, power loss density, and dose density that divided the patients into two groups with the most statistically significant difference in OS were identified.[1]

Figure 3 shows Kaplan-Meier curves for overall survival (OS) when dividing the patients into two groups according to TTFields dose. The median OS (and PFS data not shown) was significantly longer when average TTFields dose in the tumor bed was > 0.77 mW/cm^3: OS (25.2 vs 20.4 months, $p = 0.003$, HR = 0.611) and PFS (8.5 vs 6.7 months, $p = 0.02$, HR = 0.699). In similar analysis, dividing the patients according to TTFields intensity yielded that median OS and PFS were longer when average TTFields intensity at the tumor bed was >1.06 V/cm OS (24.3 vs 21.6 months, $p = 0.03$, HR = 0.705) and PFS (8.1 vs 7.9 months, $p = 0.03$, HR = 0.721).

This work sets a foundation for defining TTFields dose. It shows that TTFields dose can be defined in terms of power and usage and that delivery of higher doses of

---

[1]Defining dose for TTFields therapy is important to remember that TTFields are delivered by two sets of arrays, with the field direction switching direction every second. Thus, TTFields therapy essentially involves delivery of two incoherent electric fields to the tumor. A key question is how to meaningfully combine the two fields into a single metric defining dose. Ballo et al. established connections between local minimum field intensity (LMiFI) and local minimum power density (LMiPD) and survival. In this paper, dose was defined as LMiPD multiplied by usage.

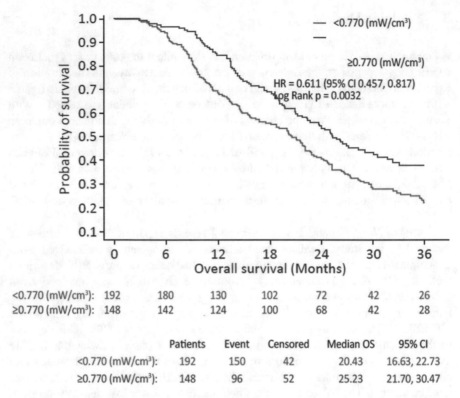

Fig. 3 Kaplan-Meier curves showing overall survival for patients treated with TTFields during the EF-14 trial. Graph shows survival curves when patients are divided into two groups based on the average dose in the tumor bed. The graphs clearly show improved survival in the group of patients who received an average dose of above 0.77 mW/cm^3 at the tumor bed. (Graph adapted from Ballo et al. [2])

TTFields to the tumor bed does indeed improve OS. Thus, a first principle for TTFields treatment planning is that treatment planning should strive to maximize average dose delivered to a region comprising the tumor and the peritumoral region, analogous to radiation therapy planning. A major difference between radiation therapy planning and TTFields treatment planning is that due to the highly toxic nature of radiation therapy; radiation therapy plans also need to account for avoidance structures, in which the radiation dose should be limited. This adds a level of complexity to the planning process. TTFields has a very low toxicity profile, with the only adverse effect reported being skin toxicity [19, 22]. Thus, there is no need to plan treatment to minimize dose in critical structures. It might be wise, however, to incorporate avoidance areas on the skin into the plan where arrays should not be placed.

Finally, it should be emphasized that TTFields treatment planning could benefit from understanding how the distribution of the field alters the progression of the tumor. This type of knowledge may help to devise more efficient strategies for planning the delivery of TTFields. These methods could be aimed at containing tumor growth by delivering higher doses to regions to which the tumor is more likely to grow or preventing tumor growth to critical structures by enhancing the field intensity in these regions.

4 Patient-Specific Model Creation

In order to properly perform numerical simulations of TTFields delivery, it is necessary to create accurate computational models that are patient-specific [11, 12, 24–26]. This involves creating a 3D volume representing the patient, in which dielectric properties (primarily conductivity) are assigned to each voxel.

Two different approaches can be used for creating these models:

Segmenting medical images of the patients to identify the various tissue types in the model and assignment of typical conductivity values to each tissue type [26]

Mapping conductivity from imaging data to assign a conductivity value to each voxel in the patient model based on some signal in the imaging data that provides information about the dielectric properties at a point [26, 27]

To date, most modelling work associated with TTFields has relied on the segmentation of patient data and assignment of conductivity values to each tissue type. Conductivity values have been assigned to the tissues based on empirical measurements that appear in the literature. There is a high degree of certainty associated with the conductivity values reported for healthy tissues of the brain, as reported measurements are relatively consistent when comparing different reports. However, little to no information exists on the electric properties of brain tumors. As response to TTFields seems to depend on dose delivered to the tumor bed, and as dose to the tumor will be influenced by the electric properties of the tumor, it is important to gain reliable data on the electric properties of tumors. At EMBC 2019, Proescholdt et al. [17] presented data on this topic. The data relied on measurements performed on tissue probes acquired from 53 patients with tumors of different histology and malignancy grades: low-grade glioma ($n = 5$), glioblastoma (GBM; $n = 16$), meningioma ($n = 19$), brain metastases ($n = 10$), and other histology types (1 craniopharyngioma, 1 lymphoma, 1 neuroma). Tissue probes were acquired from the vital and perinecrotic compartments of the tumor if present. Several probes (up to five) were sampled from each region. Immediately after acquisition, the electric properties of tissue fragments taken from the probes were determined using a parallel plate setup. The impedance of the sample was recorded at frequencies 20 Hz– 1 MHz. These measurements revealed significant differences between the conductivity observed in different tumor types, with meningiomas showing the lowest conductivity (mean conductivity [S/m]: 0.189; range: 0.327–0.113) and GBM tissue exhibiting the highest conductivity values (mean conductivity [S/m]: 0.382; range: 0.533–0.258). Consistently, the perinecrotic areas of tumors displayed lower conductivity values compared to the solid tumor compartments and also significant intratumoral heterogeneity in tumors of one specific histological diagnosis. The results of this study are summarized in Fig. 4.

This study sheds light on the dielectric properties of intracranial tumors, currently not accounted for in numerical models. An understanding of the cause of heterogeneity is needed in order to improve model quality and better predict field distributions around the tumor. In the interim, sensitivity analysis analyzing the effect of altering the electric properties of the tumor on field distributions is needed in order to complete our understanding on how best to plan TTFields therapy and the uncertainties associated with this planning.

EMBC 2019 also included a talk by Wenger et al. [26], discussing the use of water content-based electric property tomography (wEPT) in order to create patient models for TTFields-related numerical simulations. wEPT is an imaging tomography technique that models electrical conductivity, σ, and relative permittivity, ε, as monotonic functions of water content (WC) according to Maxwell's mixture theory [13]. WC maps are found via a transfer function mapping the image ratio (IR) of two T1w images with different repetition times (TR) into water content. Previously, wEPT was adapted to map WC, σ, and ε at 200 kHz in animal brain samples and tumor-bearing rats with mixed results [26]. When comparing wEPT-based predictions to empirical measurements of tissue samples using a parallel plate setup, we found a good match between wEPT-based estimations in the healthy tissue, while the quality of the match was poor within the tumors. At EMBC 2019, data were presented on the applicability of wEPT to mapping the electric properties of the human brain. The images used for wEPT mappings included, for this purpose, an image with a short TR resembling a conventional T1w MRI and a proton density (PD) image with the same parameters except for a long TR (Fig. 5).

EP maps for three patients who participated in the EF-14 trial were created using wEPT. The adapted wEPT model coefficients were found via curve fitting according to previous experiments and MRI scanner-specific parameters. Analysis of the results showed that wEPT estimates of WC, σ, and ε in healthy brain tissues (white and gray matter) appear accurate and comparable with reports in literature. The properties were also relatively homogenous throughout the tissues and did not vary much between patients. Contrary, wEPT estimates of σ and ε in tumor tissues (necrosis, enhancing and non-enhancing tumor) were highly heterogeneous with high variability between patients.

These results, combined with results of our previous study, show the potential of wEPT-like methods for mapping the electric properties of the brain. However, the results suggest that wEPT alone is insufficient to map the electric properties of the tumor as well as the heterogeneous nature of the tumor. Future studies should focus on understanding the connection between tissue microstructure and the electric properties of the tissues at 200 kHz. When these processes are well understood, then methods for accurately mapping electric properties can be devised.

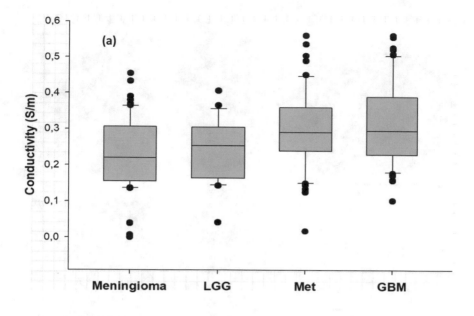

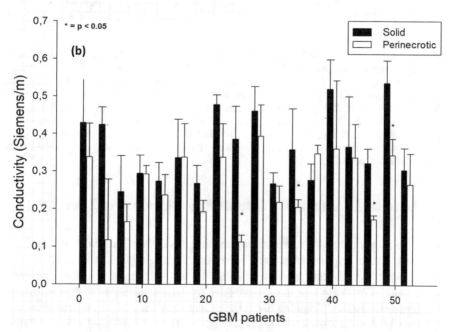

Fig. 4 (**a**) Boxplots showing the distribution of conductivities measured for four different brain tumor types: meningiomas, low-grade gliomas, brain metastases, and glioblastomas. Conductivity differs between tumor types, with the highest median conductivity measured in glioblastoma and the lowest median conductivity measured in meningioma. In all tumor types, a high heterogeneity in the electric conductivity is observed. (**b**) Bar plot showing the average conductivity measured in the solid and perinecrotic regions of the tumor for several GBM patients. Surprisingly, conductivity is consistently lower in the perinecrotic region of the tumor

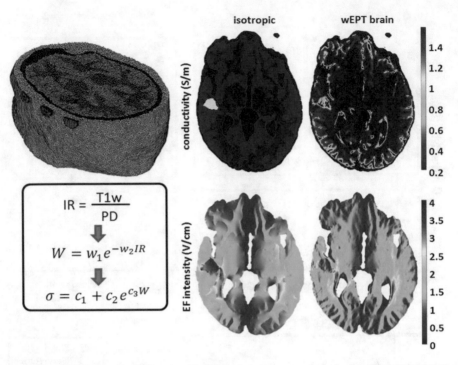

Fig. 5 (top-left) The tetrahedral mesh a of glioblastoma patient used to calculate TTFields induced electric field distributions, along with the (top row) distribution of conductivity for (top middle) a model created by segmenting a T1c image and assigning conductivity values to each tissue type and (top right) a model created using wEPT. Bottom row shows the field distribution in the (bottom middle) model created through segmentation and in the (bottom right) model created with wEPT. The flow chart in the bottom-left corner shows the wEPT scheme. First an image ration (IR) is calculated from the T1w and PD images. Next, WC is found from the IR, and conductivity derived from the WC. Note the parameters in the equations are found using curve-fitting to empirical data as detailed in Wenger et al. [26]

5   Advanced Imaging for Monitoring Response to Therapy

A key component in adaptive treatment planning is imaging. The ability to image the tumor and assess response effectively is key to mapping regions in which the tumor is responding to therapy and identifying the regions in which the disease is progressing. Treatment can then be adapted to target the regions of progression. Furthermore, advanced imaging holds the potential of early identification of molecular and biological responses occurring at the tissue and cellular level. These changes could indicate response or resistance to a specific treatment regimen, enabling the treating physician to adapt treatment early in order to improve the probability of positive outcomes. In the context of TTFields, this may mean adapting the transducer array position in order to increase the dose in regions of progression in order to suppress tumor growth in this region.

At EMBC Mohan et al. presented a study investigating the use of advanced imaging to map response to TTFields [14]. Twelve patients (both newly diagnosed and recurrent GBM patients) previously treated with standard-of-care maximal safe resection and chemoradiation received TTFields. Patients underwent baseline (prior to initiation of TTFields therapy) and two follow-up (1 and 2 months post initiation of TTFields) image acquisitions on a 3 T MRI. DTI data were acquired using 30 directions with a single-shot spin-echo EPI sequence. Motion and eddy current corrections of raw DTI data were performed, and parametric maps of mean diffusivity (MD) and fractional anisotropy (FA) generated using in-house software. Perfusion-weighted imaging (PWI) was performed using T2*-weighted gradient-echo EPI sequence which was acquired with a temporal resolution of 2.1 s. Leakage-corrected cerebral blood volume (CBV) maps were constructed. 3D-EPSI was acquired using a spin-echo-based sequence. EPSI data were processed using the Metabolic Imaging and Data Analysis System (MIDAS) package. MD, FA, EPSI [choline (Cho)/creatine(Cr)], CBV maps, and FLAIR images were co-registered to post-contrast T1-weighted images, and contrast-enhancing neoplasms were segmented using a semiautomated algorithm. Median values of MD, FA, relative CBV (rCBV), and Cho/Cr were computed at each time point, as were the 90th percentile rCBV (rCBVmax) values. Percent changes of each parameter between baseline and follow-up time points were evaluated.

Analysis of the images demonstrated an increasing trend in MD (~3%) and declining trend in FA (~8%) at the 2-month follow-up relative to baseline. Additionally, reductions in Cho/Cr and rCBV max from baseline to post-TTFields were also observed. All patients were clinically stable at 2-month follow-up. The changes in MD, FA, and Cho/Cr may indicate inhibition of cellular growth. Reduction in rCBVmax may indicate anti-angiogenic effects associated with TTFields and decreased perfusion within the tumor bed after the therapy. These preliminary results suggest that advanced MR imaging may be useful in evaluating response to TTFields in GBM patients. Further work is required to validate the findings in a larger patient cohort in which these findings could be correlated with clinical endpoints of PFS and OS. Fully utilizing the power of these findings for TTFields treatment planning would also warrant studies looking to connect voxel-based changes observed in the images with field intensity distribution patterns derived from simulations. The completion of such studies would provide physicians with valuable information about how to plan and dynamically adapt TTFields dose distributions in order to maximize their ongoing effect on the tumor.

6 Discussion and Conclusions

In this chapter, we have provided an overview on some of the TTFields-related research presented at EMBC 2019. The research presented in this chapter relates to three key areas, which together lay the foundations for the field of TTFields dosimetry and treatment planning:

- Definition of TTFields dose and the influence of dose on patient outcome
- The creation of patient-specific computational models for simulating delivery of TTFields
- Advanced imaging techniques for monitoring response to therapy

The area of TTFields dosimetry and treatment planning is very much in its infancy. The work presented at EMBC 2019 not only provides a basis for this field but also highlights the many open questions related to this field. The work presented by Bomzon et al. provides a robust and relatively intuitive definition for TTFields dose as the average power delivered by the fields. The authors clearly show a connection between TTFields dose at the tumor bed and patient survival. However, a crucial point required for effective treatment planning is to understand how dose distributions influence progression patterns. Do tumors really tend to progress to regions in which TTFields dose is lower? At a more fundamental level, given a TTFields dose distribution map, can we predict the probability that the tumor will progress in a certain region?

The work presented by Wenger et al. shows the potential of image-based electric property tomography to accurately map conductivity within patients, thereby providing a quick and accurate method for creating the patient-specific models required for TTFields treatment planning. This work also emphasizes the difficulty and knowledge gap that needs to be bridged in order to accurately map the electric properties of tissues in the vicinity of the tumor. The work by Proescholdt et al. shows that electric properties of tumors are indeed highly heterogeneous. Thus, accurate methods for modelling tumor tissue properties may be needed in order to accurately model electric field distributions in the vicinity of the tumor when performing treatment planning.

Finally, the work presented by Mohan shows the potential of advanced imaging techniques to identify metabolic and physiological changes within the tumor. These changes could be used as markers for response to therapy and could be adapted to plan therapy throughout the course of treatment.

Thus, the combination of work presented at EMBC poses key questions that need to be answered as the field of TTFields dosimetry and treatment planning evolves:

- How are progression patterns influenced by TTFields dose distributions? Do tumors progress in regions where dose is lower?
- How do we improve methods for mapping conductivity in a patient-specific method, specifically around the tumor?
- Can we utilize advanced imaging techniques to effectively monitor disease response/progression in order to better tailor therapy in an adaptive manner?

A final component required for the maturation of TTFields treatment planning is the development of a clinical quality assurance (QA) system analogous to that used for radiation therapy planning. This system should aim to establish the uncertainties associated with treatment plans and their effect on the dose distributions. This in turn could lead to the development of clinical guidelines related to the desired accuracy of the computational models used for TTFields treatment planning, the desired accuracy in the electric properties assigned to the various tissue types, and the desired accuracy in the placement of the transducer arrays on the skin when initiating treatment. Key to answering these questions are studies examining the sensitivity of numerical simulations of TTFields delivery to all of the above parameters, as well as studies aiming to experimentally validate the simulations using, for instance, suitable anthropomorphic phantoms. The development of this QA framework would enable the derivation of guidelines for best practices when performing TTFields treatment planning, thereby guiding the practicalities associated with the assimilation of sophisticated treatment planning procedures into the clinic.

As a concluding comment, we note that TTFields treatment planning could benefit enormously from emerging studies utilizing mathematical models to predict tumor progression [1, 7, 8]. These models attempt to incorporate information about factors such as tumor cell density, cell proliferation rates, and cell invasiveness into models that predict how tumors progress over time. Radiomic methods can be used to extract relevant information on the tumor, which can be fed into such models [5]. The effect of specific drugs or radiation therapy on tumor progression can then be modelled, and patients would gain potential benefit from a specific treatment quantified [15, 16]. A natural expansion to these models is to incorporate TTFields dose distributions. In fact, an attempt to do this has previously been reported [18]. The benefit of different TTFields treatment plan can be evaluated, and the optimal plan selected. As the patient is monitored, and additional imaging data collected, the treatment plan, combination of therapies, and patient-specific model could be updated to continuously provide the patient with optimal care.

In summary, TTFields are emerging as a powerful addition to a growing arsenal of tools applied in the fight against cancer. The development of effective techniques for TTFields treatment planning will help to maximize the utility of this exciting treatment modality, ultimately leading to improved patient outcomes.

References

1. Alfonso, J., Talkenberger, K., Seifert, M., Klink, B., Hawkins-Daarud, A., Swanson, K., et al. (2017). The biology and mathematical modelling of glioma invasion: A review. *Journal of the Royal Society, Interface, 14*(136), 20170490.
2. Ballo, M., Urman, N., Lavy-Shahaf, G., Grewal, J., Bomzon, Z., & Toms, S. (2019). Correlation of tumor treating fields dosimetry to survival outcomes in newly diagnosed glioblastoma: A large-scale numerical simulation-based analysis of data from the phase 3 EF-14 randomized trial. *International Journal of Radiation Oncology Biology Physics, 104*(5), 1106.
3. Bomzon, Z., Urman, N., Levi, S., Lavy-Shahaf, G., Toms, S., & Matthew, B. (2019). *Development of a framework for tumor treating fields dosimetry and treatment planning using computational phantoms (I)*. Berlin: EMBC.
4. Ceresoli, G., Aerts, J., Dziadziuszko, R., Ramlau, R., Cedres, S., van Meerbeeck, J., et al. (2019). Tumour treating fields in combination with pemetrexed and cisplatin or carboplatin as first-line treatment for unresectable malignant pleural mesothelioma (STELLAR): A multicentre, single-arm phase 2 trial. *The Lancet Oncology, 20*(12), 1702–1709.

5. Gaw, N., Hawkins-Daarud, A., Hu, L., Yoon, H., Wang, L., Xu, Y., et al. (2019). Integration of machine learning and mechanistic models accurately predicts variation in cell density of glioblastoma using multiparametric MRI. *Scientific Reports, 9*(1), 10063.

6. Giladi, M., Schneiderman, R., Voloshin, T., Porat, Y., Munster, M., Blat, R., et al. (2015). Mitotic spindle disruption by alternating electric fields leads to improper chromosome segregation and mitotic catastrophe in cancer cells. *Scientific Reports, 5*, 18046.

7. Hawkins-Daarud, A., Johnston, S., & Swanson, K. (2019). Quantifying uncertainty and robustness in a biomathematical model–based patient-specific response metric for glioblastoma. *JCO Clinical Cancer Informatics, 3*, 1–8.

8. Hu, L., Yoon, H., Eschbacher, J., Baxter, L., Dueck, A., Nespodzany, A., et al. (2019). Accurate patient-specific machine learning models of glioblastoma invasion using transfer learning. *AJNR. American Journal of Neuroradiology, 40*(3), 418–425.

9. Kirson, E., Dbalý, V., Tovaryš, F., Vymazal, J., Soustiel, J., Itzhaki, A., et al. (2007). Alternating electric fields arrest cell proliferation in animal tumor models and human brain tumors. *Proceedings of the National Academy of Sciences of the United States of America, 104* (24), 10152–10157.

10. Kirson, E., Gurvich, Z., Schneiderman, R., Dekel, E., Itzhaki, A., Wasserman, Y., et al. (2004). Disruption of cancer cell replication by alternating electric fields. *Cancer Research, 64*(9), 3288–3295.

11. Korshoej, A., Hansen, F., Mikic, N., von Oettingen, G., Sørensen, J., & Thielscher, A. (2018). Importance of electrode position for the distribution of tumor treating fields (TTFields) in a human brain. Identification of effective layouts through systematic analysis of array positions for multiple tumor locations. *PLoS ONE, 13*(8), e0201957.

12. Korshoej, A., Hansen, F., Thielscher, A., Von Oettingen, G., & Sørensen, J. (2017). Impact of tumor position, conductivity distribution and tissue homogeneity on the distribution of tumor treating fields in a human brain: A computer modeling study. *PLoS ONE, 12*(6), e0179214.

13. Michel, E., Hernandez, D., & Lee, S. (2017). Electrical conductivity and permittivity maps of brain tissues derived from water content based on T1-weighted acquisition. *Magnetic Resonance in Medicine, 77*(3), 1094–1103.

14. Mohan, S. (2019). *Advanced imaging for monitoring response to TTFields in glioblastoma patients*. Berlin: EMBC.

15. Neal, M., Trister, A., Ahn, S., Baldock, A., Bridge, C., Guyman, L., et al. (2013a). Response classification based on a minimal model of glioblastoma growth is prognostic for clinical outcomes and distinguishes progression from pseudoprogression. *Cancer Research, 73*(10), 2976–2986.

16. Neal, M., Trister, A., Cloke, T., Sodt, R., Ahn, S., Baldock, A., et al. (2013b). Discriminating survival outcomes in patients with glioblastoma using a simulation-based, patient-specific response metric. *PLoS ONE, 8*(1), e51951.

17. Proescholdt, M., Haj, A., Lohmeier, A., Stoerr Eva-Maria, E.-M., Eberl, P., Brawanski, A., et al. (2019). *The dielectric properties of brain tumor tissue*. Berlin: EMBC.

18. Raman, F., Scribner, E., Saut, O., Wenger, C., Colin, T., Fathallah-Shaykh, H.M. (2016). Computational Trials: Unraveling Motility Phenotypes, Progression Patterns, and Treatment Options for Glioblastoma Multiforme. PLoS One. *11*(1), e0146617. https://doi.org/10.1371/journal.pone.0146617PMCID: PMC4710507

19. Stupp, R., Taillibert, S., Kanner, A., Kesari, S., Steinberg, D., Toms, S., et al. (2015). Maintenance therapy with tumor-treating fields plus temozolomide vs temozolomide alone for glioblastoma a randomized clinical trial. *JAMA: The Journal of the American Medical Association, 314*(23), 2535–2543.

20. Stupp, R., Taillibert, S., Kanner, A., Read, W., Steinberg, D., Lhermitte, B., et al. (2017). Effect of tumor-treating fields plus maintenance temozolomide vs maintenance temozolomide alone on survival in patients with glioblastoma: A randomized clinical trial. *JAMA, 318*(23), 2306–2316.

21. Stupp, R., Wong, E., Kanner, A., Steinberg, D., Engelhard, H., Heidecke, V., et al. (2012). NovoTTF-100A versus physician's choice chemotherapy in recurrent glioblastoma: A randomised phase III trial of a novel treatment modality. *European Journal of Cancer, 48* (14), 2192–2202.
22. Taphoorn, M., Dirven, L., Kanner, A., Lavy-Shahaf, G., Weinberg, U., Taillibert, S., et al. (2018). Influence of treatment with tumor-treating fields on health-related quality of life of patients with newly diagnosed glioblastoma a secondary analysis of a randomized clinical trial. *JAMA Oncology, 4*(4), 495–504.
23. Toms, S.A., Kim, C.Y., Nicholas, G., Ram, Z. (2019) Increased compliance with tumor treating fields therapy is prognostic for improved survival in the treatment of glioblastoma: a subgroup analysis of the EF-14 phase III trial. J Neurooncol. *141*(2):467–473. Published online 2018 Dec 1. https://doi.org/10.1007/s11060-018-03057-zPMCID: PMC6342854.
24. Urman, N., Levy, S., Frenkel, A., Manzur, D., Hershkovich, H., Naveh, A., et al. (2019). Investigating the connection between tumor-treating fields distribution in the brain and glioblastoma patient outcomes. A simulation-based study utilizing a novel model creation technique. In N. Urman, S. Levy, A. Frenkel, D. Manzur, H. Hershkovich, A. Naveh, et al. (Eds.), *Brain and human body modeling* (pp. 139–154). Cham: Springer.
25. Wenger, C., Hershkovich, H., Tempel-Brami, C., Giladi, M., & Bomzon, Z. (2019). Water-content electrical property tomography (wEPT) for mapping brain tissue conductivity in the 200–1000 kHz range: Results of an animal study. In C. Wenger, H. Hershkovich, C. Tempel-Brami, M. Giladi, & Z. Bomzon (Eds.), *Brain and human body modeling* (pp. 367–393). Cham: Springer.
26. Wenger, C., Miranda, P., Salvador, R., Thielscher, A., Bomzon, Z., Giladi, M., et al. (2018). A review on tumor-treating fields (TTFields): Clinical implications inferred from computational modeling. *IEEE Reviews in Biomedical Engineering, 11*, 195–207.
27. Wenger, C., Salvador, R., Basser, P., & Miranda, P. (2016). Improving tumor treating fields treatment efficacy in patients with glioblastoma using personalized array layouts. *International Journal of Radiation Oncology Biology Physics, 94*, 1137.

How Do Tumor-Treating Fields Work?

Kristen W. Carlson, Jack A. Tuszynski, Socrates Dokos, Nirmal Paudel, Thomas Dreeben, and Ze'ev Bomzon

1 Introduction

Since approved by the FDA for the treatment of glioblastoma brain cancer in 2015, tumor-treating fields (TTFields) have rapidly become the fourth modality to treat cancer, along with surgery, chemotherapy, and radiation [1]. TTFields are now in clinical trials for a variety of cancer types. While efficacy has been proven in the clinic, higher efficacy is demonstrated in vitro and in animal models, which indicates much greater clinical efficacy is possible. To attain the great promise of TTFields, uncovering the mechanisms of action (MoA) is necessary.

TTFields are 200 kHz AC electric fields directed transcranially or transdermally to a tumor site with target field strength of 2–4 V/cm. Through unknown MoA, TTFields kill cancer cells, extending survival for victims of brain cancer. Empirical studies show TTFields exert a variety of effects on cell processes [2–5]. And while the cause-effect chain is under investigation, they ultimately disrupt mitosis, the

K. W. Carlson (✉)
Department of Neurosurgery, Beth Israel Deaconess Medical Center/Harvard Medical School, Boston, MA, USA
e-mail: kwcarlso@bidmc.harvard.edu

J. A. Tuszynski
Department of Physics, University of Alberta, Edmonton, AB, Canada

S. Dokos
Department of Biomedical Engineering, University of New South Wales, Sydney, Australia

N. Paudel
Independent Consultant, Greenville, NC, USA

T. Dreeben
OSRAM Opto Semiconductors, Exeter, NH, USA

Z. Bomzon
Novocure Ltd., Haifa, Israel

© The Author(s) 2021
S. N. Makarov et al. (eds.), *Brain and Human Body Modeling 2020*,
https://doi.org/10.1007/978-3-030-45623-8_2

delicately orchestrated process of cell division, which occurs more often in cancer cells than in healthy cells. This article will review the clues and hypotheses about TTFields MoA that have been uncovered to date. We present the novel hypothesis that the intrinsic, mitochondrial apoptotic pathway upregulated by Bcl-2 and inhibited by BAX is a key component of TTFields MoA.

1.1 TTFields Affect Large, Polar Molecules

Early hypotheses on TTFields MoA pointed to disruptive effects on subcellular structures such as individual dipoles, e.g., tubulin, which are common in the cell [2, 3]. However, detailed subsequent calculations show that TTFields' effects on individual dipoles seem insufficient to disrupt their function (Fig. 1) [6]. Using parameters at the high end of their ranges, Tuszynski et al. calculated that the force on a free, unpolymerized 3000 debye tubulin dimer by 1 V/cm TTFields is 10^{-24} J, several orders of magnitude less than the cell's background thermal energy, and therefore too low to torque it (Table 1). On the other hand, larger molecules composed of many dipoles, such as microtubules (MT), septin, and organelles, may accumulate enough dipole moment in their structure to be significantly affected.

Further, it is likely that TTFields' energy is amplified at various locations within the cell. Identifying those amplifying locations is an important part of our research program [7].

TTFields could also modulate ion channel gates in the mitochondrial membranes, which in turn regulate cellular processes such as apoptosis (programmed cell death).

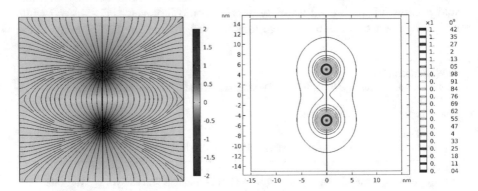

Fig. 1 COMSOL (Burlington, MA, USA) 2D axisymmetric finite element model of a dipole approximating a tubulin dimer. Left: color scale, electric potential in volts. Streamlines: electric field. Right: electric field norm contours in V/m; cf. to TTFields imposed field strength of ~ 2×10^2 V/m. Model bears out calculations by Tuszynski et al. predicting the TTFields' direct effect on comparable subcellular dipoles is orders of magnitude too weak to have a disruptive effect on cell structures [6]. Thus, workers look for locations in the cell where TTFields' effect is amplified, such as in highly conductive microtubules composed of tubulin dimers, or at the cell furrow during late mitosis [7]

Table 1 Three ubiquitous cellular energy levels, which, if exceeded via an exogenous source, would disrupt processes

Disruption metric	Energy
Background thermal energy	$kT = 4 \times 10^{-21}$ joules
Energy of ions crossing cell membrane (e.g. K+, Na+)	20 – 100 meV ~1 – 4 kT
Adenosine Triphosphate (ATP)	~25 kT $\approx 10^{-19}$ joules

Table 2 Signal-to-noise ratio guidance in signal theory

Signal/noise (dB)	Quality
1–10	Unreliable
20	Intelligible but noisy
60	High fidelity

Scale is logarithmic: 1 dB $= 10 \log_{10} (P_1/P_2)$ where P is power (watts)

In general, cellular processes evolved to send signals in the presence of background thermal noise, whose value is given by Boltzmann's constant k times absolute temperature T [6]. For practical purposes, we assume energy required for disruption must be at least 1–2 orders of magnitude above kT, since the signal/noise ratio in a reliable system is never 1:1 (Table 2) [8]. Another ubiquitous energy threshold in cellular activities, for instance, in MT depolymerization or motor protein transport, is ATP energy (Table 1). Since ATP energy is the actual energy used by the cell for numerous binding or unbinding processes, and not a probabilistic floor as is kT, TTFields energy need not substantially exceed it to interfere with an ATP-driven process. While kT and ATP energies are ubiquitous in the cell, other more specialized energy levels exist as well. Any cellular process in which TTFields' energy exceeds the threshold energy driving that process is a candidate for TTFields MoA.

Tuszynski et al. propose an "energetic constraint from above," 3×10^{-12} W, approximately the power generated by a cell, based on the hypothesis that TTFields cannot disrupt many normal and mitotic cellular processes, which would occur if the energy absorbed by the cell appreciably heated it [6], and empirically no significant heating effects have been observed [2]. However, while much of a cell's energy is devoted to heating it to a temperature where its biochemical reactions are possible, energy absorbed from TTFields may be concentrated at key structures and have nonthermal effects higher up on the entropic food chain.

1.2 The Need for a "Complete" TTFields Theory

We advocate constructing a "complete" theory of TTFields tying together the underlying mechanisms of TTFields' efficacy with its unobserved effects on

Table 3 Ingredients of a complete TTFields theory

Components of TTFields theory	Examples
Predicted electromagnetic effects on subcellular structures	Dielectrophoresis of large molecules near the cellular furrow during late mitosis [9, 10]
Predicted electromagnetic effects on cell-signaling pathways	Triggering of intrinsic, Bcl2/BAX-mediated apoptosis by direct effect on mitochondrial outer membrane (MOM) or indirect effect via excess free tubulin obstructing MOM voltage-gated ion channels
Observed effects of TTFields on cellular structures	Excess free tubulin [4] and decrease of septin at cell midline [5], both correlated with mitotic spindle deformities and other mitotic aberrations, such as cell blebbing, multiple nuclei, aneuploidy, and apoptosis in interphase
Verifiable predictions leading toward maximum efficacy	Apply direction changes from multiple directions so as to increase the incidence of field aligned and orthogonal to cell axis of randomly oriented cells in vivo To treat edema in brain cancer, replace dexamethasone (dex) with celecoxib to eliminate dex inhibiting TTFields' triggering of intrinsic apoptosis [11]

subcellular structures, cell signaling pathways, and observed effects (Table 3). A complete theory will sort out what is causal versus epiphenomenal or downstream effects. For instance, it is unknown whether the decrease of the key motor protein septin at the cell midline [5] or the 25% increase in free tubulin in relative terms [4] in TTFields-treated cells are causing disruption of mitosis or are just side effects. Importantly, a complete theory imposes the maximum constraints on its formulation, which is helpful conceptually and in modeling TTFields effects. The several hypotheses on TTFields MoA (Sect. 3) conflict with each other and with empirical evidence in varying respects, which must be sorted out. Examining isolated TTFields phenomena results in under-constrained models.

2   Empirical Clues to TTFields MoA

In this section, we compile key clues informing and constraining a theory of how TTFields work.

2.1   TTFields Only Kill Fast-Dividing Cells

TTFields do not affect normal cells and only affect cells that divide more often than normal cells, i.e., cancer cells. Thus, MoA ideas focus on subcellular structures that differ in the cell division stages (mitosis) versus between cell division (interphase). For example, MTs become much more dynamic in their length during mitosis, and during mitosis some septin structures align with the cell axis, while other septin structures rotate orthogonal to it, while in interphase they are randomly aligned in general.

2.2 TTFields Require 2–4 V/cm Field Strength

Figure 2 (left) shows a plot of killed versus live cells according to TTFields ambient field strength in vitro [2]. Based on this and other early works, numerous head and body modeling studies use 2 V/cm as a threshold for efficacy required in TTFields. However, considering the weakened effect of TTFields when not aligned with the cell axis (Sect. 2.4) and the need to kill all cells in cancer to eliminate recurrence and metastases (tumor cells that spread to other organs from their original site), we suggest 4 V/cm as an efficacy threshold.

2.3 TTFields Are Frequency-Sensitive and Effective Only in the 100–300 KHz Range

TTFields have proven effective at killing a variety of tumor cell types, but in all cases, the frequency range is confined to 100–300 KHz, equivalent to a period range of 3.3–10 µs, and the specific frequency varies with tumor cell type (Fig. 2, right) [2, 3]. For instance, TTFields' efficacy for glioblastoma and astrocytoma brain cancers, ovarian cancer, and mesothelioma lung cancer is maximal at 200 KHz, while for various types of lung, pancreatic, cervical, and mammary cancers, the peak effect is at 150 KHz [4, 12]. We believe that part of the underlying mechanism is that the cell membrane becomes translucent to electromagnetic radiation in this

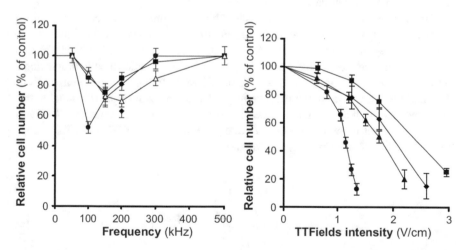

Fig. 2 Frequency- and electric field intensity-dependent TTFields' effects on proliferation of several cancer cell types. Left: relative change in number of cells after 24-h treatment of different frequencies at 1.75 V/cm. Right: effect of 24-h exposure to TTFields of increasing intensities (at optimal frequencies). ● B16F1 (mouse melanoma). ■ MDA-MB-231 (human breast adenocarcinoma). ▲△ F-98 (human glioma brain cancer). H1299 (small lung cell carcinoma). (From Kirson et al. [3], with permission, copyright (2007) National Academy of Sciences)

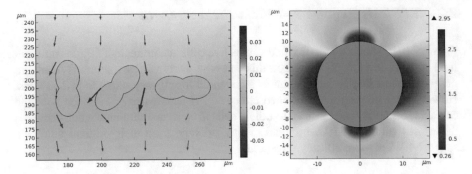

Fig. 3 At 200 KHz, a 10 μm radius cell becomes "translucent" to an exogenous electric field (2 V/cm) due to complex permittivity of its membrane, while at lower frequencies it is opaque and at higher frequencies (e.g., MHz-GHz) it is transparent [9]. The field is largely, but not entirely shunted around the cell, with a fraction of its energy penetrating the cell to affect subcellular structures. Left: dividing cell at 0, 45°, and 90° (left to right) to the cell's axis. Electric potential (color bar, V) and current flow (arrows). Arrows: current density. Right: color bar: Electric field strength (V/cm)

frequency range, allowing enough field strength to enter the cell and affect subcellular mitotic structures (Fig. 3) [9]. The mechanism underlying the upper limit of the range and different frequency responses is unknown, which could be, for instance, tied to relaxation time constants in DEP effects, the duration of sustained current flow at MT+ end required to disrupt polymerization, or the duration required to charge the MT counter-ion/depleted ion layer/C-termini capacitor which in turn could disrupt MT functions.

The efficacy versus field strength relationship is nonlinear for all cell types (Fig. 2 right) and suggests a power law, for which there are at least two candidates. In general, energy and power in an electromagnetic wave are related to the square of field strength produced by the wave (Eq. 1). Dielectrophoretic force is proportional to the square of field strength (Eq. 2) [9]; see Zhao & Zhang for a different analysis [13], and there are other possibly relevant power law relationships. It is notable that while the electric field strength squared is also proportional to frequency, TTFields efficacy does not increase with their frequency outside of the 100–300 KHz range (Sect. 2.2). The underlying reason is unknown.

$$I_{\text{ave}} = \frac{c\epsilon_0 E_0^2}{2} \tag{1}$$

$$\langle \mathbf{F}_{\text{DEP}} \rangle \sim \nabla \left[\operatorname{Re}\left[\widetilde{\boldsymbol{E}}\right]^2 + \operatorname{Im}\left[\widetilde{\boldsymbol{E}}\right]^2 \right] \tag{2}$$

2.4   TTFields Are Highly Directional

One in vitro experiment examined the proportion of cells killed in relation to the angle of the cell axis to the ambient TTFields direction [2]. This study found that TTFields are most lethal when aligned with the cell axis and have a weaker, secondary effect when orthogonal to the axis, but no significant effect at 45° to the axis. We interpret these results to indicate there are at least two MoA, and they are not additive.

Following the conclusions of the experiments just described, and assuming cells are randomly oriented when they begin mitosis, it is logical to hypothesize that changing the direction of TTFields would subject more cells to field alignment or orthogonality than applying the treatment from one direction only. A possible trade-off is the reduced duty cycle—the proportion of time spent in a given direction. It is possible that the field must be applied for a minimum duration in order to produce its disruptive effect on subcellular structures.

Thus, another study looked at TTFields' efficacy with no change of direction versus one change of direction twice per second [2]. The latter protocol killed 20% more cells than TTFields applied with no change of direction.

A third set of experiments was published in US patent [14]. Figure 4, from the patent, compares the changes in growth rate of glioma F98 cells (brain cancer) subjected to 200 KHz TTFields imposed from two orthogonal directions at different change-of-direction frequencies.

The several sets of experiments support each other, while the mechanisms underlying the increased tumor cell-killing effect with change-of-direction duty cycle remain to be understood.

2.5   TTFields Have Their Strongest Effect in Prophase
and Metaphase

Kirson et al. also empirically examined TTFields efficacy at different points in the cell division cycle. They found that the strongest effects occur in metaphase, i.e., early in mitosis, with the field direction aligned with the cell axis [2]. Secondarily, when TTFields are orthogonal to the cell axis, they have a lesser effect, also in metaphase, specifically in prophase at the start of mitosis.

2.6   TTFields Increase Free Tubulin and Decrease
Polymerized Tubulin in the Mitotic Spindle Region

Giladi et al., through careful experimentation, demonstrated that free tubulin is increased by 12.8% in absolute terms, which is ~25% in relative terms, and

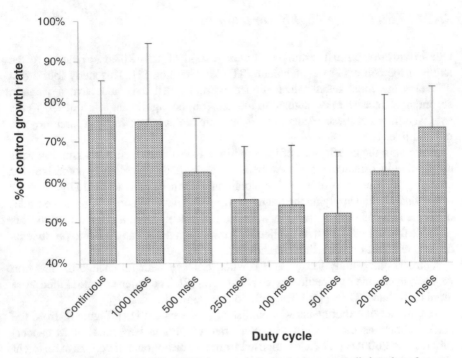

Fig. 4 (From Palti [14]). 200 KHz TTFields were applied to brain cancer cells in culture from two orthogonal directions at different duty cycle frequencies (i.e., at different frequencies of change between the two directions). The greatest efficacy occurred when the changes of direction occurred every 50 ms. The 50 ms optimal duty cycle may be a significant clue as to the underlying mechanism of action, for instance, a dielectrophoretic relaxation time in a large intracellular macromolecule

polymerized tubulin (i.e., in microtubules) is decreased concomitantly in TTFields-treated cells [4]. The effect was similar to, but not as pronounced as, the increase of free versus polymerized tubulin due to a vinca-alkaloid chemotherapeutic agent, vinorelbine, and opposite to effects of the paclitaxel class of chemotherapeutic agents that stabilize MT polymerization, thereby decreasing the ratio of free versus polymerized tubulin. Giladi et al. also showed phenomena such as aberrant spindle formation, mitotic "slippage" (delayed mitotic phases), aneuploidy (abnormal number of chromosomes), and other effects that can trigger the intrinsic form of apoptosis or are emblematic of it.

These results raise the possibility that TTFields act like the vinca alkaloids to disrupt MT polymerization, but without the noted deleterious side effects of chemotherapy. In one study, TTFields were shown to be as effective as chemotherapy in treating brain cancer [15] and in others were shown to be synergistic with chemotherapy for lung cancer, decreasing the dosage required to achieve equal efficacy [16] and increasing efficacy when used in conjunction with several chemo agents [12].

3 Candidate Mechanisms of Action (MoA)

3.1 Dielectrophoretic (DEP) Effects

Several modeling studies have shown that as the mother and daughter cells elongate during late mitosis and enter cytokinesis (division into the two daughter cells), field strength is significantly amplified near the furrow joining the two cells [9, 10, 13]. Since DEP effects are proportional to the electric field amplitude squared (Eq. 2), DEP forces increase dramatically in late mitosis when cells are maximally elongated and field lines are most concentrated near the cell furrow. Further, it is easy to see that DEP effects are enhanced when TTFields are aligned with the cell axis, in accord with experimental data (Sect. 2.4).

However, DEP effects do not correspond to empirical data showing TTFields have no effect when oriented at 45° to the cell axis and a moderate effect at 90° (Sect. 2.4 and Fig. 5). Also arguing against the DEP MoA theory is their weak effect early in the mitotic cycle and strong effect in the late mitotic cycle, in contradiction to experimental data (Sect. 2.5).

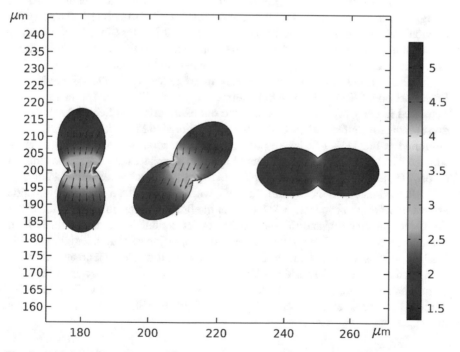

Fig. 5 A 10 μm radius dividing cell in late mitosis approaching cytokinesis is shown with an exogenous 200 kHz TTFields of 2 V/cm imposed at 0, 45°, and 90° (left to right) to the cell's axis. Field strength (color bar, V/cm) and current flow (arrows) are amplified most when the field is aligned with the cell axis, less so at 45°, and not at all at 90°. (*COMSOL* model courtesy Wenger et al. [9])

3.2   Microtubule Effects

Several hypotheses have been proposed for TTFields disruption of MTs, including direct effects on C-termini state transitions, stall of the motor protein kinesin which transports cargo along the MT, and interference with MT polymerization. Since MTs become more active and increase in length and a proportion of them align with the cell axis as mitosis progresses, hypotheses that they are involved in TTFields MoA fit the constraints of acting on mitotic but not interphase cells and efficacy when the field is aligned or orthogonal to the cell axis (Sects. 2.1 and 2.4).

We calculate the force exerted by TTFields on elementary charges as follows. For each 100 V/m $= 100$ N/C $= 100$ N/(6.2415×10^{18} e), or 1.60218×10^{-17} N/e (where N is newtons and e is an elementary charge). For 8 C-termini charges and TTFields target strength of 200 V/m, this yields 2.56×10^{-16} N, orders of magnitude less than the minimum estimated 10^{-10}–10^{-12} N required to perturb C-termini state transitions.

On the other hand, estimates vary for the stall force on the motor protein kinesin's "walk" along a MT, 8 nm dimer by dimer. One group separates each step of the walk into two phases: the first driven by the mechanical tension inherent in the kinesin "neck," which requires at least pN level force to stall, while the second, thermal energy-driven step of the walk which requires just 10^{-16} N to stall [17]. Thus, by the previous calculation and our modeling, disruption by TTFields of the kinesin walk is possible, and under the "complete" theory approach, attention is directed to correlating predicted effects of the disruption with what is observed.

Several possibilities exist for TTFields interference with MT polymerization. Direct action of TTFields on tubulin dimers at unamplified field strength, while suspected in early MoA proposals, is ruled out as described in Sect. 1.1. However, due to capacitive effects when TTFields are orthogonal to MTs, field strength may be amplified at the MT+ end where polymerization occurs. Second, Santelices et al. showed that MT conductivity is orders of magnitude greater than that of the ambient cytosol [18]. This quality alone implies that MTs will act as electrical shunts of TTFields-induced currents within the cell, which would have field amplification effects. Further, the conductive MT layer is likely the counter-ion layer attracted to the C-termini charges, and the mechanism of its conduction is uncertain [19]. If counter-ion resistance is non-Ohmic, e.g., not proportionate to MT length, possibly lossless, then as MT length increases in early mitosis, it would pick up an increasing field gradient, and if the resulting voltage, current, energy, or power exceeded a disruption threshold, e.g., for polymerization, that could explain why TTFields affect mitotic but not interphase cells (Sect. 2.1). This hypothesis is currently under investigation.

3.3 Septin Effects

As mentioned, Gera et al. showed decreased septin concentration at the cell midline during mitosis, specifically as the cell enters anaphase, when errors in mitotic spindle formation become uncorrectable and ineluctably result in aberrant mitotic exit and/or programmed cell death [5]. Septin is, in fact, an ideal candidate for TTFields effects since septin family members self-assemble into various structures—hexameric and octameric quaternary structures, filaments, rings, and gauzes—that form rapidly and are precisely aligned or orthogonal to the cell axis during various mitotic phases, notably including prophase (Fig. 6; Sects. 2.4 and 2.5) [20–22]. Gera et al. estimate the dipole moment of the septin 2-6-7 complex at 2711 debye and note a higher value is possible when the complex is aggregated into higher-order septin structures.

They suggest that TTFields' effects on septin correlate with results showing that septin depletion leads to impairment of its cytoskeletal role to prevent improper

Cell Cycle Phase	Septin Component & Alignment	Schematic
Interphase	Single fibers are present with no alignment	
Prophase	Double octamer filaments begin to align in cell furrow along mother-daughter axis	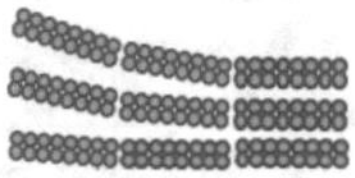
Metaphase, anaphase	Single fibers are added orthogonal to mother-daughter axis cross-linking the double filaments at regular intervals	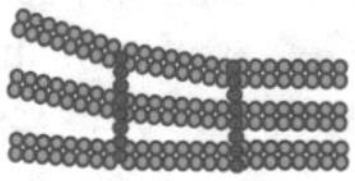
Telophase	Double filaments disassemble	
Cytokinesis	Double filaments re-align with single fibers into a circumferential double-ring structure	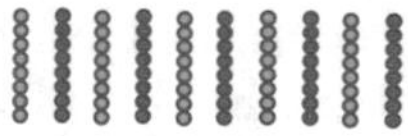

Septin octamer	ᴏᴏᴏᴏᴏᴏ	Single filament	ᴏᴏᴏᴏᴏᴏ

Fig. 6 Septin in the cell cycle (After Ong et al., [21]). For cell division to succeed, members of the cytoskeletal protein septin family are required to precisely assemble into various structures that are either aligned with or orthogonal to the cell axis. Some septins are large polarizable molecules, e.g., septin complex 2-6-7, which Gera et al. estimated its dipole moment at 2711 debye, and perhaps greater in higher-order structures [5]. Since TTFields can affect large polarized molecules and have their strongest effects when aligned or orthogonal to the cell axis, septins are a possible TTFields mechanism target.

membrane shape, e.g., to stabilize aberrantly protruding membranes ("blebbing") against internal hydrostatic pressure as the cell changes shape drastically during later mitosis [20, 23], and which they and others have observed in TTFields-affected cells [24]. Note, though, that Gilden et al. found that septin rapidly accumulates in conjunction with a second cytoskeletal protein, actin, at the site of the blebs, and so depletion of septin at the cell midline has no obvious connection with accumulation of septin at the cell membrane, while it could be tied to effects observed at the cell cleavage furrow [25]. Another possibility is that depletion of any septin at the midline could trigger septin depletion at other cell locations since septin depletion across family members is correlated, perhaps since septin complex formation is dependent on multiple septins [26]. Further, Gera et al. observed no septin depletion early in mitosis (cf. Sect. 2.4). In contrast, Estey et al. showed that depletion of septins 2 and 7 early in cytokinesis causes binucleation, which has been observed in TTFields-treated cells [26].

Gera et al. hypothesize that (1) TTFields torque the septin 2-6-7 complex and interfere with its self-assembly into a lattice as the cell enters anaphase; (2) anillin is then unable to recruit the septin complex to perform its role in preserving the structural integrity of the cell as it elongates toward cytokinesis, and (3) membrane blebbing triggers cell death via the immune system. While they speculate that p53-induced extrinsic, immune-driven apoptosis may be the ultimate cause of cell death and found evidence of p53-dependent $G_{0/1}$ cell cycle block, they were unable to find greater expression of p53 in TTFields-treated cells. Further, p53 has a complex role in cell signaling, cell cycle checkpoints, and apoptosis [27], and non-p53-dependent apoptotic pathways are possible, as is p53 involvement in intrinsic, mitochondrial apoptosis versus extrinsic apoptosis (Sect. 3.4) [28, 29]. While Gera et al. conclude that the role of p53 in TTFields MoA is unclear, their focus on septin and their results merit further investigation.

3.4 Is Intrinsic Apoptosis the Key Signaling Pathway Triggered by TTFields?

Wong et al. hypothesized that TTFields may induce an immune system response [24] and noted that the MoA of dexamethasone (dex), routinely prescribed to reduce inflammation and edema in brain cancer patients, is steroidal-driven reduction of immune system function [11]. They created two cohorts: one with dex dose reduced to <4.1 mg/day, and result was a significant extension of overall survival among the low-dex cohort, perhaps supporting the hypothesis that the immune system is involved in TTFields MoA.

An alternate hypothesis is possible. First, note that immune system activity was not implicated in the original in vitro studies where TTFields were shown to have higher efficacy than in human trials [2, 3]. Second, immune response correlated with

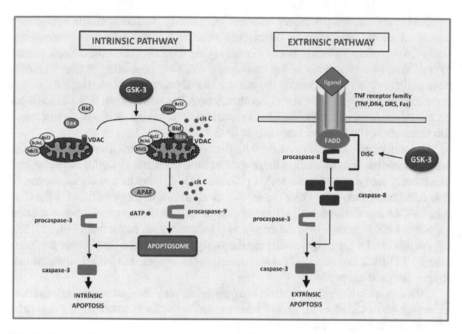

Fig. 7 Diagram showing the intrinsic, mitochondrial (left), and extrinsic, immune-related (right) apoptotic pathways. In the intrinsic pathway, the steroid dexamethasone (dex) is *anti-apoptotic*, while in the extrinsic pathway, dex is *pro-apoptotic*. Since dex interferes with TTFields efficacy, this suggests that intrinsic apoptosis is a key signaling pathway of its mechanism of action (MoA). Since dex also interferes with radiotherapy and it is known that intrinsic apoptosis is a main MoA, it supports the intrinsic apoptosis theory of TTFields MoA and suggests that TTFields act like radiotherapy and vinca alkaloid chemotherapy, but notably without their deleterious side effects. (Source, with permission via ResearchGate: [39])

better response in human patients that occurred months into treatment [24], which again was not a factor in the in vitro efficacy occurring in a few days.

Third, in reviewing the literature on dex and apoptosis, we found that dex has pleiotropic effects on two different apoptotic pathways. In some tissues, dex *enhances* immune system, p53-driven ("extrinsic") apoptosis [30, 31], while in others, dex *inhibits* Bcl-2/BAX-regulated ("intrinsic" or "mitochondrial") apoptosis (Fig. 7) [32, 33]. Fourth, in fact, dex has been shown to inhibit intrinsic apoptosis in glioblastoma and is suspected to interfere with apoptosis in chemo- and radiation therapy [34–37]. Fifth, a principal MoA of radiotherapy, which induces DNA damage in target cells, is *intrinsic* apoptosis [38]. Thus, it seems likely that dex interferes with TTFields and radiotherapy for the same reason: inhibition of intrinsic apoptosis.

We hypothesize that TTFields trigger the intrinsic apoptotic pathway as a key component of its MoA and that the MoA of TTFields, vinca alkaloids, temozolomide, and radiotherapy overlap. It is notable that TTFields produce no side effects while chemo- and radiotherapy have serious deleterious effects on healthy, non-tumor cells and decrease quality of life for patients.

The intrinsic apoptosis theory does not necessarily conflict with the hypotheses described above, which focus on electric field effects on subcellular structures. Rather, a complete TTFields theory requires tying field effects on subcellular targets to a downstream triggering of intrinsic apoptosis. One possibility is that TTFields have a direct or indirect effect on the mitochondrial outer membrane (MOM), which regulates intrinsic apoptosis via its ion channels. It has been shown that TTFields act on the ion channels directly, similar to electric field modulation of axon and neuron ion channels at much lower frequencies (0–30 kHz) [40].

An indirect effect is also possible. Studies have shown that free tubulin has the correct conformation to block voltage-gated anion channels (VDAC) in MOM and that tubulin subtypes modulate MOM potential [41–43]. An excess concentration of free tubulin, as shown by Giladi et al. [4], could result in modulation of VDAC or other MOM ion channels and affect regulation of intrinsic apoptosis. Since tubulin regulates MOM channels via the degree to which subtypes block them [44, 45], this hypothesis can be tested by identifying the subtype(s) rendered free versus polymerized by TTFields and seeing if the same subtype(s) regulate the MOM channels that trigger intrinsic apoptosis.

Cells in which intrinsic apoptosis is triggered display similarities to the empirical observations of Giladi et al. of TTFields-treated cells, for instance, that many cells exit mitosis and die in interphase [4, 28].

The intrinsic apoptosis theory predicts that decreasing the Bcl-2/BAX ratio would enhance TTFields' efficacy as it does for radiotherapy and, concomitantly, increased Bcl-2/BAX ratio would explain TTFields resistance as it does for radiotherapy, notably in glioma cells [46–48].

Thus, the intrinsic apoptosis hypothesis has the appeal of possibly meeting the requirements of a "complete" TTFields theory (Sect. 1.2) tying together electromagnetic effects on intracellular structures, cell signaling pathways, the observed empirical effects of TTFields, and predictions leading to match clinical efficacy with the high in vitro and animal model in vivo efficacy.

4 Conclusion

Since 2004, a significant body of evidence has accumulated from which increasingly specific theories of how TTFields work have been proposed. Such theories must ultimately tie together the effects of electric fields on subcellular structures, cell-signaling pathways, empirically observed effects on cells, and predictions leading to clinical efficacy matching the high efficacy proven in vitro and in animal models. We have presented the leading candidates and highlighted the theory that TTFields trigger the intrinsic, mitochondrial-regulated, apoptotic pathway as a key component of their MoA. The intrinsic apoptosis theory suggests that TTFields' MoA is similar to that of vinca alkaloids and radiotherapy, but lacks their deleterious side effects, and predicts methods to enhance TTFields clinical efficacy.

Acknowledgements Eduard Fedorov, Novocure Ltd., helped with dimensionless modeling. Magnus Olsson and Kiran Uppalapati, COMSOL, offered techniques for modeling on a nanometer scale. We thank Cornelia Wenger for sharing her COMSOL model. Jeffrey E. Arle, BIDMC/ Harvard Medical School, introduced the lead author to tumor-treating fields. Partial funding was provided by Novocure Ltd.

References

1. Mun, E. J., et al. (2017). Tumor-treating fields: A fourth modality in cancer treatment. *Clinical Cancer Research, 24*, 266–275.
2. Kirson, E. D., et al. (2004). Disruption of cancer cell replication by alternating electric fields. *Cancer Research, 64*, 3288–3295.
3. Kirson, E. D., et al. (2007). Alternating electric fields arrest cell proliferation in animal tumor models and human brain tumors. *Proceedings of the National Academy of Sciences of the United States of America, 104*, 10152–10157.
4. Giladi, M., et al. (2015). Mitotic spindle disruption by alternating electric fields leads to improper chromosome segregation and mitotic catastrophe in cancer cells. *Scientific Reports, 5*, 18046.
5. Gera, N., et al. (2015). Tumor treating fields perturb the localization of septins and cause aberrant mitotic exit. *PLoS One, 10*, e0125269.
6. Tuszynski, J. A., et al. (2016). An overview of sub-cellular mechanisms involved in the action of TTFields. *International Journal of Environmental Research and Public Health, 13*(11), 1128.
7. Tuszynski, J. A. (2019). The bioelectric circuitry of the cell. In S. Makarov, M. Horner, & G. Noetscher (Eds.), *Brain and human body modeling: Computational human modeling at EMBC 2018* (pp. 195–208). Cham: Springer. https://doi.org/10.1007/978-3-030-21293-3_11.
8. Pierce, J. R. (1981). *Signals: The telephone and beyond*. San Francisco: W.H. Freeman.
9. Wenger, C., et al. (2015). Modeling Tumor Treating Fields (TTFields) application in single cells during metaphase and telophase. *Conference Proceedings: Annual International Conference of the IEEE Engineering in Medicine and Biology Society, 2015*, 6892–6895.
10. Berkelmann, L., et al. (2019). Tumour-treating fields (TTFields): Investigations on the mechanism of action by electromagnetic exposure of cells in telophase/cytokinesis. *Scientific Reports, 9*, 7362.
11. Wong, E. T., et al. (2015). Dexamethasone exerts profound immunologic interference on treatment efficacy for recurrent glioblastoma. *British Journal of Cancer, 113*, 232–241.
12. Giladi, M., et al. (2014). Alternating electric fields (tumor-treating fields therapy) can improve chemotherapy treatment efficacy in non-small cell lung cancer both in vitro and in vivo. *Seminars in Oncology, 41*(Suppl 6), S35–S41.
13. Zhao, Y., & Zhang, G. (2018). *Elucidating the mechanism of 200 kHz tumor treating fields with a modified DEP theory*. IEEE International Symposium on Signal Processing and Information Technology (ISSPIT).
14. Palti, Y. (2011). *Optimizing characteristics of an electric field to increase the Field's effect on proliferating cells*. US patent 7,917,227 B2 US Application 11/537,026.
15. Stupp, R., et al. (2012). NovoTTF-100A versus physician's choice chemotherapy in recurrent glioblastoma: A randomised phase III trial of a novel treatment modality. *European Journal of Cancer, 48*, 2192–2202.
16. Kirson, E. D., et al. (2009). Chemotherapeutic treatment efficacy and sensitivity are increased by adjuvant alternating electric fields (TTFields). *BMC Medical Physics, 9*, 1.
17. Sozanski, K., et al. (2015). Small crowders slow down kinesin-1 stepping by hindering motor domain diffusion. *Physical Review Letters, 115*, 218102.

18. Santelices, I. B., et al. (2017). Response to alternating electric fields of tubulin dimers and microtubule ensembles in electrolytic solutions. *Scientific Reports, 7*, 9594.
19. Priel, A., et al. (2005). Transitions in microtubule C-termini conformations as a possible dendritic signaling phenomenon. *European Biophysics Journal, 35*, 40–52.
20. Gilden, J. K., et al. (2012). The septin cytoskeleton facilitates membrane retraction during motility and blebbing. *The Journal of Cell Biology, 196*, 103–114.
21. Ong, K., et al. (2014). Architecture and dynamic remodelling of the septin cytoskeleton during the cell cycle. *Nature Communications, 5*, 5698.
22. Ong, K., et al. (2016). Visualization of in vivo septin ultrastructures by platinum replica electron microscopy. *Methods in Cell Biology, 136*, 73–97.
23. Stewart, M. P., et al. (2011). Hydrostatic pressure and the actomyosin cortex drive mitotic cell rounding. *Nature, 469*, 226–230.
24. Swanson, K. D., et al. (2016). An overview of alternating electric fields therapy (NovoTTF therapy) for the treatment of malignant glioma. *Current Neurology and Neuroscience Reports, 16*, 8.
25. Normand, G., & King, R. W. (2010). Understanding cytokinesis failure. *Advances in Experimental Medicine and Biology, 676*, 27–55.
26. Estey, M. P., et al. (2010). Distinct roles of septins in cytokinesis: SEPT9 mediates midbody abscission. *The Journal of Cell Biology, 191*, 741–749.
27. Ozaki, T., & Nakagawara, A. (2011). Role of p53 in cell death and human cancers. *Cancers (Basel), 3*, 994–1013.
28. Varmark, H., et al. (2009). DNA damage-induced cell death is enhanced by progression through mitosis. *Cell Cycle, 8*, 2951–2963.
29. Luna-Vargas, M. P., & Chipuk, J. E. (2016). The deadly landscape of pro-apoptotic BCL-2 proteins in the outer mitochondrial membrane. *The FEBS Journal, 283*, 2676–2689.
30. Price, L. C., et al. (2015). Dexamethasone induces apoptosis in pulmonary arterial smooth muscle cells. *Respiratory Research, 16*, 114.
31. Li, H., et al. (2012). Glucocorticoid receptor and sequential P53 activation by dexamethasone mediates apoptosis and cell cycle arrest of osteoblastic MC3T3-E1 cells. *PLoS One, 7*, e37030.
32. Tsai, H. C., et al. (2015). Dexamethasone inhibits brain apoptosis in mice with eosinophilic meningitis caused by *Angiostrongylus cantonensis* infection. *Parasites & Vectors, 8*, 200.
33. Lee, I. N., et al. (2015). Dexamethasone reduces brain cell apoptosis and inhibits inflammatory response in rats with intracerebral hemorrhage. *Journal of Neuroscience Research, 93*, 178–188.
34. Das, A., et al. (2004). Dexamethasone protected human glioblastoma U87MG cells from temozolomide induced apoptosis by maintaining Bax:Bcl-2 ratio and preventing proteolytic activities. *Molecular Cancer, 3*, 36.
35. Das, A., et al. (2008). Modulatory effects of acetazolomide and dexamethasone on temozolomide-mediated apoptosis in human glioblastoma T98G and U87MG cells. *Cancer Investigation, 26*, 352–358.
36. Sur, P., et al. (2005). Dexamethasone decreases temozolomide-induced apoptosis in human glioblastoma T98G cells. *Glia, 50*, 160–167.
37. Pitter, K. L., et al. (2016). Corticosteroids compromise survival in glioblastoma. *Brain, 139*, 1458–1471.
38. Rupnow, B. A., et al. (1998). Direct evidence that apoptosis enhances tumor responses to fractionated radiotherapy. *Cancer Research, 58*, 1779–1784.
39. Gomez-Sintes, R., et al. (2011). GSK-3 mouse models to study neuronal apoptosis and neurodegeneration. *Frontiers in Molecular Neuroscience, 4*, 45.
40. Neuhaus, E., et al. (2019). Alternating electric fields (TTFields) activate Cav1.2 channels in human glioblastoma cells. *Cancers (Basel), 11*, 110.
41. Rostovtseva, T. K., & Bezrukov, S. M. (2012). VDAC inhibition by tubulin and its physiological implications. *Biochimica et Biophysica Acta, 1818*, 1526–1535.

42. Maldonado, E. N., et al. (2010). Free tubulin modulates mitochondrial membrane potential in cancer cells. *Cancer Research, 70*, 10192–10201.
43. Carre, M., et al. (2002). Tubulin is an inherent component of mitochondrial membranes that interacts with the voltage-dependent anion channel. *The Journal of Biological Chemistry, 277*, 33664–33669.
44. Sheldon, K. L., et al. (2015). Tubulin tail sequences and post-translational modifications regulate closure of mitochondrial voltage-dependent anion channel (VDAC). *The Journal of Biological Chemistry, 290*, 26784–26789.
45. Rostovtseva, T. K., et al. (2018). Sequence diversity of tubulin isotypess in regulation of the mitochondrial voltage-dependent anion channel. *The Journal of Biological Chemistry, 293*, 10949–10962.
46. Azimian, H., et al. (2018). Bax/Bcl-2 expression ratio in prediction of response to breast cancer radiotherapy. *Iranian Journal of Basic Medical Sciences, 21*, 325–332.
47. Sun, D., et al. (2018). MicroRNA-153-3p enhances cell radiosensitivity by targeting BCL2 in human glioma. *Biological Research, 51*, 56.
48. Liu, J. J., et al. (2005). Expression of survivin and bax/bcl-2 in peroxisome proliferator activated receptor-gamma ligands induces apoptosis on human myeloid leukemia cells in vitro. *Annals of Oncology, 16*, 455–459.

A Thermal Study of Tumor-Treating Fields for Glioblastoma Therapy

Nichal Gentilal, Ricardo Salvador, and Pedro Cavaleiro Miranda

1 Introduction

1.1 Electromagnetic Radiation and Matter

The interaction between electromagnetic (EM) radiation and biological tissues has been a subject of study for a long time. It is known that the effects of these interactions depend on factors such as intensity, frequency, and duration of the field. Due to the wide range of possible outcomes, EM radiation has been used in very different areas.

In medicine, techniques that work at a low frequency range (< 10 kHz) such as transcranial magnetic stimulation (TMS), transcranial alternating current stimulation (tACS) and transcranial/transcutaneous spinal direct current stimulation (tDCS/tsDCS) are employed in diagnosis and treatment of some neurological and psychiatric conditions. As the frequency increases, the main effect of these fields at the cellular level is no longer membrane depolarization, and thus stimulation does not occur.

As the wavelength gets shorter, the medical applications start to change. For instance, frequencies of a few MHz and higher are typically used to produce anatomical images of the human body in techniques such as magnetic resonance imaging (MRI) and microwave imaging (MWI). In other medical procedures, such as hyperthermia and ablation techniques, the main goal is not to diagnose but rather to kill tumoral cells by overheating them.

N. Gentilal (✉) · P. C. Miranda
Instituto de Biofísica e Engenharia Biomédica, Faculdade de Ciências da Universidade de Lisboa, Campo Grande, Lisboa, Portugal
e-mail: ngentilal@fc.ul.pt

R. Salvador
Neuroelectrics, Avinguda Tibidabo, 47 bis, Barcelona, Spain

© The Author(s) 2021
S. N. Makarov et al. (eds.), *Brain and Human Body Modeling 2020*,
https://doi.org/10.1007/978-3-030-45623-8_3

For frequencies higher than that of the visible light, EM radiation is strong enough to ionize cells and affect chemical bonds. This type of radiation finds its most valuable use in cancer treatment where techniques such as radiotherapy have been developed based on this type of interaction. Another common application of such type of radiation is in diagnosis (e.g., X-ray imaging, mammography, computed tomography (CT), etc.).

There is a frequency range, in the hundreds kHz region, that is too high to stimulate tissues but too low to cause significant temperature increases. Up until the beginning of the century, it was commonly believed that fields with this frequency did not have any biological effect on tissues, and thus, they were practically neglected for medical purposes. However, in 2004 the in vitro studies of Kirson et al. [1] revealed for the first time a potential application for this type of radiation: disruption of cancer cell replication. These findings led to a new cancer treatment technique named tumor-treating fields (TTFields).

1.2   Tumor-Treating Fields

TTFields consist in applying an alternating electric field (EF) with a frequency between 100 and 500 kHz. The mechanisms of action of this technique are still not fully understood. The first hypotheses suggested a two-stage action: during metaphase, these fields could disrupt mitotic spindle formation by acting on cells' highly polar structures, and during cytokinesis, they could affect correct cell division by leading to membrane blebbing [1, 2]. However, more recent studies presented some calculations that indicate that these hypotheses might not fully explain how TTFields affect the mitotic process [3], and new possible mechanisms have been suggested since [4, 5]. An example of the latter are the calculations done by Berkelmann et al. [5] that propose that the high energy deposited at the cleavage furrow might kill tumoral cells by overheating them during telophase and cytokinesis. Despite these uncertainties, the minimum EF intensity that should be induced at the tumor bed to achieve an antimitotic effect is well-established to be 1 V/cm. Furthermore, application of these fields in two perpendicular directions alternately also proved to increase the number of cells targeted [1].

TTFields are FDA-approved for the treatment of recurrent and newly diagnosed cases of glioblastoma multiforme (GBM), following the outcomes of the EF-11 [6] and EF-14 [7, 8] clinical trials, respectively. More recently, TTFields were also approved for the treatment of malignant pleural mesothelioma after the results from the STELLAR clinical trial [9] showed a treatment improvement when this technique was applied jointly with chemotherapy.

Post hoc analysis of the results from these trials allowed to identify some of the most significant factors that could affect treatment outcome. A detailed analysis on the relation between TTFields inhibitory effect and the intensity of the electric field at the tumor site was performed by Ballo et al. [10] using data from the EF-14 trial. For this study, the median overall survival (OS) and the progression-free survival

(PFS) were significantly higher for the patients in which the EF was greater at the tumor bed. This corroborates what was already seen in in vitro cell cultures in the study of Kirson et al. [1]. On the other hand, other studies investigated the correlation between treatment time and OS and PFS. Kanner et al. [11] showed that in patients with recurrent GBM, a daily compliance of at least 75% leads to significantly better treatment outcomes. In newly diagnosed GBM cases, where TTFields are applied in combination with temozolomide, this value drops down to 50% as discussed in [12]. The idea of an extended treatment time is further supported by the study of Giladi et al. [13] in which it was shown that slowly dividing cells are more likely to be affected by the electric fields the longer they are applied. Despite this clear need of maximizing patient's exposure time to TTFields, it is also important to address how this technique could affect biological tissues due to the temperature increases as a result of the Joule effect. The purpose of this work is to quantify these rises and predict the thermal impact using a realistic head model. Additionally, we suggest new possible ways to apply the fields based on the results we obtained.

1.3 The Optune Device

Optune (https://www.optune.com) is the device developed by Novocure (https://www.novocure.com) to treat GBM with TTFields. It is composed of an EF generator connected to four transducer arrays that work in pairs. The arrays are strategically placed on patient's shaved scalp to increase the electric field at the tumor using the NovoTAL System. This FDA-approved software uses MR images to create personalized treatment maps by adapting the configuration of the paired arrays. The importance of individualized planning was clearly shown in the work of Wenger et al. [14]. In this computational study, the maximum electric field at the tumor almost doubled when the arrays were specifically adapted for each patient. These conclusions were further corroborated by Korshoej et al. [15] in which the authors evaluated the best array positioning for different tumor locations.

Each array consists of a matrix of 3×3 interconnected transducers with temperature sensors. Each transducer is a ceramic disk with a diameter of 18 mm and a thickness of 1 mm. The Optune device works at 200 kHz and injects around 900 mA of current amplitude in two perpendicular directions alternately, with a switching time of 1 s. To allow current to flow from the capacitive transducers to the scalp, a thin layer of conductive hydrogel is used between them.

As noted in several works [2, 6–8] and summarized by Lacoutre et al. [16], the occurrence of dermatologic effects is the main side effect reported so far during TTFields therapy. These might be explained by chemical irritation in the skin due to the use of the hydrogel, moisture, mechanical trauma from shaving, or array removal, for example. To avoid burns, Optune controls the injected current to keep transducers' temperature below 41 °C.

2 Methods

2.1 The Realistic Human Head Model

A realistic human head model was used to perform the studies here presented. Creation of this model to study the electric field in the cortex during tCS is described in detail in [17]. Briefly, MR images of the Colin27 template with a resolution of $1 \times 1 \times 1$ mm^3 were retrieved from BrainWeb. Segmentation was performed using BrainSuite, and as a result five different surface meshes were created delimiting different tissues: scalp, skull, cerebrospinal fluid (CSF), gray matter (GM), and white matter (WM). Mimics was then used for some corrections and model improvement. The latter included representation of the superior orbital fissures and optical foramina and addition of the lateral ventricles that were considered to be filled with CSF.

The adaptation of this model for TTFields studies is described in [18]. Shortly, during model manipulation in Mimics, a virtual lesion was added in the right hemisphere of the brain, near the lateral ventricle, more or less at the same distance to the anterior and posterior regions of the head. This lesion intended to represent a GBM tumor, and it consisted in two concentric spheres: one that represented a necrotic core, with a diameter of 1.4 cm, and the other that represented an active shell, with a diameter of 2 cm. At this stage, the four transducer arrays were also added to the model. One pair was placed over the anterior and posterior regions of the head (AP configuration) and the other pair over the left and right temporal and parietal regions (LR configuration). Although this array positioning is not optimized for this virtual lesion, it represents a possible layout for a real patient. To fill in the gap between the transducers and the scalp, a layer of gel with variable thickness (0.7 mm median) was added to the model. After these modifications were made, a volume mesh was generated (Fig. 1), which consisted of around 2.3×10^6 tetrahedral elements with an average element quality of 0.50. This mesh was then imported to COMSOL Multiphysics to run the simulations.

2.2 Heat Transfer in TTFields: Relevant Mechanisms

Under normal physiological conditions, the human body core temperature is kept between 36 and 37.5 °C over a wide range of environmental circumstances [19]. In contrast, skin temperature can vary significantly as a function of the surroundings. In fact, heat flows from the most internal organs and tissues to the skin, where it is dissipated to the air [20].

The interplay between heat production and heat loss is a complex system that allows the body to regulate itself in order to maintain its activity. In this section, an overview of the most significant heat transfer mechanisms for a patient undergoing TTFields therapy is discussed. In this context, the main cooling mechanisms include conduction, convection, sweat, radiation, and blood perfusion. On the other hand,

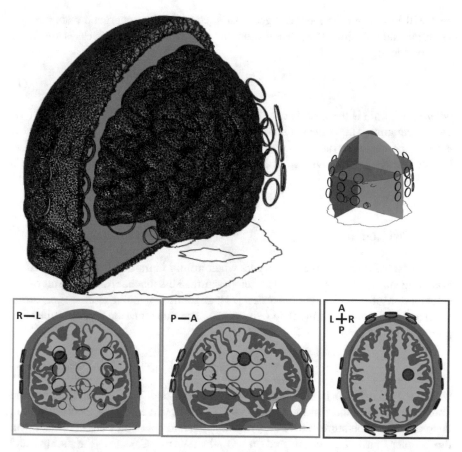

Fig. 1 Top: Partial representation of the mesh for the realistic human head model. Bottom: coronal (left), sagittal (middle), and axial (right) cuts through the center mass of the virtual lesion. Each color represents a different tissue: scalp (orange), skull (blue), CSF (yellow), gray matter (gray), white matter (green), tumor's outer shell (brown), and tumor's necrotic core (red). The eyes and the optical foramina are not colored

there are two main mechanisms that are responsible for increasing tissues' temperature: their metabolic activity and the Joule effect. As noted previously, the significance of the latter is a consequence of the high daily compliance needed for this therapy. The possible thermal harm is thus predicted by considering all these mechanisms and evaluating their relative contribution to the temperature distribution.

2.2.1 Conduction

Thermal conduction can be seen as a process that involves atomic and molecular motion [21]. When there is a temperature difference between two bodies in contact, heat will flow from the most energetic particles to the less energetic ones through vibration and collision. The rate equation that characterizes this process is known as Fourier's law:

$$F = -k\nabla T \tag{1}$$

where F (W/m^2) is the heat flux, k (W/(m$\cdot\,^{\circ}$C)) the thermal conductivity, and T ($^{\circ}$C) the temperature. This is the main process through which energy flows from the most internal tissues to the most superficial ones. For the head tissues we considered, the lowest thermal conductivity is scalp's (0.34 W/(m$\cdot\,^{\circ}$C)), and the highest is skull's (1.16 W/(m$\cdot\,^{\circ}$C)).

2.2.2 Convection

Heat transfer through convection occurs when a fluid in motion is in contact with a bounding surface that is at a different temperature. This mechanism is a result of the contribution of two different processes [21]: random molecular motion (i.e., diffusion) and bulk motion of the fluid (advection). The rate equation that describes this phenomenon is:

$$F = h(T_{\text{surface}} - T_{\text{fluid}}) \tag{2}$$

where h (W/(m$^2\cdot\,^{\circ}$C)) is the convection coefficient and T_{surface} and T_{fluid} are the bounding surface and fluid temperatures ($^{\circ}$C), respectively. This mechanism can decrease the temperature of the scalp, transducers, and gel as they are in contact with the cooler air. We considered $h = 4$ (W/(m$^2\cdot\,^{\circ}$C)) for the scalp and $h = 0.25$ (W/(m$^2\cdot\,^{\circ}$C)) for the medical tape covering the arrays. There are two other regions where convection can occur: between the blood and each perfused tissue, which is discussed below, and between the CSF and adjacent tissues (GM and skull). Heat exchange between the brain and CSF is still under investigation [22]. The very large surface contact between these two tissues plays an important role in maintaining brain's thermal safety as well as in providing structural and biochemical support [23]. Nonetheless, diffusion of molecules between the brain and the CSF is very restrained [24], and there is barely any study that describes the role of CSF advection in brain's temperature control. Further studies are also needed to characterize the importance of convection between skull and CSF.

2.2.3 Radiation

Heat transfer through radiation occurs whenever a body is at a nonzero temperature. For a given room temperature, the net radiation loss, in W/m^2, is given by:

$$F = \varepsilon\sigma\left(T_{surface}^4 - T_{room}^4\right) \tag{3}$$

where ε represents the surface emissivity (unitless), σ is the Stefan-Boltzmann constant (5.67×10^{-8}W/(m^2 · K^4)), and T are the temperatures in K. According to Wien's displacement law, loss of energy through radiation at body temperature falls within the infrared region. Radiation is significant for all surfaces exposed to the environment (scalp, gel, and transducers).

2.2.4 Sweat

In hot environments when ambient temperature is higher than the skin's temperature, sweat is the only way for the body to dissipate heat [20]. Under normal conditions, sweat rate is around 600–700 mL per day [19], though this value can change depending on the environment, physical activity, and physiological conditions.

An equation that could realistically model sweat losses has been sought for a long time. The first attempts could not satisfactorily predict these losses for a practical range of conditions, either environmental or considering all the remaining significant energy processes. Shapiro's equation [25] is typically considered to be the first mathematical expression to correctly model whole body sweat losses in response to exercise, environment, and clothing. In their study, 34 heat-acclimatized subjects walked in a treadmill in different environments and wearing distinct outfits. The data obtained allowed the authors to deduce the following equation for sweat losses:

$$\text{Sweating rate } \left(\text{W/m}^2\right) = 18.7 \times E_{req} \times \left(E_{max}\right)^{-0.455} \tag{4}$$

where E_{req} (W/m^2) is the required evaporative cooling necessary to maintain thermal balance and E_{max} (W/m^2) is the maximum evaporative capacity of the environment. Both these quantities are defined in [26]. For a cooling effect to occur, E_{max} should be higher than E_{req}, which is normally the case. In some situations, when air vapor pressure is too high, such as in humid conditions, an increase in the skin vapor pressure and wetted area is not enough to cool down the body, and thus heat storage occurs. Mathematically, E_{req} is defined as:

$$E_{req} \left(\text{W/m}^2\right) = [M_{net} + (R + C)]/1.8 \tag{5}$$

M_{net} (W) is the metabolic heat load given by the difference between body's total metabolic rate (M, in W) and the external work performed by the subject walking in a treadmill. The factor 1.8 represents the average surface of the human body (in m^2).

$$M_{net} \left(W\right) = M - (0.098 \, m_t \times v \times G) \tag{6}$$

In the previous equation, m_t is the mass of the subject (in kg), v the walking speed (in m/s), and G the grade (in %) of the treadmill during the experiments. The value 0.098 is a conversion factor (in m/s). The authors calculated the metabolic rate as:

$$M\ (W) = \ m_t\left[\left(2.7 + 3.2(v - 0.7)^{1.65}\right) + G(0.23 + 0.29(v - 0.7))\right] \quad (7)$$

At rest (i.e., when $v = 0$ m/s), M is assumed to be 105 W.

The environmental heat load due to radiation and convection $(R + C)$, in W, is defined as:

$$(R + C)\ (W) = \left(\frac{11.6}{clo}\right)(T_{air} - T_{skin}) \quad (8)$$

In the former expression, clo represents the total thermal resistance of subject's clothes, which was a factor of interest in this experiment. A value of zero would correspond to a naked person, whereas clo $= 1$ is the insulating value of clothing needed to maintain a person at rest in comfort in a room at 21 °C, with air movement of 0.1 m/s and humidity less than 50%. Typical values for this parameter and for different outfits can be found in [27].

On the other hand, E_{max} in Eq. (4) is given by:

$$E_{max}\left(W/m^2\right) = \frac{25.5}{1.8}\left(\frac{i_m}{clo}\right)(P_{skin} - P_{air}) \quad (9)$$

P_{skin} and P_{air} are the vapor pressure in the skin and air (in mmHg), respectively. Similarly to the clo factor, the permeability index of clothing, i_m, also depends on the subject's clothes [26]. A value of zero means that the fabric is impermeable, whereas a value of 1 means that all the moisture passes through the material.

The equation for sweat deduced by Shapiro and colleagues is valid for ambient temperatures between 20 and 54 °C and a relative humidity between 10% and 90%. Furthermore, E_{req} is limited to a range of values between 50 and 360 W/m^2, while E_{max} range of validity is between 20 and 525 W/m^2. This equation was largely used during more than 20 years to predict sweat losses. However, a more recent study [28] showed that it might be overestimating the predicted values. Additional experimental tests were made for a wider range of conditions, and a correction equation was deduced by Gonzalez et al. [28]:

$$\text{Corrected Shapiro's equation }\left(g/\left(m^2 \cdot h\right)\right) = 147\ \exp\ (0.0012 \times OSE) \quad (10)$$

where h means hours and OSE stands for original Shapiro equation (in g/(m^2 · h)). To convert the result to W/m^2, it is necessary to multiply the value obtained from the previous equation by water's latent heat (2410 J/g at 37 °C) and by 1/3600 h/s. This correction proved to increase the accuracy of the results by 58%. Additionally, a new sweat equation was also purposed:

$$\text{Gonzalez's sweating equation } \left(\text{g}/\left(\text{m}^2 \cdot \text{h}\right)\right)$$
$$= 147 + 1.527 \times E_{\text{req}} - 0.87 \times E_{\text{max}} \tag{11}$$

The last equation proved to be even more accurate (65%) in predicting sweat losses. The last two equations are valid for ambient temperatures between 15 and 46 °C, water vapor pressure between 0.27 and 4.45 kPa, and wind speed of 0.4–2.5 m/s.

Occurrence of sweat in TTFields patients is not an atypical situation [16]. It is observed when the environment is hot and humid or when the patient wears a wig to cover the arrays. Sweat represents an additional way to cool down the head. Although the number of eccrine glands varies depending on the region of the scalp, it is commonly believed that the forehead has a higher density of glands [29]. Thus, this is also the region that might cool down the most when sweating occurs. However, due to other aspects discussed further ahead, not all sweat is directly transferred to the environment by evaporation.

2.2.5 Metabolism

Metabolism can be defined as the necessary chemical reactions that occur in cells for the body to maintain its function and respond to external stimuli. As a result of this process, heat is released. The sum of the basal metabolic rate of cells with the rate from physical exercise and extra chemical reactions results in the final metabolic rate of the body [19]. For head tissues in particular, this mechanism can contribute significantly to temperature increases due to the brain's activity, which is a highly metabolic demanding organ. It accounts for 25% of the body's total glucose usage and around 20% of the total oxygen consumption [23]. Not surprisingly, in our model, the GM had the highest metabolic rate (16,229 W/m^3), whereas the CSF had the lowest (0 W/m^3).

2.2.6 Blood Perfusion

When body's temperature is too high, vasodilation of blood vessels is one of the main mechanisms to dissipate heat along with sweating and a decrease in the heat production [19]. For instance, vasodilation can increase heat exchange through the skin by a factor of 8 [19]. As previously mentioned, the physical process behind this cooling mechanism is convection between the blood and perfused tissues. A mathematical expression was deduced by Pennes [30] to express blood perfusion for each tissue:

$$\text{Blood perfusion } \left(W/m^3\right) = \omega_b \rho_b c_b (T_b - T) \tag{12}$$

where ω is tissue's blood perfusion rate (s^{-1}), ρ is the density (kg/m^3), c the specific heat $(J/(kg \cdot {}^{\circ}C))$, and T the temperature $({}^{\circ}C)$. The subscript b stands for blood. As the brain is the organ that needs more energy, it is also the one which has the highest blood perfusion. We considered GM and WM's perfusion rate to be $0.0133 \ s^{-1}$ and $0.0037 \ s^{-1}$, respectively.

2.2.7 Joule Heating

Joule heating, also known as Ohmic heating, occurs when there is a potential difference between two regions of a conductive medium. Electrons collide with the surroundings and energy transfer in the form of heat occurs. Mathematically, this term can be expressed as:

$$\text{Joule heating} \ (W/m^3) = \boldsymbol{J} \cdot \boldsymbol{E} = \sigma ||\boldsymbol{E}||^2 \tag{13}$$

where $\boldsymbol{J}$ is the current density vector (A/m^2), $\boldsymbol{E}$ is the electric field vector (V/m), and σ is the electric conductivity (S/m). In TTFields the need to maximize the EF intensity at the tumor bed and the time that the patient is subjected to the field lead inevitably to temperature increases which may have a significant impact on tissues.

2.3 Heat Transfer in TTFields: Pennes' Equation

Pennes' equation [30] was used to obtain the temperature distribution as a function of space and time. This equation was derived based on experimental measurements of the temperature on the human forearm, and it is defined as:

$$\rho c \frac{\partial T}{\partial t} = \underbrace{\nabla \cdot (k \nabla T)}_{\text{Conduction}} + \underbrace{\omega_b \rho_b c_b (T_b - T)}_{\text{Blood perfusion}} + \underbrace{Q_m}_{\text{Metabolic heat}} + \underbrace{\boldsymbol{J} \cdot \boldsymbol{E}}_{\text{Joule heating}} \tag{14}$$

Some of the heat transfer mechanisms previously discussed are already considered in this equation. The remaining heat processes (convection, radiation, and sweat) were considered at the boundaries. More specifically, we considered energy exchanges through radiation and convection to occur on the outer boundaries of the scalp, transducers, and gel.

In our model, the medical tape that covers the arrays is not represented. In a real patient, this adhesive covers a large part of the scalp, and thus it changes the rate at which energy is exchanged with the environment. To minimize the error associated with this lack of representation, we chose emissivity and convection factor values for the transducers and the gel that were close to that of the medical tape. As previously mentioned, TTFields patients might sweat to reduce the temperature increases that occur on the scalp, but some of this energy might not be transferred to the environment. The fabric of the medical tape can retain some of the sweat, and consequently changes in the hydrogel composition could occur. Due to all these uncertainties and the fact that there is no equation that predicts cooling by sweating just for the head, we opted to use a constant value based on Gonzalez's sweat equation for the whole scalp (125 W/m^2) except where covered by the gel and transducers.

2.4 Simulations' Conditions

The frequency used in TTFields therapy, 200 kHz, falls within the frequency range where the electromagnetic wavelength is significantly larger than the size of the human head which allowed us to use the electroquasistatic approximation of Maxwell's equations. Under this approximation, the electric potential could be calculated using Laplace's equation. The AC/DC module (electric currents, frequency domain study) was used in COMSOL Multiphysics to obtain the electric field distribution when the AP configuration was on. The potential difference between the anterior and posterior arrays (V_{AP}) was defined in such a way that a current with an amplitude of 900 mA passed between the two arrays. In our model, this corresponded to having $V_{AP} = 91.8$ V. This study took around 8 h in a workstation with 2×4 core CPU's (Intel Xeon W5580 @ 3.2 GHz) and 48 GB RAM. Solver's relative tolerance was set to 10^{-6}. At the end of this simulation, we obtained a total power density spatial map that represented the contribution of this configuration to the Joule effect.

The Heat Transfer Module (electromagnetic heating, Joule heating, frequency-transient study) was then used to calculate the electric field distribution when the LR configuration was applied and also to predict the temperature increases. Similarly to the previous study, a fixed potential difference was defined between the left and right arrays ($V_{LR} = 68.2$ V) to ensure that 900 mA were injected.

Regarding the Joule heating term, the last term of Eq. (14), the contribution from the AP and LR configurations was taken alternately every second after both studies were coupled. Each simulation of 360 s took around 22 h in the aforementioned workstation. We also assumed that all the physical properties were isotropic and uniform. The values assigned to each tissue and material were chosen after an extensive literature review. Additional details about these values and how the studies were performed can be found in [31].

3 Results

3.1 Duty Cycle and Effective Electric Field at the Tumor

In our first work [31], we predicted the thermal impact of TTFields and quantified the duty cycle of Optune. We defined the latter as the time that each configuration is on

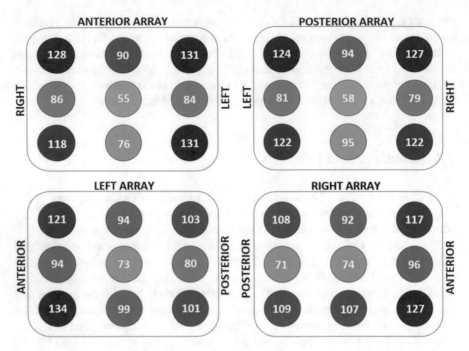

Fig. 2 Absolute current injected in each transducer. Each array can be seen as an isopotential surface where the edge effects are evident as more current is injected in the outer transducers. All values are in mA

divided by the maximum time it could be delivering the fields and assumed that there was a complete current shutdown whenever the average temperature of any transducer reached a cutoff temperature. We considered this critical value to be 40.4 °C because it corresponded to having around 5% of the volume of the transducer at 41 °C, which we assumed to be the volume occupied by the temperature sensor. In this work, we did not consider heat losses through sweat. Our simulations showed that there was one transducer whose temperature reached the threshold temperature quicker than the other 35 (the Most Significant Transducer or MST). Consequently, it was the one that controlled if current was injected or not. In our head model in particular, it was the transducer located in the anterior array, superior row, left side (ASL). Not surprisingly, it was also one of the transducers where the most current was injected (around 131 mA, see Fig. 2). Each array can be seen an isopotential surface in which more current is injected through the outer transducers, a phenomenon commonly known as the edge effect [32].

According to our results, Optune's way of operating can be summarized as follows: at the beginning of treatment, current is injected in two perpendicular directions alternately. Transducers' temperature starts to increase at different rates and biological tissues' temperature also rises. After around two and a half minutes, the temperature of one transducer (the MST) exceeds the safety threshold of 40.4 °C, and the fields are completely shut down in both directions. It then takes around 2–4 s

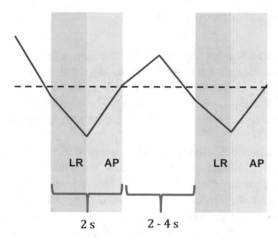

Fig. 3 Temperature variation of the Most Significant Transducer during intermittent operation. The dashed line represents the critical temperature (40.4 °C). After each shutdown, it takes around 2–4 s for the temperature of the transducer to decrease below the 40.4 °C. After current starts to be applied again, it takes about 2 s to surpass once again the threshold. These heating and cooling rates lead to different duty cycles for each configuration

to decrease the temperature of the MST to values below the critical limit and consequently to start injecting current again. After 2 s of alternate application of the fields, the temperature will once again surpass the threshold, and current shutdown will occur once more. This represents an intermittent operating mode of Optune (Fig. 3).

This led to different duty cycles for each configuration. In our model, current was injected in the AP direction 47% of the maximum time, while this value decreased to around 30% for the LR configuration. This means that, on average, Optune is used for GBM treatment around 39% of the time that the patient is using the device.

To know the effective electric field in the tumor, we started by calculating the EF magnitude induced by each configuration in this region. We defined ATV1 as the percentage of the tumor shell volume above the therapeutic threshold of 1 V/cm. As it can be seen in Fig. 1, the tumor is much closer to the right array and more or less at the same distance to the anterior and posterior arrays. This explains why the ATV1 value for the AP configuration is just 13%, while for the LR is 72%. Following this rationale, we can define the effective ATV1 (ATV1$_{\text{eff}}$) at the tumor shell as the combined contribution of the duty cycle and the induced electric field as:

$$\text{ATV1}_{\text{eff}}^{\text{std}} = \text{On}_{\text{AP}} \times \%\text{ATV1}_{\text{AP}} + \text{On}_{\text{LR}} \times \%\text{ATV1}_{\text{LR}}$$
$$= 0.47 \times 13\% + 0.30 \times 72\% = 27.7\%$$

where On$_x$ is the fraction of time that configuration x is on and ATV1$_x$ the percentage of the tumor volume above 1 V/cm when this configuration is applied.

3.2   Improving the Duty Cycle

There are two possible ways to increase the effective ATV1: by rising the electric field magnitude at the tumor shell or by increasing the duty cycle. We investigated these possibilities by performing different studies (studies 1–4) where we slightly changed how Optune worked. Here we discuss the operating modes that improved the results. Note that these results do not consider heat losses through sweat. For more details, see [31].

We first hypothesized that decreasing the amount of injected current by a factor of $1/\sqrt{2}$, but activating both arrays simultaneously, might allow to reduce the temperature maxima by reducing the Joule effect by half and thus increasing the time that the device is on (study 1). This operating mode affected both the electric field magnitude and direction and resulted in a duty cycle of 70%. However, the ATV1 decreased to around 45% of the tumor shell. Even so, this results in an effective ATV1 of 31.5%, which is more than the 27.7% observed in the standard configuration. Additionally, in this mode, current is injected only in one direction because the EF is a vector quantity, and thus the net field is oriented at approximately 45° relative to the standard AP and LR directions. However, it is possible to intentionally change the phase of the injected current so that more electric field directions are achieved. The latter approach might increase the number of affected cells and thus enhance treatment outcome [1].

We then changed the switching time between configurations. Currently, Optune alternates the direction of the applied field every second as this is the optimal time to affect cell division according to in vitro studies. We investigated the impact of increasing the switching time to 2 s (study 2). Our results showed that although this leads to higher temperatures, the cooling rate is also augmented because it depends on the temperature difference with the environment. As a consequence, the duty cycle improved for this case, reaching 47% and 40% for the AP and LR configurations, respectively. Thus, the effective ATV1 also increased to 34.9%. It is important to note that switching times different from 1 s might not be the most suitable to affect the mitotic process of GBM cells. For instance, a recent in vitro study of Berkelmann et al. [21] concluded that for the BT4Ca cell line, the optimal switching time that maximizes the antiproliferative effect of TTFields is 60 s.

In all the studies we performed, there was always one transducer that controlled if Optune was injecting current or not. Due to the cooling time and the location of the MST on the anterior array, the LR configuration was applied less time than the AP's. We hypothesized that if temperature control was made independently for each configuration, then the duty cycle would increase. To apply this feature, we considered that there could be two MST's, one for each configuration. In practice, this means that if a transducer of one configuration reaches the threshold temperature (40.4 °C), we would only shut down that specific configuration, while the other could continue to operate alternately until a transducer of that configuration reached the critical temperature (Study 3). In these conditions, the duty cycle for the AP

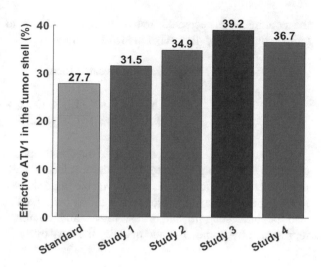

Fig. 4 Effective volume of the tumor shell above 1 V/cm. Study 1: less current injected, but all arrays were simultaneously activated. Study 2: switching time between configurations increased to 2 s. Study 3: temperature control for each configuration is made independently. Study 4: each transducer gets exactly 100 mA

configuration was 47%, while current was injected in the LR direction 46% of the time, leading to an ATV1 effective of 39.2%.

Lastly, we tried to increase the electric field in the tumor. We investigated what would happen if we avoided the edge effects by injecting exactly 100 mA in each transducer (Study 4). We noticed that this changed the MST and also that it improved the duty cycle (AP: 46%; LR: 42%), although the ATV1 at the tumor bed did not change significantly compared to the standard operation mode. In this case, the ATV1$^{std}_{eff}$ increased to 36.7%.

The effective ATV1 for each study is presented in Fig. 4.

Our results point out that the best way to improve TTFields therapy, among the hypotheses we tested, is by controlling the current into the array pairs independently but still alternately. This allows to compensate the fact that, in our model, there was always one transducer that controlled if the device was applying the fields or not. Including this feature in Optune might imply changing the device's hardware and software.

As the duty cycle increases, so does the Joule effect, which can lead to significant temperature increases compared to the standard case. In Sects. 3.4 and 3.5, we analyze temperature changes and make predictions about the thermal impact of this technique.

3.3 The Effect of Sweat

Before predicting the thermal impact, we started by studying the main differences in the temperature distributions and duty cycle if we considered heat losses through sweat. As already stated, we considered a constant value of 125 W/m^2 for the whole scalp except at the regions where the transducers were placed. In these conditions, the cooling time decreased, and it took around 2 s to cool down the MST to temperatures below 40.4 °C. This led to a duty cycle of around 50% for both the AP and LR configurations.

The effective ATV1 considering sweat is then:

$$ATV1_{eff}^{sweat} = On_{AP} \times \%ATV1_{AP} + On_{LR} \times \%ATV1_{LR}$$
$$= 0.50 \times 13\% + 0.50 \times 72\% = 42.5\%$$

Comparing the latter value with the ones presented in Fig. 4, it is possible to conclude that adding sweat to the model has led to the largest improvement in the predicted outcome for this therapy. Given the importance of this cooling mechanism, it would be important to improve and validate the model used. Additionally, it would also be interesting to add sweat losses and combine the different conditions represented by studies 1–4 and quantify the duty cycle.

3.4   Temperature Increases

The temperature distribution on each tissue surface for the standard operating mode and without considering sweat can be seen in Fig. 5. The temperature distribution on each tissue surface with this additional cooling mechanism is shown in Fig. 6.

The two figures clearly show localized temperature increases, mainly underneath the regions where the transducers were placed. In Fig. 5, temperature on the left and right sides is lower compared to the values presented in Fig. 6 because the duty cycle for the LR configuration is also lower. Even though the duty cycle was increased to nearly 50% for both configurations when sweat was considered, the maximum temperatures reached by each tissue were practically the same in both situations because current shutdown occurred at the same temperature. Given that there are electric field hotspots at tissue interfaces due to accumulation of charges [18], the Joule effect will also be more significant in these regions. Consequently, it is expected that the temperature maxima in each tissue occur on the surfaces. In Fig. 7 this superficial heating is well seen on the brain's surface. Despite the fact that both configurations are applied for the same time, the temperature increases are higher in the anterior-posterior direction as it is clearly shown in Fig. 6. This is a result of the array positioning in our model. The distance between the central transducers of the arrays pairs is around 20 cm for the AP configuration and 17 cm for the LR. Under these circumstances, the potential difference between each pair is higher for the first to ensure that 900 mA are injected (91.8 V vs. 68.2 V, both values are in amplitude), and thus the electric field and Joule heating will also be higher near this array pair. This explains why the MST is located in the anterior array. For other head model and another array layout, the most significant transducer might not be the same.

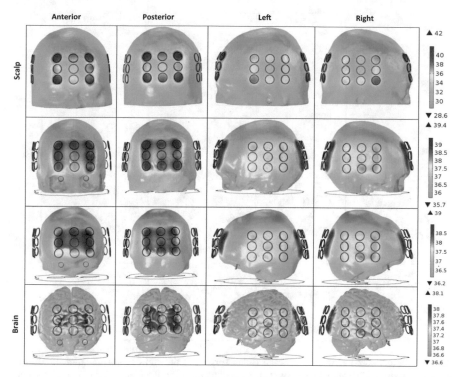

Fig. 5 Temperature distribution for Optune's standard operating mode without considering sweat and for each tissue surface at the end of the simulation (360 s). First row, scalp; second row, skull; third row, CSF; fourth row, the brain. Note the different scales for each row. All values are in °C

3.5 Prediction of the Thermal Impact

We used two different metrics to predict the thermal impact of TTFields: the maximum temperature reached by each tissue and the cumulative number of equivalent minutes at 43 °C (CEM 43 °C). The latter was defined by Sapareto and Dewey [33] as:

$$\text{CEM } 43^\circ\text{C} = \sum_{i=1}^{n} t_i R^{43-T_i} \tag{15}$$

In this equation, t_i(s) and T_i (°C) are the duration of the *i-th* time interval and temperature during that period, respectively, n is the total number of intervals, and R is a constant related with the thermotolerance acquired by the cells. It has a value of 0.5 for temperatures higher than 43 °C and 0.25 otherwise. Thresholds for thermal effects based on CEM 43 °C values are available in the literature [34, 35].

Given that the temperature increases are very localized (Figs. 6 and 7) and occur mainly underneath the arrays, we chose to calculate the CEM 43 °C for each tissue

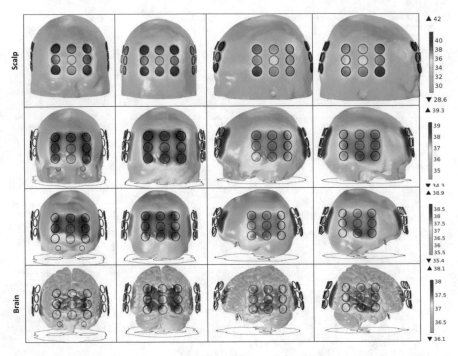

Fig. 6 Temperature distribution on each tissue surface at the end of the simulation (360 s) for Optune's standard operating mode when sweat is considered for the scalp. First row, scalp; second row, skull; third row, CSF; fourth row, the brain. Note the different scales for each row. All values are in °C

within a cylinder defined by the MST and for the results represented in Fig. 6. The basis of this cylinder has the area of the transducer, and the cylinder's axis is parallel to the transducer surface's normal. To make our predictions, we only considered thresholds for thermal impact for the five tissues that were represented in our model. Additionally, because we only simulated the first 6 min of therapy, we assumed that the temperature at the last time step would remain the same to calculate the CEM 43 °C for a typical treatment day (18 h). This approach is feasible because the temperature in each tissue had practically reached a steady-state distribution [31].

Our analysis shows that the CEM 43 °C thresholds for the scalp and skull were not surpassed, and thus thermal damage at histopathological and functional levels is not expected when the patient is subjected to the action of the fields. Regarding thermal safety for CSF, there are no thresholds defined in [34, 35] to compare our measurements to. We observed maximum increases of around 1.8 °C in this tissue. Given that CSF is in contact with the interstitial fluid (ISF) and that the latter helps to dissipate heat from the brain [36], the actual temperature increase might be even lower. For the brain, we obtained a local CEM 43 °C value of 0.30 min underneath the transducers assuming 18 h of treatment. Some studies [34, 35] reported that values higher than 0.03 min are enough to increase the permeability of the blood-brain barrier (BBB). Other possible effects include changes in the cerebral blood

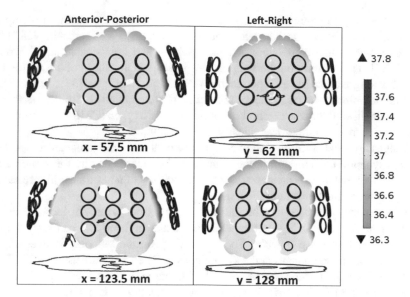

Fig. 7 Temperature distribution in the brain in four different slices under the arrays at $t = 360$ s when sweat is considered. Note that the temperature maxima occur at the surface under the transducers and that there is a quick drop to a constant value (37 °C). Values are in °C

flow (CBF) and variation in the concentration of some neurotransmitters. Regarding the last two, whether there is a rise or a decrease is not consensual among different studies. Some researchers observed a higher irrigation of cerebral tissue for CEM 43 °C values of 0.03 and 16 min, whereas other showed a decreased blood flow for values around 1 and 34 min. Neurotransmitter concentration was also seen to increase for values above 0.115 min, but when CEM 43 °C surpassed 1.29 min, there was a drop below the baseline concentration.

It is important to note, however, that there are a lot of uncertainties regarding the CEM 43 °C thresholds, and these conclusions might not hold when these limits are used to predict the thermal impact of tumor-treating fields. This metric was created with the intention of having a way to calculate a thermal isoeffect dose for cancer therapy. In other words, the rationale behind it is that it is possible to convert the time that a tissue is at a given temperature to an equivalent time at 43 °C. However, this is not necessarily accurate. As pointed out by Sapareto and Dewey [33], the heating regime is an important parameter as well as when the damage is assessed, tissue's pH, and the thermotolerance to heat of different cells. Not surprisingly, the lack of experimental protocols leads to different conclusions regarding variations of the same physiological process.

Due to all these uncertainties, it is first necessary to investigate if CEM 43 °C can be really applied to TTFields by measuring sensitive biomarkers to the changes here reported and confirm or disprove these hypotheses. In case that this metric proves to be a good measurement for thermal impact, specific thresholds for long and repetitive exposure times are needed. Most of the thresholds defined in the literature concern whole-body heating and were not specifically designed for the human head.

Another way to predict thermal impact is by evaluating the maximum temperature reached by each tissue. The temperature maxima (see Fig. 6) are still well below 45 °C, which is the temperature at which protein and lipid denaturation starts to occur [23]. Temperature on the scalp is also below the threshold of 44 °C for skin burn for long exposures, as reported by Moritz and Henriques [37]. In the brain, maximum temperature variations of around 2 °C were obtained. Studies performed in animals showed that the cerebral metabolic rate of oxygen ($CMRO_2$) [38, 39] and CBF [39] increase more or less linearly with the temperature.

Volgushev et al. [40] also showed that temperature variations can temporarily change synaptic transmission in the neocortex. Further studies by the same author [41] revealed that there is an increase of the probability of neurotransmitter release at the synapses as the temperature increases.

Whether these changes can be harmful or not is not clear. An increase on the $CMRO_2$ might lead to higher temperatures, but a higher CBF can also help to dissipate the heat more quickly. The uncertainty on the outcome is even higher when it comes up to changes at the synaptic level. The frequency used for TTFields treatment is too high to be capable of cell stimulation. Evidence from clinical trials [6, 8] shows no additional significant nervous system disorders when TTFields are used as a part of the standard of care which might indicate the absence or the low impact of these variations. As none of these effects were observed [6–8], it is necessary to understand why the theory does not match with the practice. The main explanations might be:

1. These changes do not occur during TTFields therapy: the conclusions drawn from the literature only apply to the conditions and technique for which those specific studies were performed. This implies that thresholds defined based on animal studies must be reviewed for humans.
2. These changes do occur but are negligible: given that the temperature variations are not very high, these physiological alterations are not meaningful, and thus they are not observed as side effects.
3. These changes occur but were not measured: the sensitive biomarkers for the changes here reported were not acquired.
4. These are long-term changes and are masked by patient's condition: these possible side effects might not occur on the first months of treatment, and only patients who survive longer manifest these alterations.
5. The high daily compliance leads to development of thermotolerance by the cells: given that the daily minimum recommended time for recurrent cases is 18 h and that treatment should be performed every day, cells might become less sensitive to temperature variations, and thus the thresholds should be higher in these conditions.
6. The predictions made are not correct: the model does not realistically represent an actual treatment, and consequently the temperature distribution is being overestimated.

The first step to exclude some of these hypotheses would be to acquire sensitive biomarkers during TTFields treatment. If these changes do occur, it is necessary to investigate further if they can be harmful or beneficial for the patient. On the other hand, if the predictions are not correct, further improvements should be made to our model to make it more reliable.

Regarding the tumor, the maximum temperature variation also occurred at the surface, although it was not higher than 0.1 °C. It is known that significant temperature increases can make cells more likely to be affected by radiation or some drugs, which explains why hyperthermia is sometimes used in cancer treatment. In our studies the maximum temperature reached by the tumor was 37.1 °C, which suggests that TTFields mechanisms of action are not expected to be temperature-related. However, a recent work by Berkelmann et al. [5] opened up new perspectives on the possible role of heat in this therapy. The authors predict a very high SAR value at the cleavage furrow during telophase and cytokinesis when the direction of the electric field is parallel to the longitudinal axis of the hourglass morphology of the dividing cells.

3.6   Continuous Versus Intermittent Application of the Fields

As seen in Sect. 3.1. due to the necessary thermal restrictions, Optune might work intermittently with duty cycle lower than 100%. In our next study, we investigated by how much it was necessary to decrease the injected current so that the average temperature of the MST would not exceed the shutdown temperature (40.4 °C). In other words, this means that if the injected current is below a critical value (the critical current or I_C), then it is possible to have an uninterrupted application of the fields and consequently a duty cycle of 100%. Figure 8 shows the average temperature of the MST as a function of the injected current for a 360 s simulation when sweat is considered. Current was applied alternately between the two configurations with a switching time of 1 s and when no current shutdown is considered. Each simulation took around 24 h in the aforementioned workstation. An analytical function (Eq. 16) was fitted to each curve represented in this figure.

$$T = C_1 + C_2 * (1 - \exp(-t/C_3)) \tag{16}$$

where C_1 (°C), C_2 (°C), and C_3 (s) are constants. The maximum average temperature (T_{max}, in °C) reached by this transducer can be predicted by calculating the limit of T when $t \to \infty$:

$$T_{max} = C_1 + C_2 \tag{17}$$

Table 1 summarizes C_1, C_2, and C_3 values, as well as T_{max}, as a function of the injected current.

The physical meaning of each parameter is as follows: C_1 values represent the initial average temperature of the transducer; C_2 can be seen as the maximum contribution of the injected current to temperature increases. C_3 is a time constant

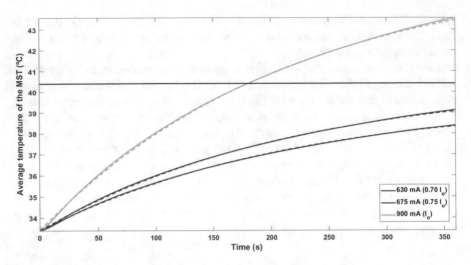

Fig. 8 Average temperature variation of the MST when no current shutdown is implemented. The dashed lines represent the best fit for each case based on the expression shown in Eq. (16)

Table 1 Curve fitting parameters as a function of the current injected into the MST

I (mA)	C_1(°C)	C_2(°C)	C_3 (s)	R^2	T_{max} (°C)
900 (I_0)	33.45	12.50	222.2	0.9998	45.95
675 (0.75I_0)	33.42	7.05	222.2	0.9998	40.47
630 (0.70I_0)	33.42	6.15	222.2	0.9998	39.57

The values of R-squared suggest a high correlation between current and the average temperature reached by the MST

related with the heat transfer with the environment and the injected current. When $t = C_3$, the temperature of the MST reached around 63.2% of its maximum temperature. Figure 8 also shows in dashed lines the best curve fitting based on the parameters of Table 1 for each amount of injected current. Note that the fitted equation might underestimate the temperatures for long periods of time.

According to our calculations, the critical current is 674 mA which represents a decrease of around 25% compared to the current that is typically injected in a real patient. Naturally, this decrease is followed by a lower induced electric field in tissues and more specifically in the tumor. The ATV1 for each configuration in the tumor shell is given in Table 2.

The ATV1 values for the AP configuration reduce to 0% for current values near I_C, which, as mentioned previously, has to do with tumor's location. In terms of treatment efficacy, decreasing the current to achieve a duty cycle of 100% leads to an effective ATV1 of:

$$\text{ATV1}_{\text{eff}}^{\text{No\_shut}} = 0\% \times 1.00 + 34 \times 1.00 = 34.0\%$$

Table 2 Percentage of the tumor shell volume above 1 V/cm as a function of the injected current

I (mA)	$ATV1_{AP}(\%)$	$ATV1_{LR}(\%)$
900 (I_0)	13	72
675 (0.75I_0)	0	39
674 (I_C)	0	34
630 (0.70I_0)	0	31

The latter value is still lower than the 42.5% obtained when current shutdown was implemented, which is mainly influenced by the AP configuration not producing a significant electric field (i.e., higher than 1 V/cm) at the tumor bed. Nonetheless, injecting less current to avoid the need of shutting down the fields might be a good solution to increase the effective ATV1 for more superficial tumors.

As expected, temperature maxima in the deepest tissues decreased for this new current injection mode because the power density dissipated in tissues was 44% lower compared to what was deposited when 900 mA were applied. The temperature distribution considering heat losses through sweat when 674 mA are injected continuously is shown in Fig. 9. Regardless of the injected current, the spatial distribution of the temperature on tissues' surface remains practically the same (compare Figs. 6 and 9). Despite the small decreases in CEM 43 °C and in the maximum temperature, the same physiological changes predicted in Sect. 3.5 apply to the results here presented.

4 Limitations and Future Work

One of the major drawbacks when modelling tumor-treating fields is the lack of a metric that allows to simultaneously evaluate the electric field, the duty cycle, and the temperature increases and to weigh the different parameters to find, e.g., the best array placement. Treatment efficiency is commonly quantified by metrics such as the ATV1 that allow to study if the induced EF is higher than the minimum therapeutic value or not. More recently, the power density has been used as a measurement of both the electric field and temperature increases [10]. In the future, it would be useful to develop and validate a new metric that combines all the features mentioned above.

The results of our simulations predict some effects that were not reported elsewhere. Acquiring the necessary biomarkers sensitive to the physiological changes here described would allow to corroborate or disprove our conclusions. If our model turns out to be inaccurate to represent a real patient and the conclusions derived from it are not reliable, then fine-tuning the most critical parameters is necessary. The latter includes having a more realistic equation to calculate sweat losses and perform a sensitivity analysis to evaluate the impact of the uncertainty of the electric and thermal parameters of tissues in the temperature distribution. Representation of the medical tape that covers the transducers, gel, and a large surface of the scalp would also make our model more realistic. Additionally, performing this

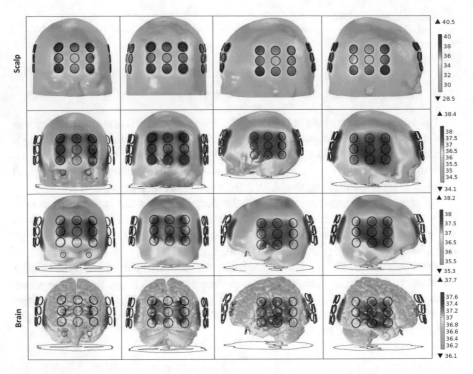

Fig. 9 Temperature distribution at the end of the simulation (360 s) on each tissue surface when current is decreased to 674 mA to avoid the need of shutting down the fields and considering sweat. First row, scalp; second row, skull; third row, CSF; fourth row, the brain. Note the different scales for each row. All values are in °C

type of thermal studies in different head models and doing an experimental validation using phantoms would be useful to support our work.

Acknowledgements Instituto de Biofísica e Engenharia Biomédica is supported by Fundação para a Ciência e Tecnologia (FCT), Portugal, under grant n° UID/BIO/00645/2013.

Conflict of Interests Pedro Cavaleiro Miranda has a research agreement funded by Novocure. Nichal Gentilal holds a PhD grant from the research agreement with Novocure.

References

1. Kirson, E. D., et al. (2004). Disruption of cancer cell replication by alternating electric fields. *Cancer Research, 64*, 3288–3295.
2. Kirson, E. D., et al. (2007). Alternating electric fields arrest cell proliferation in animal tumor models and human brain tumors. *Proceedings of the National Academy of Sciences of the United States of America, 104*, 10152–10157.
3. Tuszynski, J. A., et al. (2016). An overview of sub-cellular mechanisms involved in the action of TTFields. *International Journal of Environmental Research and Public Health, 13*, 1128.

4. Carlson, K. W., et al. (2019). Simulating the effect of 200 kHz AC electric fields on tumour cell structures to uncover the mechanism of a cancer therapy. In S. Makarov, M. Horner, & G. Noetscher (Eds.), *Brain and human body modeling*. Cham: Springer.

5. Berkelmann, L., et al. (2019). Tumour-treating fields (TTFields): Investigations on the mechanism of action by electromagnetic exposure of cells in telophase/cytokinesis. *Scientific Reports, 9*, 7362.

6. Stupp, R., et al. (2012). NovoTTF-100A versus physician's choice chemotherapy in recurrent glioblastoma: A randomised phase III trial of a novel treatment modality. *European Journal of Cancer, 48*, 2192–2202.

7. Stupp, R., et al. (2015). Maintenance therapy with tumor-treating fields plus temozolomide vs temozolomide alone for glioblastoma: A randomized clinical trial. *JAMA, 314*, 2535–2543.

8. Stupp, R., et al. (2017). Effect of tumor-treating fields plus maintenance temozolomide vs maintenance temozolomide alone on survival in patients with glioblastoma: A randomized clinical trial. *JAMA, 318*, 2306–2316.

9. Ceresoli, G., et al. (2018). MA12.06 STELLAR – Final results of a phase 2 trial of TTFields with chemotherapy for first-line treatment of malignant pleural mesothelioma. *Journal of Thoracic Oncology, 13*, S397–S398.

10. Ballo, M. T., et al. (2019). Correlation of tumor treating fields dosimetry to survival outcomes in newly diagnosed glioblastoma: A large-scale numerical simulation-based analysis of data from the phase 3 EF-14 randomized trial. *International Journal of Radiation Oncology, Biology, Physics, 104*, 1106–1113.

11. Kanner, A. A., et al. (2014). Post Hoc analyses of intention-to-treat population in phase III comparison of NovoTTF-100A™ system versus best physician's choice chemotherapy. *Seminars in Oncology, 5*, S25–S34.

12. Toms, S. A., et al. (2019). Increased compliance with tumor treating fields therapy is prognostic for improved survival in the treatment of glioblastoma: A subgroup analysis of the EF-14 phase III trial. *Journal of Neuro-Oncology, 141*, 467–473.

13. Giladi, M., et al. (2015). Mitotic spindle disruption by alternating electric fields leads to improper chromosome segregation and mitotic catastrophe in cancer cells. *Scientific Reports, 5*, 18046.

14. Wenger, C., et al. (2016). Improving tumor treating fields treatment efficacy in patients with glioblastoma using personalized array layouts. *International Journal of Radiation Oncology, Biology, Physics, 94*, 1137–1143.

15. Korshoej, A., et al. (2018). Importance of electrode position for the distribution of tumor treating fields (TTFields) in a human brain. Identification of effective layouts through systematic analysis of array positions for multiple tumor locations. *PLoS One, 13*, e0201957.

16. Lacoutre, M. E., et al. (2014). Characterization and management of dermatologic adverse events with the NovoTTF-100A System, a novel anti-mitotic electric field device for the treatment of recurrent glioblastoma. *Seminars in Oncology, 41*, S1–S14.

17. Miranda, P. C., et al. (2013). The electric field in the cortex during transcranial current stimulation. *NeuroImage, 70*, 48–58.

18. Miranda, P. C., et al. (2014). Predicting the electric field distribution in the brain for the treatment of glioblastoma. *Physics in Medicine and Biology, 59*, 4137–4147.

19. Guyton, A., & Hall, J. E. (2006). *Text book of medical physiology* (12th ed.). Philadelphia: Elsevier Saunders. Chapter 73.

20. Lim, C. L., Byrne, C., & Lee, J. K. (2008). Human thermoregulation and measurement of body temperature in exercise and clinical settings. *Annals of the Academy of Medicine, Singapore, 37*, 347–353.

21. Bergman, T. L., et al. (2011). *Fundamentals of heat and mass transfer* (7th ed.). New York: Wiley.

22. Hladky, S. B., & Barrand, M. A. (2014). Mechanisms of fluid movement into, through and out of the brain: Evaluation of the evidence. *Fluids Barriers CNS, 11*, 26.

23. Wang, H., et al. (2014). Brain temperature and its fundamental properties: A review for clinical neuroscientists. *Frontiers in Neuroscience, 8*, 307.

24. Pardridge, W. M. (2011). Drug transport in brain via the cerebrospinal fluid. *Fluids Barriers CNS, 8*, 7.
25. Shapiro, Y., Pandolf, K. B., & Goldman, R. F. (1982). Predicting sweat loss response to exercise, environment and clothing. *European Journal of Applied Physiology, 48*, 83–96.
26. Givoni, B., & Goldman, R. F. (1972). Predicting rectal temperature response to work, environment, and clothing. *Journal of Applied Physiology, 32*, 812–822.
27. ASHRAE. (2013). *Fundamentals handbook*. Atlanta: ASHRAE.
28. Gonzalez, R. R., et al. (2009). Expanded prediction equations of human sweat loss and water needs. *Journal of Applied Physiology, 107*, 379–388.
29. Harker, M. (2013). Psychological sweating: A systematic review focused on aetiology and cutaneous response. *Skin Pharmacology and Physiology, 26*, 92–100.
30. Pennes, H. (1948). Analysis of tissue and arterial blood temperatures in resting human forearm. *Journal of Applied Physiology, 1*, 93–133.
31. Gentilal, N., Salvador, R., & Miranda, P. C. (2019). Temperature control in TTFields therapy of GBM: Impact on the duty cycle and tissue temperature. *Physics in Medicine and Biology, 64*, 225008.
32. Nathan, S. S., et al. (1993). Determination of current density distributions generated by electrical stimulation of the human cerebral cortex. *Electroencephalography and Clinical Neurophysiology, 86*, 183–192.
33. Sapareto, S. A., & Dewey, W. C. (1984). Thermal dose determination in cancer therapy. *International Journal of Radiation Oncology Biology Physics, 10*, 787–800.
34. Dewhirst, M. W., et al. (2003). Basic principles of thermal dosimetry and thermal thresholds for tissue damage from hyperthermia. *International Journal of Hyperthermia, 19*, 267–294.
35. Yarmolenko, P., et al. (2011). Thresholds for thermal damage to normal tissues: An update. *International Journal of Hyperthermia, 27*, 320–343.
36. Cserr, H. F. (1986). The neuronal microenvironment. *Annals of the New York Academy of Sciences, 481*, 123–134.
37. Moritz, A. R., & Henriques, F. C. (1947). Studies of thermal injury: II. The relative importance of time and surface temperature in the causation of cutaneous burns. *The American Journal of Pathology, 23*, 695–720.
38. Lanier, W. L. (1995). Cerebral metabolic rate and hypothermia: Their relationship with ischemic neurologic injury. *Journal of Neurosurgical Anesthesiology, 7*, 216–221.
39. Rosomoff, H. L., & Holaday, D. A. (1954). Cerebral blood flow and cerebral oxygen consumption during hypothermia. *American Journal of Physiology, 179*, 85–88.
40. Volgushev, M., et al. (2000). Synaptic transmission in the neocortex during reversible cooling. *Neuroscience, 98*, 9–22.
41. Volgushev, M., et al. (2004). Probability of transmitter release at neocortical synapses at different temperatures. *Journal of Neurophysiology, 92*, 212–220.

Improving Tumor-Treating Fields with Skull Remodeling Surgery, Surgery Planning, and Treatment Evaluation with Finite Element Methods

Nikola Mikic and Anders R. Korshoej

1 Introduction

Tumor-treating fields (TTFields) are alternating fields (200 kHz) used to treat glioblastoma (GBM), which is one of the deadliest cancer diseases of all. Glioblastoma is a type of malignant brain cancer, which causes significant neurological deterioration and reduced quality of life, and for which there is currently no curative treatment. TTFields were recently introduced as a novel treatment modality in addition to surgery, radiation therapy, and chemotherapy. The fields are induced noninvasively using two pairs of electrode arrays placed on the scalp. Due to low electrical conductivity, significant currents are shielded from the intracranial space, potentially compromising treatment efficacy. Recently, skull remodeling surgery (SR-surgery) was proposed to address this issue. SR-surgery comprises the formation of skull defects or thinning of the skull over the tumor to redirect currents toward the pathology and focally enhance the field intensity. Safety and feasibility of this concept were validated in a clinical phase 1 trial (OptimalTTF-1), which also indicated promising survival benefits. This chapter describes the FE methods used in the OptimalTTF-1 trial to plan SR-surgery and assess treatment efficacy. We will not present detailed modeling results from the trial but rather general concepts of

N. Mikic
Department of Neurosurgery, Aarhus University Hospital, Aarhus, Denmark

Department of Clinical Medicine, Aarhus University, Aarhus, Denmark

Department of Neurosurgery, Aalborg University Hospital, Aalborg, Denmark

A. R. Korshoej (✉)
Department of Neurosurgery, Aarhus University Hospital, Aarhus, Denmark

Department of Clinical Medicine, Aarhus University, Aarhus, Denmark
e-mail: andekors@rm.dk

© The Author(s) 2021
S. N. Makarov et al. (eds.), *Brain and Human Body Modeling 2020*,
https://doi.org/10.1007/978-3-030-45623-8_4

model development and field calculations. Readers are kindly referred to Wenger et al. [1] for a more general overview of the clinical implications and applications of TTFields modeling.

2 Glioblastoma

GBM is the most common and one of the most aggressive primary malignant tumors in the central nervous system [2]. GBM is a WHO grade IV glial tumor characterized by invasive growth and significant anaplasia. The age-standardized incidence rate of GBM in Denmark is 6.3/100,000 person-years for males and 3.9/100,000 person-years for females with a median age of 66 years and a median overall survival of 11.2 months [3], which corresponds well with survival estimates from other Western countries [4]. Today standard therapy consists of maximal surgical resection followed by radiotherapy with concomitant and adjuvant temozolomide chemotherapy [5].

3 Tumor Treating Fields

In the search for new treatment options for GBM, TTFields have recently been introduced as a fourth and supplementary treatment modality applied in parallel with adjuvant temozolomide. TTFields are alternating electric fields of low intensity (100–500 V/m) and intermediate frequency (200 kHz) that are transmitted through the head and brain between electrodes placed noninvasively in an individualized pattern on the patient's scalp (Fig. 1). The electric fields affect dividing cells in particular and hereby primarily cancer cells. The therapeutic effect of TTFields is explained by two physical principles, dielectrophoresis and dipole alignment. In combination, the two principles affect the normal movement of charged and polarizable structures, including septin and tubulin, which is highly responsible for successful mitosis. Thus the disruption of these mechanisms leads to cell death [1]. In patients with newly diagnosed GBM, TTField therapy in combination with chemotherapy has been proved to have a significant effect on median overall survival (OS) and median progression-free survival (PFS) compared to chemotherapy alone [6]. A recent meta-analysis of studies on TTField treatment of GBM patients further concludes that TTFields are an efficient and safe treatment modality [7]. The positive effects of TTFields, recently, led to the introduction of TTFields as a category 1 recommendation of TTFields for a selected population of patients with newly diagnosed GBM by the National Comprehensive Cancer Network in the USA [8].

Fig. 1 TTField therapy. Two pairs of electrode arrays are connected to a TTField generator carried by the patient in a bag (**a**). The arrays are placed on the patient's head (**b**). Each array pair induces alternating TTFields in sequence (**c**) using a 50:50 duty cycle. (The patient photograph is published with permission from the patient (Courtesy of Novocure). The figure is adapted from Korshoej et al. [10])

In regard to the practical use of TTFields, patients are recommended to wear the active device as much as possible – designated as the level of compliance. A compliance threshold above 50% correlates positively with improved outcome, but maximal effect on survival rates is attained with a compliance of >90% [9], and therefore continuous treatment is recommended whenever possible.

4 TTFields Dosimetry

In recent years, finite element (FE) methods have been used to estimate the distribution of TTFields intensity in the patient's head and tumor with the objective of improving technology design and treatment implementation. The rationale behind this approach is that high field intensities correlate positively with longer overall survival [11] and increased tumor kill rate in vitro [12, 13], so field estimation can be considered an approach to TTField dosimetry with potential applications for individual treatment planning as well as identification of expected responders to therapy and prediction of the expected treatment prognosis and topographical patterns of recurrence in the brain. Although previous studies have established that field intensity is a highly relevant surrogate dose parameter, it is well-known that other factors such as field frequency, treatment duration, and spatial correlation also affect the efficacy of TTFields [14–16]. Ongoing work is being conducted to refine the dosimetry methods and establish a golden standard with a strong correlation to clinical outcome.

5   Skull Remodeling Surgery and the Utility of FE Modeling

As an example of FE modeling utility, we recently demonstrated that the high resistivity of the skull causes significant amounts of currents to be shielded from the intracranial regions of interest, which may compromise treatment efficacy. To overcome the obstacle, we proposed a surgical skull remodeling procedure (SR-surgery) aiming to introduce localized skull defects (with reduced skull resistivity) and thereby redirect the tumor inhibiting currents toward the underlying regions of interest (Fig. 2) [17]. SR-surgery encompasses thinning of the skull or formation of burr holes or larger skull defects (craniectomies) over the tumor region, which causes the intensity of the field (i.e., treatment dose) to increase in these regions (Fig. 3) and further reduces the amount of wasted electrical energy deposited in the skin (Fig. 2b).

In search of a feasible approach for clinical implementation, we previously explored a number of different configurations of craniectomy and found that the field intensity in the underlying tumor increases with craniectomy diameter, until the skull defect is approximately the same size as the underlying region of interest. When the defect area exceeds the size of the underlying pathology, it causes currents to be shunted around and pass the intended target and therefore does not contribute to further dose enhancement in the desired area (Fig. 4). In addition, we found that it was more effective to use multiple smaller burr holes distributed over the region of interest, rather than a single craniectomy. With this approach it was possible to achieve higher field enhancement per skull defect area, which made the approach favorable from a clinical safety perspective.

Recently, we demonstrated the safety and feasibility of the SR-surgery concept in a clinical phase 1 trial (OptimalTTF-1, clinicaltrials.gov ID: NCT02893137). We found that SR-surgery combined with TTFields was not associated with serious adverse events related to the intervention, and adverse events observed could be attributed to medical therapy or TTField treatment alone. In addition, the trial further indicated a promising treatment efficacy with prolonged overall survival and progression-free survival compared to historical data from comparable patient cohorts [18].

6   The Aim and Motivation of Field Modeling in SR-Surgery Planning and Evaluation

In the OptimalTTF-1 trial, we used field modeling for a number of purposes. The most important motivation was the need for a method to ensure that enrolled patients would gain an expected benefit from the participation in the trial. Since all enrolled patients underwent SR-surgery, and thereby had to accept the potential risks of the surgery itself in addition to the risks associated with reduced skull protection in the operated region, we required the expected dose enhancement to be considerable for

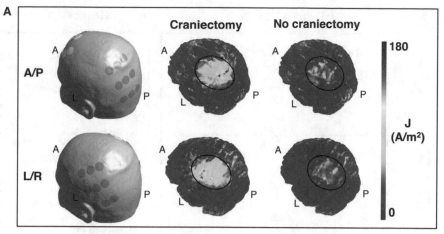

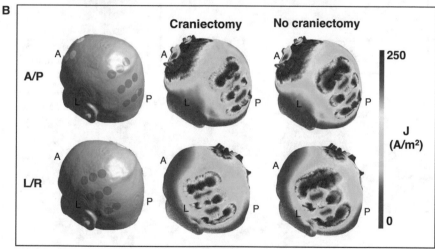

Fig. 2 Effect of craniectomy on the field and current distribution in a human head model. (Reproduced from Korshoej et al. [17]). (**a**) Surface representations of a patient's head with the left/right (L/R) and anterior/posterior (A/P) array pairs positioned on the scalp. The middle panel shows the current density distribution on the brain surface induced by the corresponding array configurations in the presence of a craniectomy (encircled) above the tumor region. Compared to a situation with no craniectomy (right-most panels), it is clear that craniectomy causes a significant amount of current to flow through the craniectomy and toward the underlying brain region. (**b**) This panel shows results similar to (**a**), but with the current distributions shown for the skin and electrode surfaces, respectively. The craniectomy redistributes how the impressed currents flow through the electrodes, and more importantly it causes a lower amount of current to flow through the skin between the electrodes and rather redirects the current toward the brain region underneath the hole in the skull

ethical reasons. Therefore, we set the threshold to an average expected field enhancement of >25% in the region of pathology, i.e., the remnant tumor or the peritumoral border zone. This was assessed using a reasonably quick and flexible modeling

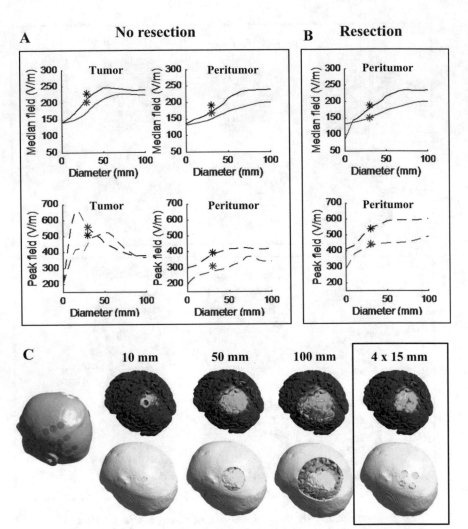

Fig. 3 Effect of different craniectomy configurations. (a) This panel shows the peak field and median field values in the tumor region and peritumoral region (2 cm around the tumor), when no resection is performed, for different sizes of circular craniectomies. The red line represents the anterior/posterior array pair, and the black line represents the left/right pair. The enhancing effect of the craniectomies tends to plateau around a diameter of 5–7 cm, equivalent to the size of the underlying tumor. The asterisks represent the equivalent values for a configuration with four 15 mm burr holes distributed over the tumor. This configuration is equally effective as a 5-cm-diameter craniectomy. (b) This panel shows results similar to panel **a**, but following resection. The same conclusions apply although the plateau tendency is less pronounced in the given case. (c) This panel shows examples of the investigated craniectomy configurations with the underlying brain, tumor, and peritumoral region and field intensity in the brain surface. (The figure is reproduced from Korshoej et al. [17])

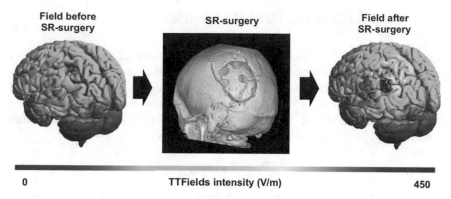

Field before SR-surgery **SR-surgery** **Field after SR-surgery**

0 TTFields intensity (V/m) 450

Fig. 4 The effect of skull remodeling in a single-trial case. Surface representation of the field intensity distribution in the brain and resection cavity from a patient in the OptimalTTF-1 trial. Furthermore, the middle panel shows the SR-surgery configuration applied for the given patient, and the right panel shows the field distribution after the craniectomy. SR-surgery caused the field in the peritumoral region around the resection cavity to be enhanced by approximately 50% in the given case. (This figure is adapted from Korshoej et al. [10])

approach, in which a tumor mimicking the actual patient case were introduced virtually in a preexisting computational head model based on MRI data from a healthy individual (see below). The reason for adopting the approach was that we needed a technique for quick evaluation and exploration of SR-surgery benefit in various configurations. In Denmark, there is a legal requirement to initialize treatment of cancer patients (i.e., operate in this case) within 2 weeks of suspected tumor diagnosis or establishment of disease progression. Therefore, it was not possible to construct detailed and personalized head models for each enrolled patient prior to surgery, as this procedure is very time-consuming. Instead, we used the flexible approach, with which model creation and surgery planning could be completed within approximately 2 days. The computations were initiated immediately upon patient enrollment. We used the model to explore different SR-surgery configurations and identify the optimal configuration with the highest field gain possible for each patient. This configuration was then used to guide the surgery. As a predefined rule, the total skull defect area had to be <30 cm^2.

In addition to validating treatment benefit, an important motivation was to be able to correlate topographical patterns of disease recurrence on MRI with detailed individual assessments of the TTField distribution in treated patients. This work is exploratory in nature and requires accurate computational models based on MRI data from individual patients. Moreover, these more accurate models would serve to validate the estimates obtained in the preliminary preoperative simulations. This work is still ongoing and beyond the scope of the present paper, but the concept illustrates how FE modeling may be used to address and explore many clinically relevant aspects of TTField therapy. The following sections will focus on describing the basic framework of the quick and flexible modeling technique that was used for the assessment of treatment benefit upon patient enrollment.

7   Physical Basis of the Field Calculations

Before we continue to discuss the construction of the head models, we will briefly present the physical framework assumed for the calculations. Given the dielectric properties of biological tissues, the low to intermediate frequency of TTFields (200 kHz), and the small width of the head (approximately 20 cm) [19], we can assume TTFields to behave in a quasi-stationary fashion. Therefore, the electric potential φ can be approximated with Laplace's equation

$$\nabla \cdot (\sigma \nabla \varphi) = 0, \tag{1}$$

where $\nabla \cdot$ is the divergence operator and σ is the real-valued conductivity [20]. In our calculations, we used the FE approach to obtain an approximate numerical solution to Laplace's equation of the electrostatic potential. The field distribution was then initially derived by taking the gradient of the potential distribution and the current density subsequently from Ohm's law and using the derived field and the scalar conductivity assigned to the element. All distributions were calculated separately for each of the electrode pairs, as they are activated sequentially in the real treatment scenario. In addition, calculations were performed both before and after introducing a virtually planned SR-surgery procedure into the model. This allowed us to calculate the absolute and relative changes in the average field intensity in the respective regions of interest, including the tumor and peritumoral border zone, and thereby to quantify the expected field enhancement caused by the intervention.

8   Creating the Head Models

The head models used for computations were constructed from the dataset "almi5," which was created using SimNIBS [21] and which is available from simnibs.org. The model was initially composed of five volumes, namely, skin, skull, cerebrospinal fluid (CSF), gray matter (GM), and white matter (WM). To incorporate the tumor, necrotic regions, and resection cavities, we post-processed the surface mesh STL files of the model for every patient. The post-processing was based on morphological measurements of the pathology regions on preoperative MRI images of the patient, including gadolinium-enhanced T1 sequences. The tumor was incorporated into the GM volume, the necrotic region into the tumor interior, and the resection cavity into the CSF volume. The edited surface meshes were "cleaned" for self-intersections and triangle degenerations using MeshFix. Subsequently, all volumes encapsulated by neighboring surfaces were tessellated with Gmsh (gmsh. info) to construct a tetrahedral computational mesh. The skull defects, i.e., virtual SR-surgeries, were initially outlined in MeshMixer by producing closed (often spherical or cylindrical) compact surface files traversing the exterior and interior boundaries of the skull in a desired geometrical configuration and location. These

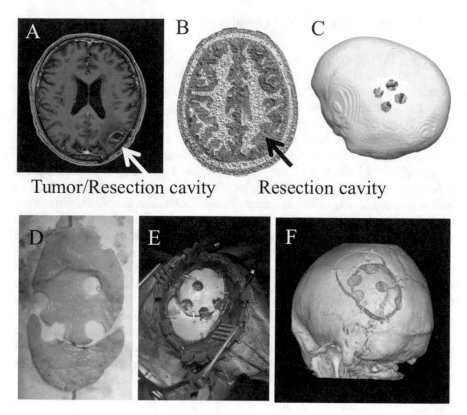

Tumor/Resection cavity   Resection cavity

Fig. 5 SR-surgery planning for the patient shown in Fig. 4. (**a**) Contrast-enhanced T1 MRI showing the tumor/resection. (**b**) Patient head model showing approximation of intended resection cavity. (**c**) Outline of the SR-surgery plan. (**d** and **e**) Images of the remodeled bone plate. Four burr holes (15 mm diameter) were created and the interior plate thinned out in a 5-cm-diameter area over the tumor. (**f**) CT scan of the SR-surgery result. (The figure and legend is reproduced from Korshoej et al. [10])

volumes were then used to define binary volume masks used to select the elements to be contained in the surgical skull defects. These elements were then assigned a uniform isotropic conductivity equal to the skin, based on the assumption that the removed skull tissue would be replaced with a better-conducting skin tissue. The holes in the skull were typically placed directly above the tumor and resection border. A number of configurations were then tested in a trial-and-error fashion, and the model selected for SR-surgery was then visualized using Gmsh and used as a guiding framework for surgery in combination with neuronavigation technologies (Fig. 5).

9   Placement of TTField Transducer Arrays

The 3 × 3 TTField electrode arrays were positioned to maximize TTField intensity for each patient and portray the clinical treatment scenario planned for the individual patient. In a normal clinical setting, the array layout is determined using the NovoTAL® software (Novocure™). NovoTAL® uses individual measurements of the head size and tumor size/position to design a layout for each treated individual, which maximizes the field intensity in the tumor. However, the alteration and redistribution of the current density and electric field caused by SR-surgery arguably invalidate this approach, and we therefore planned the array layouts using the guiding principles of optimized and individualized array placement outlined in Korshoej et al. [22, 23] as well as generalized principles determining the distribution of TTFields [24, 25]. Basically, the arrays were placed so that a row of edge transducers from one array in each pair overlaid the tumor (Fig. 6) and the remodeled region of the skull, while the other array in the same pair was placed on opposite side of the skull, ensuring that currents would flow through the holes in the skull and toward the opposite side of the head and thereby induce high fields in the tumor. This approach is based on the observation that stronger fields are induced in tissues underlying the periphery of the electrode arrays ("edge effect"). Hence, it is not desirable to have the skull holes located under the central parts of the array or in a far distance from the array, as this would reduce the amount of current likely to pass through the holes. The virtual placement of electrodes was performed using the SimNIBS GUI and a custom Matlab script (Mathworks, Inc.). For further details, see [23].

10   Boundary Conditions and Tissue Conductivities

Computations were conducted using the Dirichlet boundary conditions defined by the anatomical boundaries of the head and fixed electrical potentials at the top of the array transducers. Particularly, the potential was set to 1 V in the transducers of one array in a pair, while the potential in the electrodes of the other array were set to −1 V. Numerical approximation was obtained using a conjugate gradient solver with a defined tolerance of 1 E-9. All potentials, fields, and current densities were then rescaled to obtain a total current of 1.8A through the arrays equivalent to the amount of current delivered by the Optune™ device. This allowed us to model the actual scenario that all electrodes in an array were connected to the same electrical source. In all calculations, a uniform isotropic scalar conductivity value σ was assigned to all nodes in a volume based on previous measurements from in vitro and in vivo studies (skin 0.25 S/m, bone 0.010 S/m, CSF 1.654 S/m, tumor 0.24 S/m, and necrosis 1.00 S/m [23]). All transducers were modeled with an underlying layer of conductive gel with 0.5 mm thickness and 1.0 S/m conductivity.

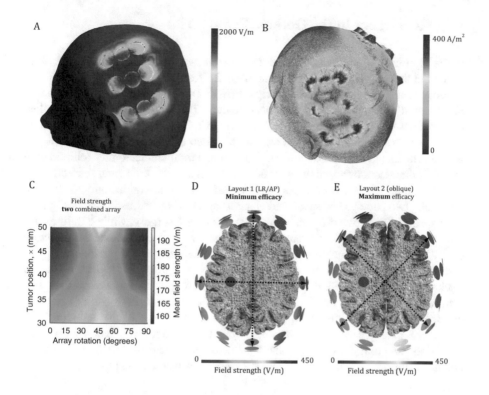

Fig. 6 The edge effect and principles of electrode array positioning. The panels **a** and **b**, respectively, show the skin surface representations of the current density (**a**) and field intensity (**b**) induced by the left/right array pair of a participant in the trial. Both panels illustrate that the stronger fields and currents are present near the periphery of the array. Panels **c–e** illustrate the underlying principle adopted when placing the arrays on the head of the patient. Panel c shows the mean field intensity in a virtually introduced 2-cm-diameter tumor with a 1.4-cm-diameter central necrotic core for different tumor positions and array positions. Specifically, we tested how the field was affect by 15-degree stepwise rotations of an orthogonal configuration of two array pairs in the same horizontal plane (**b** and **c**). This rotation was conducted around a central craniocaudal axis. Eleven tumors were investigated for all rotations. Particularly, the tumors were translated along an axis in the coronal plane from deep positions (30 mm from the median plane) to superficial positions (50 mm from the median plane). The tumors were located in the plane of the central transducers of the arrays. For all tumors, the maximum average field intensity was achieved when the array pairs were oriented both at 45 degrees to the sagittal plane, i.e., obliquely (panel **e**). The default layout (i.e., anterior/posterior and left/right, panel **d**) were the least efficient for these tumors. These results are further elaborated in Korshoej et al. [23] from which this figure has been adapted. The conclusions of these investigations are that arrays should be placed such that the edge of one array from each pair is placed in close vicinity to the tumor (and the introduced skull defects) and the other array in the same pair on the opposite side of the head. This approach was also adopted when positioning the arrays in the OptimalTTF-1 trial. (Panels **c–e** of this figure are reproduced from Korshoej et al. [23])

11   SR-Surgery in the OptimalTTF-1 Trial

In the OptimalTTF-1 trial, a total of 15 subjects were enrolled. The tumors were located in the temporal ($N = 5$), parietal ($N = 2$), frontal ($N = 2$), occipital ($N = 1$), frontoparietal ($N = 3$), and parietooccipital ($N = 2$) regions, and field enhancement >25% could be obtained for all patients (median 37%, range 25–67%). The applied skull defects had a mean area of 10.5 cm^2 (range 7–24 cm^2), and the mean absolute field values in the region of interest were in the range 100–200 V/m. Ten patients had 4–6 burr holes (15–18 mm diameter), and two had total craniectomies (elliptic with semiaxis diameters of approximately 60 × 50 mm and 85 × 65 mm, respectively). One had five 15 mm burr holes and one 25 mm mini-craniectomy, while the remaining two patients had seven and eight 20 mm burr holes, respectively. Figure 7 shows examples of two different configurations of SR-surgery, while a third example is given in Fig. 5f. The remodeled regions were placed above the resection cavity/border and residual tumor. Skull thinning was performed if possible and if the resection cavity extended to regions where the overlying skull had an estimated thickness above 3 mm. Skull thinning in areas below this limit was considered less significant because the relative gain in conductivity would be too small in these cases. For patients with temporal tumors, the squamous area of the temporal bone was therefore only perforated by burr holes, and bone bridges were left to support the overlying temporal muscle and maintain cosmetic integrity. All surgeries were conducted by trained neurosurgeons. The operation was technically feasible, easy

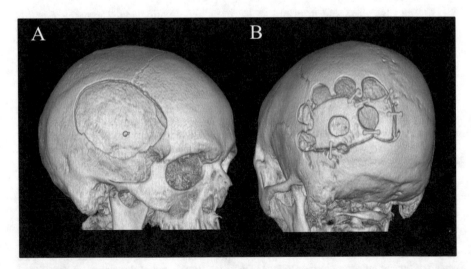

Fig. 7   CT reconstructions of two additional examples of SR-surgery configurations. (a) Total craniectomy (85 × 65 mm, elliptic) above the tumor region. This was equivalent to a standard craniotomy bone flap created during resection surgery. (b) Seven burr holes (18 mm diameter) distributed above the resection cavity and its surrounding borders, tumor region before and after SR-surgery. (This figure is reproduced from Korshoej et al. [10])

to perform, and added less than 15 min of additional surgery time. Overall survival was 15.0 months, CI95% = [9.6; 16.2], and the overall survival rate at 1 year was 64%, CI95% = [35; 85], which is promising compared to historical data.

12 Conclusion

In this chapter, we have introduced the general concept of TTFields and well as background information on the main indication of this treatment, i.e., glioblastoma. We have illustrated the technical framework and rationale for implementation FE modeling dosimetry as a method to plan and evaluate skull remodeling surgery in combination with tumor-treating field therapy of GBM. We have illustrated how SR-surgery can be used to increase the TTField dose in GBM tumors and the techniques used to quantify this enhancement. The presented framework was adopted in a phase 1 clinical trial to validate expected efficacy for patients enrolled in the trial and further to calculate the field enhancement achieved for each patient. The trial, which is concluded at this time, showed that the SR-surgery approach was safe, feasible, and potentially improved survival in patients with first recurrence of GBM [18]. Two different modeling approaches were adopted, namely, a fast but less accurate approach, in which a representative tumor or resection cavity was introduced virtually in a computational model based on a healthy individual and one based on the individual patients MRI data, which was more accurate but also too time-consuming to be used for quick preoperative calculations. Here we have mainly focused on describing the principles and workflow of the simplified framework. Although we considered this approach sufficient for the given purpose, future work is needed to improve the FE pipeline for better time-efficiency and preparation of patient-specific models as exemplified in [17]. Such models would both improve anatomical accuracy and also allow for individualized anisotropic conductivity estimation giving a more accurate and realistic basis for the calculations. In the OptimalTTF-1 trial, we conducted the necessary MRI scans for individualized modeling preoperatively, postoperatively, and at disease recurrence for most patients. Based on this data, we aim to conduct individualized and refined *post hoc* simulations to accurately reproduce the actual skull remodeling configurations including skull thinning and thereby provide more accurate estimates of the beneficial effect of SR-surgery. This will be highly valuable when exploring the dose-response relationship and effects of craniectomy enhancement of TTFields in further detail. Furthermore, efforts are being made to streamline and automate the simulation pipeline to enable quick and accurate dose estimation and treatment planning before SR-surgery. Such procedures would ideally also use automated optimization procedures as opposed to the current exploratory approach to ensure maximal dose enhancement. Finally, we are finalizing the analysis of the OptimalTTF-1 trial, which will shed important light to the clinical significance of the concept. A future clinical phase 2 trial is being planned to test treatment efficacy.

References

1. Wenger, C., Miranda, P., Salvador, R., Thielscher, A., Bomzon, Z., Giladi, M., et al. (2018). A review on Tumor Treating Fields (TTFields): Clinical implications inferred from computational modeling. *IEEE Reviews in Biomedical Engineering, 11*, 195.
2. Omuro, A., & DeAngelis, L. M. (2013). Glioblastoma and other malignant gliomas: A clinical review. *JAMA, 310*(17), 1842–1850.
3. Hansen, S., Rasmussen, B. K., Laursen, R. J., Kosteljanetz, M., Schultz, H., Nørgård, B. M., et al. (2018). Treatment and survival of glioblastoma patients in Denmark: The Danish neuro-oncology registry 2009–2014. *Journal of Neuro-Oncology, 139*(2), 479–489.
4. Ostrom, Q. T., Gittleman, H., Liao, P., Vecchione-Koval, T., Wolinsky, Y., Kruchko, C., et al. (2017). CBTRUS statistical report: Primary brain and other central nervous system tumors diagnosed in the United States in 2010–2014. *Neuro-Oncology, 19*(Suppl 5), v1–v88.
5. Weller, M., van den Bent, M., Hopkins, K., Tonn, J. C., Stupp, R., Falini, A., et al. (2014). EANO guideline for the diagnosis and treatment of anaplastic gliomas and glioblastoma. *The Lancet Oncology, 15*(9), e395–e403.
6. Stupp, R., Taillibert, S., Kanner, A., Read, W., Steinberg, D. M., Lhermitte, B., et al. (2017). Effect of tumor-treating fields plus maintenance temozolomide vs maintenance temozolomide alone on survival in patients with glioblastoma: A randomized clinical trial. *JAMA, 318*(23), 2306–2316.
7. Magouliotis, D. E., Asprodini, E. K., Svokos, K. A., Tasiopoulou, V. S., Svokos, A. A., & Toms, S. A. (2018). Tumor-treating fields as a fourth treating modality for glioblastoma: A meta-analysis. *Acta Neurochirurgica, 160*, 1–8.
8. National Comprehensive Cancer Network. (2017). NCCN guidelines version 1.2017. Sub-Committees Central Nervous System Cancers.
9. Toms, S., Kim, C., Nicholas, G., & Ram, Z. (2018). Increased compliance with tumor treating fields therapy is prognostic for improved survival in the treatment of glioblastoma: A subgroup analysis of the EF-14 phase III trial. *Journal of Neuro-Oncology, 141*, 1–7.
10. Korshoej, A. R., Mikic, N., Hansen, F. L., Thielscher, A., Saturnino, G. B., & Bomzon, Z. (2019). Enhancing tumor treating fields therapy with skull-remodeling surgery. The role of finite element methods in surgery planning. *2019 41st annual international conference of the IEEE Engineering in Medicine and Biology Society (EMBC)* (pp. 6995–6997). IEEE.
11. Ballo, M. T., Urman, N., Lavy-Shahaf, G., Grewal, J., Bomzon, Z., & Toms, S. (2019). Correlation of tumor treating fields dosimetry to survival outcomes in newly diagnosed glioblastoma: A large-scale numerical simulation-based analysis of data from the phase 3 EF-14 randomized trial. *International Journal of Radiation Oncology, Biology, Physics, 104*(5), 1106–1113.
12. Kirson, E. D., Dbaly, V., Tovarys, F., Vymazal, J., Soustiel, J. F., Itzhaki, A., et al. (2007 Jun 12). Alternating electric fields arrest cell proliferation in animal tumor models and human brain tumors. *Proceedings of the National Academy of Sciences of the United States of America, 104*(24), 10152–10157.
13. Kirson, E. D., Gurvich, Z., Schneiderman, R., Dekel, E., Itzhaki, A., Wasserman, Y., et al. (2004). Disruption of cancer cell replication by alternating electric fields. *Cancer Research, 64*(9), 3288–3295.
14. Korshoej, A. R., & Thielscher, A. (2018). Estimating the intensity and anisotropy of tumor treating fields using singular value decomposition. Towards a more comprehensive estimation of anti-tumor efficacy. *2018 40th Annual international conference of the IEEE Engineering in Medicine and Biology Society (EMBC)*. IEEE.
15. Korshoej, A. R., Sørensen, J. C. H., Von Oettingen, G., Poulsen, F. R., & Thielscher, A. (2019). Optimization of tumor treating fields using singular value decomposition and minimization of field anisotropy. *Physics in Medicine & Biology, 64*(4), 04NT03.
16. Korshoej, A. R. (2019). Estimation of TTFields intensity and anisotropy with singular value decomposition: A new and comprehensive method for dosimetry of TTFields. In *Brain and*

human body modeling: Computational human modeling at EMBC 2018 (pp. 173–193). Cham: Springer.

17. Korshoej, A. R., Saturnino, G. B., Rasmussen, L. K., von Oettingen, G., Sørensen, J. C. H., & Thielscher, A. (2016). Enhancing predicted efficacy of tumor treating fields therapy of glioblastoma using targeted surgical craniectomy: A computer modeling study. *PLoS One, 11*(10), e0164051.

18. Korshoej, A., Lukacova, S., Sørensen, J. C., Hansen, F. L., Mikic, N., Thielscher, A., et al. (2018). ACTR-43. Open-label phase 1 clinical trial testing personalized and targeted skull remodeling surgery to maximize ttfields intensity for recurrent glioblastoma–interim analysis and safety assessment (Optimalttf-1). *Neuro-Oncology, 20*(Suppl 6), vi21–vi21.

19. Wenger, C., Salvador, R., Basser, P. J., & Miranda, P. C. (2015). The electric field distribution in the brain during TTFields therapy and its dependence on tissue dielectric properties and anatomy: A computational study. *Physics in Medicine & Biology, 60*, 7339–7357.

20. Miranda, P. C., Mekonnen, A., Salvador, R., & Basser, P. J. (2014). Predicting the electric field distribution in the brain for the treatment of glioblastoma. *Physics in Medicine and Biology, 59* (15), 4137.

21. Saturnino, G. B., Antunes, A., & Thielscher, A. (2015). On the importance of electrode parameters for shaping electric field patterns generated by tDCS. *NeuroImage, 120*, 25–35.

22. Korshoej, A. R., Hansen, F. L., Mikic, N., Thielscher, A., von Oettingen, G. B., & JCH, S. (2017). Exth-04. Guiding principles for predicting the distribution of tumor treating fields in a human brain: A computer modeling study investigating the impact of tumor position, conductivity distribution and tissue homogeneity. *Neuro-Oncology, 19*(Suppl 6), vi73.

23. Korshoej, A. R., Hansen, F. L., Mikic, N., von Oettingen, G., JCH, S., & Thielscher, A. (2018). Importance of electrode position for the distribution of tumor treating fields (TTFields) in a human brain. Identification of effective layouts through systematic analysis of array positions for multiple tumor locations. *PLoS One, 13*(8), e0201957.

24. Lok, E., San, P., Hua, V., Phung, M., & Wong, E. T. (2017). Analysis of physical characteristics of Tumor Treating Fields for human glioblastoma. *Cancer Medicine, 6*, 1286.

25. Korshoej, A. R., Hansen, F. L., Thielscher, A., von Oettingen, G. B., & Sørensen, J. C. H. (2017). Impact of tumor position, conductivity distribution and tissue homogeneity on the distribution of tumor treating fields in a human brain: A computer modeling study. *PLoS One, 12*(6), e0179214.

Part II
Non-invasive Neurostimulation – Brain

A Computational Parcellated Brain Model for Electric Field Analysis in Transcranial Direct Current Stimulation

M. A. Callejón-Leblic and Pedro C. Miranda

1 Introduction

tDCS is a noninvasive brain modulation technique which produces cortical excitatory and inhibitory plastic changes by delivering small-amplitude electric currents into the brain through two or more electrodes placed on the scalp [1]. Although neuromodulation has shown promising results regarding the improvement of both cognitive and motor functions as well as the treatment of several neurological and neuropsychiatric diseases [2–4], the field still faces one major drawback: the lack of reproducibility caused by large differences among subjects and studies [5].

Realistic computational models derived from magnetic resonance imaging (MRI) have proved useful for the estimation of the EF magnitude and distribution through the brain, revealing the effect of individual anatomy as a major cause for intersubject variability [6–10]. Further to this, classic electrode placement criteria, where one electrode is usually located above the neural target and another some distance away, has been called into question when predicting similar or even higher values in both superficial and deep areas between the electrodes [10–14]. It is for this reason that the correlation between the EF magnitude and the observed neurophysiological effects of tDCS continues to be controversial [15–17]. Recent modelling studies have emphasized that the influence of tDCS-induced EFs does not only depend on

M. A. Callejón-Leblic
Biomedical Engineering Group, University of Seville, Seville, Spain

Instituto de Biofísica e Engenharia Biomédica, Faculdade de Ciências, Universidade de Lisboa, Lisboa, Portugal
e-mail: mcallejon@us.es

P. C. Miranda (✉)
Instituto de Biofísica e Engenharia Biomédica, Faculdade de Ciências, Universidade de Lisboa, Lisboa, Portugal
e-mail: pcmiranda@fc.ul.pt

© The Author(s) 2021
S. N. Makarov et al. (eds.), *Brain and Human Body Modeling 2020*,
https://doi.org/10.1007/978-3-030-45623-8_5

their magnitude but also on their relative orientation to specific cortical targets [18–20].

This chapter is structured as followed: Section 2 offers a brief review of recent works addressing the intricate relationship between the modelled EF magnitude and/or orientation and the stimulation response. Based on previous work in [21], Sect. 3 describes the modelling framework followed to obtain a computational parcellated brain model based on the finite element method (FEM) for the analysis of the distribution of tangential and normal EF components over the cortex. Section 4 presents estimates of mean and peak tangential and normal EF values for different cortical regions and four different electrode montages typically used in tDCS clinical applications. Section 5 discusses the main differences observed between the various electrode montages analyzed. Finally, Sect. 6 concludes this chapter.

2   Relation Between EF Magnitude and Orientation and tDCS-Physiological Effects

The conventional criteria for the placement of electrodes in tDCS stems from initial studies undertaken by Nitsche et al. where significant modulation of motor-evoked potentials (TMS-MEPs) was found only for the left M1-right supraorbital area (RSOA) montage [22, 23]. In apparent contradiction to this, current MRI-driven computational EF models predict considerable EF spread over intermediate and deep brain regions between the electrodes, relatively far from the presumed targeted regions [9, 11, 14, 24, 25].

As demonstrated through animal and in vitro studies, changes in membrane excitability of neurons are sensitive to orientation of EFs relative to different neuronal compartments [26–28]. However, given the complex anatomy of the highly convoluted human cortex, extrapolating these results to in-vivo studies in humans is not straightforward. Figure 1 shows an illustrative scheme of the EF vector over the human cortical sheet and the definition of both tangential and normal EF components. Recent multiscale models integrating both macroscopic and microscopic effects at both tissue and cellular levels have revealed a high correlation between soma polarization and normal EF irrespective of electrode montage [29].

At a population level, recent works have investigated the correlation between the simulated EF magnitude and/or orientation and the tDCS-induced excitatory effects, with seemingly divergent results. For instance, as reported by Fitscher et al. in [30], lower magnitude and normal EF values estimated for multifocal tDCS of the motor cortex did not explain the increased excitability changes observed when compared with standard left M1-RSOA montage. More recently in [31], Antonenko et al. applied tDCS to 24 healthy participants during eyes-closed resting-state functional resonance imaging for the standard large-pad left M1-RSOA montage under anodal, cathodal, and sham stimulation conditions. A better correlation with physiological effects was identified for the EF magnitude rather than for the normal component

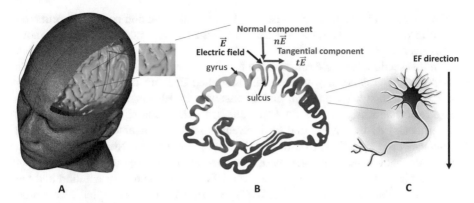

Fig. 1 Schematic illustration of the electric field and their directional components. (**a**) Realistic computational EF model at tissue level. (**b**) Description of tangential and normal EF components over a slice of the cortical sheet. (**c**) Relation between the orientation of EF and the morphology of neuronal structures at cellular level

over the targeted left precentral gyrus for both anodal and cathodal conditions. In another recent work, Foerster et al. [32] analyzed the TMS-MEP excitability response in 15 volunteers after 15 min, 1 mA anodal and cathodal stimulation with two different orientations of the 5 cm × 7 cm electrode placed over the left M1: one with the long axis of the electrode aligned with the medial-lateral direction and another with the electrode rotated 45° clockwise. The second electrode was fixed over the contralateral supraorbital area. Although modelling results in a representative brain model did not reveal significant differences regarding EF orientation over M1 for these two montages, a significant enhancement of the excitatory response for both anodal and cathodal conditions was only exhibited with the 45°-rotated electrode. This outcome was instead related to higher EF magnitude over M1 and premotor areas.

Nevertheless, other authors have recently uncovered evidence that the tDCS excitatory changes are indeed sensitive to EF orientation relative to the cortical surface. For example, in [20], Laakso et al. measured the MEPs in 28 healthy subjects before and after a 20 min sham or 1 mA anodal tDCS stimulation with large-pad electrodes over the right motor cortex. Opposite effects in modulated MEPs were observed for individuals with the strongest and weakest EFs, respectively, showing a high correlation with the normal component in the hand knob area near the TMS hotspot. In another study [18], Rawji et al. compared the tDCS-induced changes measured by TMS-MEP in 22 healthy volunteers by comparing two different montages with two small electrodes placed 7 cm around M1: one montage with the electrodes aligned so as to direct current perpendicularly to the central sulcus in the posterior-anterior (PA) direction and another with the electrodes directing current parallel to the central sulcus in the medial lateral (ML) direction. Significant after-effects of TMS-MEP for the PA montage were correlated with predicted current density normally oriented toward the hand region, whereas the inexistence of excitatory effects for the ML montage were related to a non-uniform

EF orientation in this area. Likewise in [19], Hannah et al. confirmed and furthered these results for behavioral motor learning and two PA- and ML-orientated montages over the sensorimotor cortex.

In this section, we have summarized the main results derived from recent works addressing the relationship between the modelled EFs and the tDCS neurophysiological effects. It has been seen that a consensus on EF magnitude or orientation-based mechanisms has not yet been reached. The variability observed among experimental trials may be due to different stimulation parameters and electrode montages used as well as to different modelling approaches. Nonetheless, we can conclude that computational EF models were vital to improve our understanding of the complex relationship between simulated EF magnitude and orientation and the stimulation phenomena observed in tDCS experiments. In this chapter, a computational analysis of the distribution of tangential and normal EF components over a representative brain model is presented for various common tDCS clinical montages.

3 A Computational Parcellated Brain Model in tDCS

3.1 Head Anatomy

It is common practice for the creation of a realistic computational model of the head to involve the segmentation of T1- and T2-weigthed magnetic resonance images (MRI) based on individual anatomies. The process of converting MRI to EF distribution is further explained in [33]. That said, computational models based on brain atlases have also gained interest among researchers due to their ability to represent an average human brain. Such is the case of the computational model this chapter will focus on, which is based on the 0.5 mm^3 ICBM152 V2009b symmetric template derived from the nonlinear average of MRI scans of 152 adults with high anatomical detail of inner brain tissues [34]. Here, an earlier version of SimNIBS [35] (v2.0, http://simnibs.de/) which employs Freesurfer (https://surfer.nmr.mgh.harvard.edu/), and is reported to provide higher resolution of cortical gyri and sulci morphology [36], was used to segment and to obtain 3D surface meshes of ICBM152 inner brain tissues such as gray matter (GM), white matter (WM), cerebellum, and ventricles. Regarding non-brain tissues, Huang et al. in [37] co-registered and re-sliced a prior version of the same template, ICBM152 v6, which better retained the anatomical characteristics of the scalp and the skull, to the same MRI space of ICBM152 V2009b. By artificially overlapping the resultant MRI with an average of several heads, they were able to extend the field of view (FOV) down to the neck and thus generate a full head model named ICBM-NY (New-York) head model. The ICBM-NY segmentation masks made available by Huang et al. were tessellated here to obtain the 3D triangular surface meshes for the outer head tissues such as the scalp, skull, eyeballs, air cavities, and CSF. A non-manifold assembly of these tissues as well as post-processing operations such as the smoothing, re-meshing, and cleaning of small anatomical skull details and

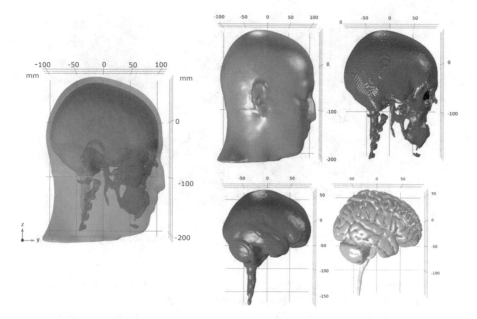

Fig. 2 View of the computational head model based on ICBM152 V2009b template and ICBM152-NY segmentations masks provided by Huang et al. in [37]. The model comprised different tissues such as the scalp, skull, vertebrae, eyeballs, air cavities, CSF, GM, WM, cerebellum, brain stem, and ventricles. Skull openings such as the eye foramina and foramen magnum were manually modelled using Mimics 3-matic

other tissue irregularities were manually carried out in Mimics 3-matic software (v16, https://www.materialise.com/es/medical/software/mimics). The resulting computational head model based on ICBM152 template can be seen in Fig. 2.

3.2   Cortex Parcellation

The use of multimodal neuroimaging studies along with EF modelling may help identify the true targeted cortical regions in tDCS. Hence, the delineation of different cortical or brain regions, often referred to as brain parcellation [38], may help provide useful quantitative comparison of the EF dose delivered to different brain substructures. Many neuroimaging and EF modelling software toolboxes include specific labelling tools for brain parcellation based on available brain atlases. The multimodal parcellation reported in [39], considered one of the most detailed in the literature so far, is that derived from the Human Connectome Project version 1.0 (HCP-MMP 1.0, https://balsa.wustl.edu/WN56). It offers a description of 180 areas per hemisphere grouped into 22 regions according to cortical anatomy, function, and connectivity criteria. In this work, we labelled these 22 regions in the ICBM152 model (see Fig. 3) based on the Freesurfer version of HCP-MMP 1.0 parcellation

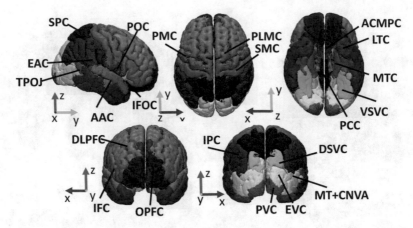

Fig. 3 Cortex parcellated into 22 cortical regions per hemisphere according to color code in Table 1

data [40]. The names of the 22 regions are listed in Table 1, where they have been grouped into 6 main sections per hemisphere, including visual cortex (VC), somato-sensory and motor cortex (SMC), auditory cortex (AC), temporal cortex (TC), parietal cortex (PC), and frontal cortex (FC).

3.3 tDCS Electrode Montages

Bipolar large-pad saline-soaked electrodes are the most commonly used in tDCS clinical trials. Both anodal (excitatory) and cathodal (inhibitory) stimulations are possible, where the only difference is the direction of the resultant EF but not the distribution [41]. As can be seen in Fig. 4, four bipolar montages consisting of 7 cm × 5 cm rectangular, 6-mm-thick electrodes were considered. This permitted the analysis of the EF distribution over various cortical areas commonly targeted in tDCS clinical applications: motor cortex (C3-right supraorbital area RSOA), dorso-lateral prefrontal cortex (F3-RSOA), visual cortex (Oz-Cz), and auditory cortex (T8-T7). The scalp coordinates for these positions were identified according to 10-20 EEG system.

3.4 The Physics of tDCS

The physics of tDCS is that of a conducting body volume with two or more electrodes attached to its surface [42]. The electric current injected through the attached electrodes causes a net flow of ions through the head tissues, which can be described in terms of an electric field, $\overrightarrow{E}$ (V/m), or a current density, $\overrightarrow{J} = \sigma\overrightarrow{E}$

Table 1 List of 22 parcellated cortical regions from HCP-MMP 1.0

	Parcelled Cortical Regions
Visual/Occipital cortex	1. Primary Visual Cortex: PVC
	2. Early Visual Cortex: EVC
	3. Dorsal Stream Visual Cortex: DSVC
	4. Ventral Stream Visual Cortex: VSVC
	5.M.T.+Complex Neighboring Visual Areas: MT+CNVA
Somatosensory and Motor Cortex	6. Somatosensory and Motor Cortex: SMC
	7. Paracentral Lobular and Midcingulate Cortex: PLMC
	8. Pre-Motor Cortex: PMC
	9. Posterior Opercular Cortex: POC
Auditory Cortex	10. Early Auditory Cortex: EAC
	11. Auditory Association Cortex: AAC
	12. Insular and Frontal Opercular Cortex: IFOC
Temporal Cortex	13. Medial Temporal Cortex: MTC
	14. Lateral Temporal Cortex: LTC
Parietal Cortex	15. Temporal-Parietal-Occipital Junction: TPOJ
	16. Superior Parietal cortex: SPC
	17. Inferior Parietal Cortex: IPC
	18. Posterior Cingulate Cortex: PCC
Frontal Cortex	19. Anterior Cingulate and Medial Prefrontal Cortex: ACMPC
	20. Orbital and Polar Frontal Cortex: OPFC
	21. Inferior Frontal Cortex: IFC
	22. Dorsolateral Prefrontal Cortex: DLPC

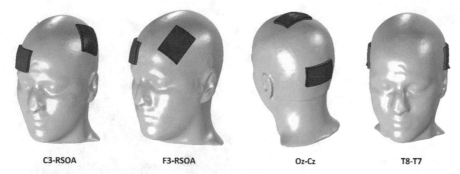

C3-RSOA   F3-RSOA   Oz-Cz   T8-T7

Fig. 4 Electrode montages commonly used in tDCS clinical applications: C3-RSOA, F3-RSOA, Oz-Cz, and T8-T7. The anode (positive) is colored red and the cathode (negative) blue

(A/m^2). In turn, these currents lead to changes in the membrane potential of excitable neurons, which are excited or inhibited depending on the stimulation applied (anodal or cathodal), the magnitude of the electric field induced, and its relative direction to the targeted neural segments.

This bioelectric problem can be mathematically modelled in terms of Laplace's equation, i.e., $\nabla \cdot (\sigma \nabla \phi) = 0$, which provides a stationary solution for the scalar potential, Φ, in the conducting head tissues. From this, the electric field $\overrightarrow{E} = -\overrightarrow{\nabla \Phi}$ and the current density, $\overrightarrow{J} = \sigma \overrightarrow{E}$, can be derived. The required boundary conditions are usually (i) the continuity of the normal current density at internal boundaries, $-\overrightarrow{n} \cdot \left(\overrightarrow{J_1} - \overrightarrow{J_2} \right) = 0$; (ii) electrical isolation at external boundaries, $-\overrightarrow{n} \cdot \overrightarrow{J} = 0$; and (iii) floating potential boundary conditions for the injection of a constant electric current I_0 through the electrodes: in this way, a constant voltage $V = V_0$ is applied on the electrode boundary such that the total normal electric current density $\overrightarrow{J}$ equals a specific current I_0, i.e., $\int_{\partial \Omega} \left(-\overrightarrow{n} \cdot \overrightarrow{J} \right) dS = I_0$.

3.5   FEM Calculation

Given the highly complex anatomy of the human head and brain tissues, an approximate solution of Laplace's equation can only be obtained through the use of numerical computational techniques such as the finite element method (FEM) [43]. Therefore, FEM software such as that used here, Comsol Multiphysics (v5.3a, www.comsol.com), provides a solution for the electric potential Φ at each node of a tetrahedral volume mesh resembling the head anatomy (see Fig. 5). Another key element of FEM calculation is the electric conductivity of head tissues, σ (S/m), often emulated as homogeneous isotropic conductor volumes. Despite the existing

Fig. 5 Volume mesh for FEM calculation

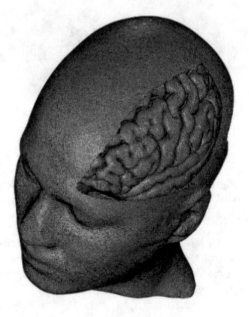

Table 2 Conductivity values in S/m for different head tissues, taken from the literature

Tissue	Conductivity value σ (S/m)
Scalp	0.33
Skull	0.008
CSF	1.79
GM	0.33
WM	0.15
Eyeballs	1.5
Air cavities	$10e^{-14}$
Sponge electrodes	2

controversy on accurate conductivity values, recent in vivo measurements [44] reveal a reasonable agreement with values commonly used in the modelling literature, especially for GM, WM, and scalp, as shown in Table 2 [41].

4   Results

4.1   Tangential and Normal EF Distribution Through the Cortex

Tangential $| t\vec{E} |$ and normal $| n\vec{E} |$ EF components calculated over the cortical surface are shown in Figs. 6 and 7 for an injected current of 1 mA and four tDCS montages commonly used in clinical trials. For the sake of comparison EF components were normalized by the peak value obtained for each electrode montage. Peak values per component, hemisphere, and electrode montage are shown in Table 3 and will be discussed in the next subsection.

Figures 6 and 7 display the coexistence of a tangential component widely spread over the gyri and a normal component mainly focalized in deeper sulci, as has previously been highlighted in the literature [41]. The low thickness and high conductivity of the CSF may partly explain this dual distribution: when the electric current reaches the CSF on its way between the two electrodes through the head, the ions are driven through this thin conductive layer. This occurs tangentially over the thinnest areas (gyri) and perpendicularly in the case of the deepest areas (sulci), which leads to this particularly interesting form of tangential and normal EF distribution over the cortex. It is also noteworthy that unlike the results derived from simplified multilayer spheres mimicking the human head, which show a tangential EF component largely distributed between the electrodes and a normal EF component strongly confined underneath them [45, 46], the irregular anatomy of the highly convoluted brain leads to tangential and normal EFs hotspots being dispersed over the cortex [13].

In addition to anatomy, the relative position of the electrodes and the inter-electrode distance also constitute key variables which determine the EF distribution

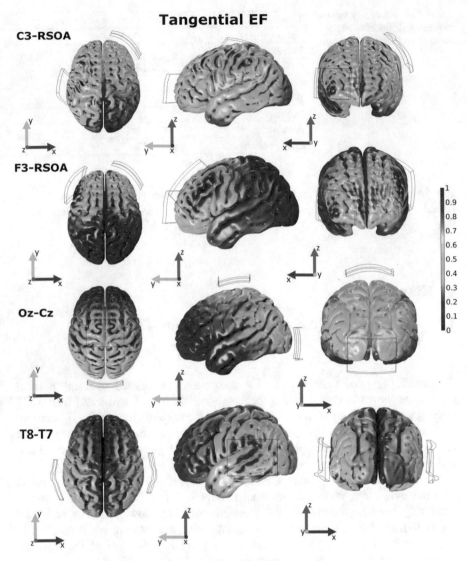

Fig. 6 Normalized magnitude of the tangential EF component for four electrode montages commonly used in clinical applications: C3-RSOA (first row), F3-RSOA (second row), Oz-Cz (third row), and T8-T7 (fourth row)

through the cortical surface [24, 47–51]. For instance, Figs. 6 and 7 show that C3-RSOA exhibits a largely distributed tangential component through the cortex over the whole left hemisphere and the right frontal cortex. This is due to the position of anode C3 in the middle of the left hemisphere, which causes a wider spread of the current lines through the head tissues toward the cathode at RSOA. On the other hand, for the second montage analyzed, F3-RSOA, reducing the inter-electrode

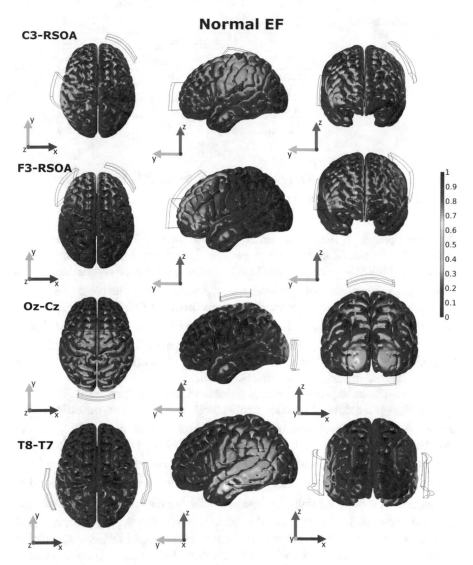

Fig. 7 Normalized magnitude of the normal EF component for four electrode montages commonly used in clinical applications: C3-RSOA (first row), F3-RSOA (second row), Oz-Cz (third row), and T8-T7 (fourth row)

distance by moving the anode from C3 to F3 enhances the focality of both tangential and normal EF in the frontal cortex between the electrodes. In the case of Oz-Cz, tangential and normal EF are more symmetrically distributed on both hemispheres, mainly over the parietal and occipital cortex, due to the alignment of both the anode and cathode on the midsagittal plane. Finally, T8-T7, with the anode and cathode placed opposite each other, yields an EF pattern mostly confined over the temporal cortex underneath the electrodes.

Table 3 Peak tangential and normal EF values

Peak EF (V/m)				
	Left hemisphere		Right hemisphere	
	$\mid t\vec{E}\mid$	$\mid n\vec{E}\mid$	$\mid t\vec{E}\mid$	$\mid n\vec{E}\mid$
C3-RSOA	0.25	0.39	0.18	0.27
F3-RSOA	0.20	0.28	0.21	0.28
Oz-Cz	0.23	0.29	0.21	0.29
T8-T7	0.29	0.31	0.19	0.31

4.2 Mean and Peak Tangential and Normal EF Values over Different Cortical Areas

As shown in Fig. 8, mean tangential $\mid t\vec{E}\mid$ and normal $\mid n\vec{E}\mid$ EF values were calculated over the 22 cortical areas per hemisphere for 4 electrode montages: C3-RSOA, F3-RSOA, Oz-Cz, and T8-T7. In the majority of cortical areas, the mean $\mid t\vec{E}\mid$ values were slightly higher than the mean $\mid n\vec{E}\mid$. The peak values over the cortex for the four electrode montages considered are listed in Table 3 and are comparatively shown in Fig. 9. It should also be noted that in the case of $\mid t\vec{E}\mid$ peak values were generally twice those of the mean, whereas for $\mid n\vec{E}\mid$ this ratio was approximately fourfold. Nevertheless, a number of hemisphere and electrode montage-specific characteristics were also found.

For instance, as expected, some noticeable differences were encountered for C3-RSOA over the right and the left hemispheres, due to its nonsymmetric configuration. As Fig. 8 shows, higher mean $\mid t\vec{E}\mid$ and $\mid n\vec{E}\mid$ values above 0.10 V/m were calculated in regions of the left hemisphere such as SMC, POC, and DLPFC. The highest value of 0.14 V/m was obtained on the left IPC near the anode for both $\mid t\vec{E}\mid$ and $\mid n\vec{E}\mid$. In contrast, lower mean $\mid t\vec{E}\mid$ and $\mid n\vec{E}\mid$ values were generally found in the majority of areas of the right hemisphere, which is explained by thicker skull and air-filled sinuses under the RSOA electrode. As shown in Fig. 9, peak $\mid t\vec{E}\mid$ values of 0.25 and 0.18 V/m were estimated over the left IPC and right OPFC, respectively, close to the anode and cathode in each hemisphere. Regarding the normal component, a peak value of about 0.4 V/m was predicted on POC just below the bottom edge of the anode on the left hemisphere. Also, a peak value of 0.27 V/m was found for $\mid n\vec{E}\mid$ over OPFC near the electrode in the right hemisphere. Peak values of 0.2 and 0.25 V/m were, calculated in areas such as SMC and PMC.

Compared with C3-RSOA, F3-RSOA resulted in lower mean $\mid t\vec{E}\mid$ and $\mid n\vec{E}\mid$ EF values below 0.06 V/m in the majority of cortical areas on both hemispheres. Higher mean values close to 0.08 V/m were only found in areas of the frontal cortex. The enhanced focality, albeit with a lower magnitude, exhibited by this montage is explained by the smaller interelectrode distance. Peak values of about 0.20 and 0.28 V/m were, respectively, calculated for $\mid t\vec{E}\mid$ and $\mid n\vec{E}\mid$ on both right and left

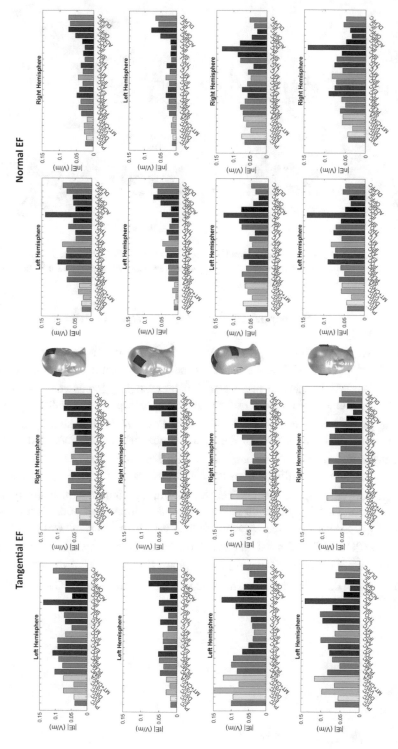

Fig. 8 Mean tangential and normal EF values (V/m) over the 22 hemispherical cortical areas for 4 electrode montages typically used in clinical applications: C3-RSOA, F3-RSOA, Oz-Cz, and T8-T7. The color code is the same as in Fig. 3

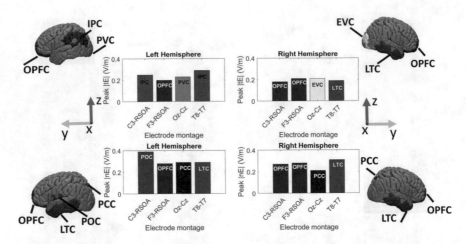

Fig. 9 Simulated peak tangential and normal EF values for the different electrode montages analyzed. The figure also depicts the cortical region where the peak was observed for both right and left hemispheres

OPFC. In addition, a peak normal EF close to 0.28 V/m was estimated on the targeted left DLPFC.

In the case of the Oz-Cz montage, higher mean $|\vec{tE}|$ values of 0.15 and 0.13 V/m were calculated for left DSVC and left IPC, respectively, near the superior edge of the anode (Oz). With respect to mean $|\vec{nE}|$ values, slightly lower values below 0.10 V/m were found in the majority of cortical areas, except on left and right IPC where a higher mean value near 0.14 V/m was seen. Peak $|\vec{tE}|$ values slightly above 0.20 V/m were calculated in areas of the visual cortex such as left PVC and right EVC. Interestingly, peak $|\vec{nE}|$ values above 0.20 V/m were seen in both left and right PCC, approximately halfway between anode and cathode.

Lastly, mean $|\vec{tE}|$ and $|\vec{nE}|$ values below 0.1 V/m were found in the majority of cortical areas for T8-T7. Due to its particular configuration with both electrodes placed above opposite ears, a higher mean value of 0.15 V/m was estimated over the left and right IPC near the stimulation electrodes. In the left hemisphere, peak $|\vec{tE}|$ and $|\vec{nE}|$ values of about 0.3 V/m were found in left IPC and LTC, respectively. A high peak value of 0.3 V/m was also estimated for $|\vec{nE}|$ on the right LTC just below the anode.

5 Summary and Discussion

A general trend has been confirmed for all the electrode montages analyzed, with a marked anatomical distribution of tangential and normal EF on cortical gyri and sulci, respectively. High mean $|\overrightarrow{tE}|$ values were estimated for the majority of cortical regions between the electrodes, which may explain the low spatial resolution often reported for standard tDCS. Instead, higher peak EF values were predicted for $|\overrightarrow{nE}|$ in sulci regions underneath and close to the electrode edges, with some electrode montage-specific characteristics.

Specifically, C3-RSOA showed the highest peak values for $|\overrightarrow{nE}|$ in areas near the somatosensory and motor cortex and the parietal cortex. This high $|\overrightarrow{nE}|$ value in areas below the anode C3 concurs with recent works that have stressed the relevance of the normal EF component over specific motor targets [20, 29]. However, it must also be noticed that other experimental studies have in fact found a greater correlation between the EF magnitude and the neurophysiological excitatory response on gyral regions such as the left precentral gyrus. For the normal component, a larger intersubject variability prevented the correlation with physiological parameters [31]. In line with these results, our results also show that the EF magnitude in gyri areas mainly accounts for the tangential EF and that a consistent higher-mean gyral tangential EF distribution is observed in the average brain modelled.

The F3-RSOA montage, which targets the left DLPC, exhibited lower $|\overrightarrow{tE}|$ and $|\overrightarrow{nE}|$ values, possibly due to the presence of air-filled sinuses and thicker skull underneath the electrodes. A more focal EF pattern confined over frontal areas between the electrodes was observed for this montage and is explained by their shorter inter-electrode distance. Peak $|\overrightarrow{tE}|$ and $|\overrightarrow{nE}|$ values were found over OPFC. These results may corroborate previous modelling studies that have suggested the medial prefrontal cortex as a new target for depression condition in tDCS [10]. Furthermore, previous modelling works analyzing intersubject variability have also reported a greater divergence of EF values over the frontal cortex among subjects for F3-RSOA compared with classic C3-RSOA montage targeting the left motor cortex [6].

Regarding Oz-Cz, this montage showed higher mean and peak $|\overrightarrow{tE}|$ values in areas of the occipital/visual cortex just underneath the cathode. Interestingly, peak $|\overrightarrow{nE}|$ values were also seen on PCC halfway between both stimulation electrodes, thus corroborating prior modelling results in [12].

In the case of T7-T8, previous computational results on auditory tDCS reported higher mean values of current density confined around the stimulation electrodes [52]. Our results here coincide with these previous outcomes showing both hemispheres being stimulated with higher mean and peak $|\overrightarrow{tE}|$ and $|\overrightarrow{nE}|$ EF values over regions of the auditory cortex and LTC near the stimulation electrodes. Finally, it is noteworthy that three of the four montages analyzed, C3-RSOA, Oz-Cz, and T8-T7, showed higher mean tangential and normal EF values over IPC, possibly explained by local CSF thinning.

6 Conclusion

In this chapter we have conducted a computational analysis of the distribution of the tangential and normal components of the electric field over a representative parcellated brain-FEM model for different electrode montages commonly used in tDCS clinical applications. The results confirmed the existence of a dual EF pattern on the cortex, based on a widely gyri-distributed tangential component and a sulci-confined normal EF component. One open question is if such a noticeable dependence of EF components on cortical morphology is the cause of the low focality reported for large-pad tDCS or if, conversely, it actually entails significant direction-based mechanisms for the stimulation outcome at distinct neural segments. Therefore, the validation of computational EF models with physiological measurements is vital in order to clarify this controversy. Our analysis could prove helpful in designing new experimental studies which allow greater understanding of the underlying modulatory mechanisms related to different EF components in tDCS.

References

1. Stagg, C. J., Antal, A., & Nitsche, M. A. (2018, September). Physiology of transcranial direct current stimulation. *The Journal of ECT, 34*(3), 144–152.
2. Truong, D. Q., & Bikson, M. (2018, September). Physics of transcranial direct current stimulation devices and their history. *The Journal of ECT, 34*(3), 137–143.
3. Morya, E., et al. (2019, November). Beyond the target area: An integrative view of tDCS-induced motor cortex modulation in patients and athletes. *Journal of Neuroengineering and Rehabilitation, 16*(1), 1–29.
4. Yavari, F., Jamil, A., Mosayebi Samani, M., Vidor, L. P., & Nitsche, M. A. (2018, February). Basic and functional effects of transcranial electrical stimulation (tES)—An introduction. *Neuroscience and Biobehavioral Reviews, 85*, 81–92.
5. Bikson, M., et al. (2018, May). Rigor and reproducibility in research with transcranial electrical stimulation: An NIMH-sponsored workshop. *Brain Stimulation, 11*(3), 465–480.
6. Laakso, I., Tanaka, S., Mikkonen, M., Koyama, S., Sadato, N., & Hirata, A. (2016, August). Electric fields of motor and frontal tDCS in a standard brain space: A computer simulation study. *NeuroImage, 137*, 140–151.
7. Ciechanski, P., Carlson, H. L., Yu, S. S., & Kirton, A. (2018, July). Modeling transcranial direct-current stimulation-induced electric fields in children and adults. *Frontiers in Human Neuroscience, 12*, 1–14.
8. Mikkonen, M., Laakso, I., Tanaka, S., & Hirata, A. (2020, January). Cost of focality in TDCS: Interindividual variability in electric fields. *Brain Stimulation, 13*(1), 117–124.
9. Evans, C., Bachmann, C., Lee, J. S. A., Gregoriou, E., Ward, N., & Bestmann, S. (2020, January). Dose-controlled tDCS reduces electric field intensity variability at a cortical target site. *Brain Stimulation, 13*(1), 125–136.

10. Csifcsák, G., Boayue, N. M., Puonti, O., Thielscher, A., & Mittner, M. (2018, July). Effects of transcranial direct current stimulation for treating depression: A modeling study. *Journal of Affective Disorders, 234*, 164–173.

11. Huang, Y., Dmochowski, J. P., Su, Y., Datta, A., Rorden, C., & Parra, L. C. (2013, December). Automated MRI segmentation for individualized modeling of current flow in the human head. *Journal of Neural Engineering, 10*(6), 1–26.

12. Rampersad, S. M., et al. (2014, May). Simulating transcranial direct current stimulation with a detailed anisotropic human head model. *IEEE Transactions on Neural Systems and Rehabilitation Engineering, 22*(3), 441–452.

13. Opitz, A., Paulus, W., Will, S., Antunes, A., & Thielscher, A. (2015, April). Determinants of the electric field during transcranial direct current stimulation. *NeuroImage, 109*, 140–150.

14. Gomez-Tames, J., Asai, A., & Hirata, A. (2019, December). Significant group-level hotspots found in deep brain regions during transcranial direct current stimulation (tDCS): A computational analysis of electric fields. *Clinical Neurophysiology, 131*, 755–765.

15. Bestmann, S., & Ward, N. (2017, July). Are current flow models for transcranial electrical stimulation fit for purpose? *Brain Stimulation, 10*(4), 865–866.

16. Peterchev, A. V. (2017, March). Transcranial electric stimulation seen from within the brain. *eLife, 6*, 1–3.

17. Polanía, R., Nitsche, M. A., & Ruff, C. C. (2018, February). Studying and modifying brain function with non-invasive brain stimulation. *Nature Neuroscience, 21*(2), 174–187.

18. Rawji, V., et al. (2018, March). tDCS changes in motor excitability are specific to orientation of current flow. *Brain Stimulation Basic Translation and Clinical Research in Neuromodulation, 11*(2), 289–298.

19. Hannah, R., Iacovou, A., & Rothwell, J. C. (2019, May). Direction of TDCS current flow in human sensorimotor cortex influences behavioural learning. *Brain Stimulation, 12*(3), 684–692.

20. Laakso, I., Mikkonen, M., Koyama, S., Hirata, A., & Tanaka, S. (2019, December). Can electric fields explain inter-individual variability in transcranial direct current stimulation of the motor cortex? *Scientific Reports, 9*(1), 1–10.

21. Callejon-Leblic, M. A., & Miranda, P. C. (2019). *A computational analysis of the electric field components in transcranial direct current stimulation.* In: 41st Annual International Conference of the IEEE Engineering in Medicine and Biology Society (EMBC), Berlin, Germany, pp. 5913–5917, July 2019.

22. Nitsche, M. A., & Paulus, W. (2000, September). Excitability changes induced in the human motor cortex by weak transcranial direct current stimulation. *The Journal of Physiology, 527*(3), 633–639.

23. Nitsche, M. A., et al. (2007, April). Shaping the effects of transcranial direct current stimulation of the human motor cortex. *Journal of Neurophysiology, 97*(4), 3109–3117.

24. Opitz, A., Yeagle, E., Thielscher, A., Schroeder, C., Mehta, A. D., & Milham, M. P. (2018, November). On the importance of precise electrode placement for targeted transcranial electric stimulation. *NeuroImage, 181*, 560–567.

25. Huang, Y., & Parra, L. C. (2019, January). Can transcranial electric stimulation with multiple electrodes reach deep targets? *Brain Stimulation, 12*(1), 30–40.

26. Liu, A., et al. (2018, vs). Immediate neurophysiological effects of transcranial electrical stimulation. *Nature Communications, 9*(1), 1–10.

27. Chakraborty, D., Truong, D. Q., Bikson, M., & Kaphzan, H. (2018, August). Neuromodulation of axon terminals. *Cerebral Cortex, 28*(8), 2786–2794.

28. Rahman, A., et al. (2013, May). Cellular effects of acute direct current stimulation: Somatic and synaptic terminal effects. *The Journal of Physiology, 591*(10), 2563–2578.

29. Seo, H., & Jun, S. C. (2019, March). Relation between the electric field and activation of cortical neurons in transcranial electrical stimulation. *Brain Stimulation, 12*(2), 275–289.

30. Fischer, D. B., et al. (2017, August). Multifocal tDCS targeting the resting state motor network increases cortical excitability beyond traditional tDCS targeting unilateral motor cortex. *NeuroImage, 157*, 34–44.

31. Antonenko, D., et al. (2019, September). Towards precise brain stimulation: Is electric field simulation related to neuromodulation? *Brain Stimulation, 12*(5), 1159–1168.
32. Foerster, Á., et al. (2019, March). Effects of electrode angle-orientation on the impact of transcranial direct current stimulation on motor cortex excitability. *Brain Stimulation, 12*(2), 263–266.
33. Miranda, P. C., Callejón-Leblic, M. A., Salvador, R., & Ruffini, G. (2018, December). Realistic modeling of transcranial current stimulation: The electric field in the brain. *Current Opinion Biomedical Engineering, 8*, 20–27.
34. Fonov, V., Evans, A. C., Botteron, K., Almli, C. R., McKinstry, R. C., & Collins, D. L. (2011, January). Unbiased average age-appropriate atlases for pediatric studies. *NeuroImage, 54*(1), 313–327.
35. Windhoff, M., Opitz, A., & Thielscher, A. (2013, April). Electric field calculations in brain stimulation based on finite elements: An optimized processing pipeline for the generation and usage of accurate individual head models. *Human Brain Mapping, 34*(4), 923–935.
36. Nielsen, J. D., et al. (2018, July). Automatic skull segmentation from MR images for realistic volume conductor models of the head: Assessment of the state-of-the-art. *NeuroImage, 174*, 587–598.
37. Huang, Y., Parra, L. C., & Haufe, S. (2016, October). The New York Head-A precise standardized volume conductor model for EEG source localization and tES targeting. *NeuroImage, 140*, 150–162.
38. Eickhoff, S. B., Yeo, B. T. T., & Genon, S. (2018, November). Imaging-based parcellations of the human brain. *Nature Reviews Neuroscience, 19*(11), 672–686.
39. Glasser, M. F., et al. (2016, August). A multi-modal parcellation of human cerebral cortex. *Nature, 536*(7615), 171–178.
40. Mills, K. (2016). HCP-MMP1.0 projected on fsaverage. *FigShare (dataset)*. https://doi.org/10.6084/m9.figshare.3498446.v2. Last access: December 2019.
41. Miranda, P. C., Mekonnen, A., Salvador, R., & Ruffini, G. (2013, April). The electric field in the cortex during transcranial current stimulation. *NeuroImage, 70*, 48–58.
42. Makarov, S. N., Noetscher, G. M., & Nazarian, A. (2016). *Low-frequency electromagnetic modeling for electrical and biological systems using MATLAB*. Hoboken: Wiley., ISBN: 978-1-119-05256-2.
43. Makarov, S. N., et al. (2017, June). Virtual human models for electromagnetic studies and their applications. *IEEE Reviews in Biomedical Engineering, 10*, 95–121.
44. Koessler, L., et al. (2017, February). In-vivo measurements of human brain tissue conductivity using focal electrical current injection through intracerebral multicontact electrodes. *Human Brain Mapping, 38*(2), 974–986.
45. Faria, P., Hallett, M., & Miranda, P. C. (2011 December). A finite element analysis of the effect of electrode area and inter-electrode distance on the spatial distribution of the current density in tDCS. *Journal of Neural Engineering, 8*(6), 066017.
46. Truong, D. Q., et al. (2014, July). Clinician accessible tools for GUI computational models of transcranial electrical stimulation: BONSAI and SPHERES. *Brain Stimulation, 7*(4), 521–524.
47. Ramaraju, S., Roula, M. A., & McCarthy, P. W. (2018, February). Modelling the effect of electrode displacement on transcranial direct current stimulation (tDCS). *Journal of Neural Engineering, 15*(1), 1–7.
48. Saturnino, G. B., Antunes, A., & Thielscher, A. (2015, October). On the importance of electrode parameters for shaping electric field patterns generated by tDCS. *NeuroImage, 120*, 25–35.
49. Salvador, R., Wenger, C., Nitsche, M. A., & Miranda, P. C. (2015, November). How electrode montage affects transcranial direct current stimulation of the human motor cortex. In: *Proceedings of the Annual International Conference of the IEEE Engineering in Medicine and Biology Society, EMBS*, pp. 6924–6927.
50. Bai, S., Dokos, S., Ho, K. A., & Loo, C. (2014, February). A computational modelling study of transcranial direct current stimulation montages used in depression. *NeuroImage, 87*, 332–344.

51. Bikson, M., Datta, A., Rahman, A., & Scaturro, J. (2010, December). Electrode montages for tDCS and weak transcranial electrical stimulation: Role of 'return' electrode's position and size. *Clinical Neurophysiology, 121*(12), 1976–1978.
52. Wagner, S., et al. (2014, February). Investigation of tDCS volume conduction effects in a highly realistic head model. *Journal of Neural Engineering, 11*(1), 1–14.

Computational Models of Brain Stimulation with Tractography Analysis

Stefanie Riel, Mohammad Bashiri, Werner Hemmert, and Siwei Bai

1 Introduction

Computational head models have been used extensively in electrophysiological studies to locate dipole sources in electroencephalography (EEG) analysis and to investigate the current distribution profile in brain stimulation, such as the electroconvulsive therapy (ECT). They have been able to provide useful information that can't be acquired or difficult to acquire from experimental or imaging studies [1]. However, most of these head models are volume conductor models, in which the electric activity of neurons is assumed to be spatially fixed and temporally independent of activity arising from other sources. In reality, the brain is a system of coupled networks that constantly interact with each other by synaptic inputs, with electrical signals travelling within neurons. Therefore, in order to utilise the head

S. Riel
Department of Mechanical Engineering, Technical University of Munich, Garching, Germany

M. Bashiri
Department of Electrical and Computer Engineering, Technical University of Munich, Munich, Germany

W. Hemmert
Department of Electrical and Computer Engineering, Technical University of Munich, Munich, Germany

Munich School of BioEngineering, Technical University of Munich, Garching, Germany

S. Bai (✉)
Department of Electrical and Computer Engineering, Technical University of Munich, Munich, Germany

Munich School of BioEngineering, Technical University of Munich, Garching, Germany

Graduate School of Biomedical Engineering, University of New South Wales, Sydney, NSW, Australia
e-mail: siwei.bai@tum.de

S. N. Makarov et al. (eds.), *Brain and Human Body Modeling 2020*,
https://doi.org/10.1007/978-3-030-45623-8_6

models for a better understanding of the mechanisms underlying brain stimulation, a passive model unable to mimic active membrane dynamics is of limited utility. Bai et al. presented a finite element (FE) whole head model with the incorporation of a Hodgkin-Huxley-based continuum excitable neural description in the brain, which was able to simulate the dynamic changes of brain activation directly elicited by ECT [2]. Nevertheless, the computation was rather lengthy. In addition, the intra-cellular potential in the model was assumed to be resistively tied to a remote fixed potential, whose physiological meaning was difficult to find in reality. This con-straint also led to the missing of the spread of excitation through neural networks in the brain (seizure). McIntyre's group has over the years introduced a representation of the white matter (WM) fibres in the vicinity of the subthalamic nucleus (STN), combined with a volume conductor model of deep brain stimulation (DBS) [3, 4]. After the electric potential induced by the DBS device was calculated by the FE solver, the time-dependent transmembrane potential was solved in the neuron using a Hodgkin-Huxley-type model based on the interpolated potential distribution along the length of each axon. The model was able to predict the activation in STN neurons and internal capsule fibres, and the degree of activation matched well with animal experimental data [3]. This chapter describes the steps necessary to imple-ment computational modelling of the human head, including a white matter tractography analysis to determine the voltage, electric field and activation function distribution for two prescribed setups of ECT. Neural activation is analysed by the neural activation function, which is in turn approximated by the second spatial derivative of the electrical potential, assuming that the axonal diameter, axoplasmic resistivity and capacity were constant.

2 Methods

To evaluate the electric field and the neural activation function along neural fibres, an FE volume conductor model of electrical brain stimulation (ECT in this case) was combined with a white matter fibre tractography model of the brain. Figure 1 details all the necessary steps for a combined tractography analysis of electrical brain stimulation.

2.1 Image Preprocessing

The FE volume conductor model and the white matter tractography model were both reconstructed from magnetic resonance image (MRI) data from the same subject. MRI data were taken from subject MGH1010 (gender: female, age: 25–29) provided by the NIH Human Connectome Project (HCP) [5]. The dataset comprised a structural T1-weighted scan, a high-resolution T2-weighted scan and a set of high-resolution, high *b*-value diffusion-weighted (DWI) scans. These MR scans provided

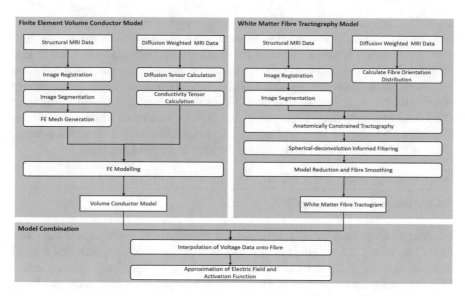

Fig. 1 Flowchart describing the workflow to generate a finite element volume conductor model and a white matter fibre tractography of the head. Information on the electrical field and the activation function can be computed by numerical approximation of the spatial derivatives of the voltage along the fibre, which in turn can be extracted by combining both finite element head model and the tractogram

by HCP had been preprocessed, and therefore image distortions/artefacts currents caused by eddy, gradient nonlinearities and motion had been removed [6]. To ensure the same frame of reference for the individually acquired image datasets, it is necessary to perform image registration. The structural image data with a resolution of 1 mm was resampled to a resolution of 1.5 mm using trilinear interpolation in order to match the dimension of DWI data. Subsequently, a rigid-body registration with six degrees of freedom was performed between the fixed DWI data and the floating T1-weighted image. Since the transformation was performed on the structural data (aligned to the DWI data), no further processing of auxiliary diffusion data (gradient tables of DWI scans) was necessary. These operations were implemented using FLIRT (FMRIB's Linear Image Registration Tool) provided in the FMRIB Software Library (FSL) which is developed by the Oxford Centre of Functional MRI of the Brain (https://fsl.fmrib.ox.ac.uk/fsl/fslwiki/FLIRT) [7–9].

2.2 White Matter Fibre Tractography

White matter consists of myelinated neuronal tracts, interconnecting different parts of the brain. This neural fibre network causes a highly anisotropic diffusion behaviour of water molecules inside the brain. Diffusion is higher along the fibres than in the transverse direction [10]. This characteristic is utilised by DWI scanning, which applies a magnetic gradient in different directions during image acquisition and thus encodes the sensitivity of diffusion to each gradient direction in the image data. Similar anisotropic behaviour is found for the electrical conductivity of white matter fibres, and a linear relationship between electrical conductivity and water diffusion has been experimentally validated [11, 12]. Therefore using anatomically constrained tractography, we extracted a geometrical model of neural fibres from DWI data, which provided information about the trajectories of white matter fibres. Structural T1-weighted MRI data containing information on the structural composition of the brain was used to determine the start and end of each fibre.

2.2.1 Image Segmentation

In order to adopt anatomical constraints for tractography, the structural data was firstly segmented. T1-weighted image data is typically used as it provides a good contrast between head structures, especially between grey and white matter [13]. Image segmentation was performed in two steps: brain extraction and brain tissue segmentation. The brain extraction procedure was implemented in the FSL Brain Extraction Tool (BET) (https://fsl.fmrib.ox.ac.uk/fsl/fslwiki/BET/UserGuide). An intensity histogram of the structural images was used to find the threshold to differentiate the brain- and non-brain regions. Triangular tessellation of a spherical surface was initialised and iteratively deformed – based on the brain/non-brain intensity threshold – to wrap the whole-brain volume [14]. The extracted brain volume was then segmented into cortical grey matter, subcortical grey matter, white matter, cerebrospinal fluid and pathological tissue. The procedure was executed in the FLS automated segmentation tools: FAST (https://fsl.fmrib.ox.ac.uk/fsl/fslwiki/FAST) and FIRST (https://fsl.fmrib.ox.ac.uk/fsl/fslwiki/FIRST). FSL FAST makes use of a hidden Markov random field method to categorise grey matter, white matter and CSF [15]. Due to poor contrast, an intensity-based segmentation procedure failed to segment subcortical grey matter, but FSL FIRST, which utilises a priori knowledge on the subcortical grey matter, improved segmentation results [16].

2.2.2 Fibre Orientation Distribution

The basis for every white matter tractography is the identification of fibre orientations from DWI data. The fibre orientation distribution (FOD) holds information on the probability of certain fibre directions within each voxel. The measured signal S (θ, ϕ) at a direction (θ, ϕ) is linked to FOD via the axially symmetric response function $R(\theta)$ that satisfies

$$S(\theta, \phi) = FOD(\theta, \phi) \bigotimes R(\theta) \tag{1}$$

where $R(\theta)$ models the signal expected for a voxel containing a single, coherently oriented fibre population.

FOD was obtained through constrained spherical deconvolution using spherical and rotational harmonics that formed a complete orthonormal basis set of functions over the sphere and the space of pure rotations. A constraint on positive fibre directions was then imposed [17–19]. The procedure was performed in the MRtrix3 software package (https://www.mrtrix.org/), which provides tools for the analysis of DWI data [20].

2.2.3   Anatomically Constrained Tractography

FOD estimates the probability that white matter fibres are aligned in a given direction. When deriving a probability density function (PDF) from the FOD, the trajectory of a neural fibre can then be constructed.

Tractography started with seeds being randomly scattered within a defined seeding area. From each seed point, the direction of a step was calculated by randomly sampling the PDF derived from a second-order integration over FODs [21]. The step size was predefined. Instead of stepping along straight-line segments defined by the step size, the connection between two points was realised by an arc of a circle of fixed length (step size), tangent to the current direction of tracking at the current point. The probability of each path was calculated as a joint probability of each infinitesimal step, thus making up the path. This joint probability was approximated by computing the product of the amplitude of the FOD, evaluated at four points (on the arc segment) along the tangent to the path. The sampling of the PDF was achieved using rejection sampling, a simple Monte Carlo method. Streamlines were iteratively generated by stepping along a path determined by the step size, the orientation derived from the PDF and a constraint on the maximum curvature of the streamline. The procedure was repeated until a stopping criterion was met.

Since reconstructed streamlines represent white matter fibres connecting spatially distant areas of grey matter, the start and end points should occur within the grey matter. Some fibres may project to the spinal cord, but fibres should never end within the white matter or fluid-filled regions. This a priori information was utilised to define the seeding area and to derive a termination criterion. Only fibres starting and terminating within cortical or sub-cortical grey matter were accepted. Fibres that ended in fluid-filled regions or inside the white matter were rejected [22].

As the density of reconstructed streamlines does not necessarily represent the true density of anatomical connections, the raw tractogram requires a filter. To find a subset of streamlines that best matched the diffusion signal, a spherical-deconvolution informed filtering procedure was used [23]. The FOD was proportional to the DWI signal and to the volume of fibres within a volume pixel. The FOD was firstly sampled using a large set of basis direction of a unit hemisphere. The sampled FOD was then segmented into individual lobes representing FOD peak orientations. The tractogram was subsequently mapped to the voxel grid of the DWI scans, and for every voxel, a FOD lobe look-up table was generated. Streamlines

Table 1 Parameters for probabilistic tractography

Step size	0.75 mm
Maximum curvature	15°
Seeding area	Grey matter-white matter interface
Max. number of seeds	500,000,000
Number of streamlines (raw tractogram)	500,000
Number of streamlines (After filtering)	207,485
Length of streamlines	3–150 mm
Track termination	Anatomical constraints

within every voxel were assigned to a lobe. For every voxel, the amount of streamlines in a specific orientation (of unit length) needed to match the amplitude of the FOD peak orientations. Therefore, a cost function, based on the FOD lobe amplitudes over all voxels and the total track density, was introduced and minimised by removing individual streamlines from the tractogram. Key characteristics of the tractogram and input settings are listed in Table 1.

2.2.4 Post-Processing

Two more processing steps are necessary before the white matter geometry can be combined with the volume conductor FE model: model reduction and fibre smoothing. The filtered tractogram still contained over 200,000 individual white matter fibres that ranged from 3 to 150 mm. In order to optimise the computation time for a preliminary analysis, the total number of fibres in the tractogram was reduced. Figure 2 shows a white matter tractogram for 1000, 2000, 5000 and 10,000 fibres ranging from 50 to 150 mm. While 1000 streamlines appeared too sparse for a proper visual inspection of the activating function, 2000 and more fibres covered the whole-brain volume in an adequate density. To limit the computational expense, a white matter model of 2000 fibres was selected. The extracted streamlines did not follow a smooth trajectory since the tractography algorithm allowed an abrupt change in direction after every step (within the fibre curvature constraint). To reduce the risk of potential errors and noise due to spatial discontinuity, the trajectory of every streamline was smoothed and subsequently oversampled. A cubic spline was used to fit the streamline data for all three dimensions separately passing through all provided points on the streamline. Every streamline was oversampled by a factor of 10, increasing the number of points and reducing the step size from 0.75 mm to 0.075 mm. This was done using the univariate spline interpolation function provided by the Python SciPy package (https://www.scipy.org/).

Fig. 2 Tractography model for different numbers of streamlines

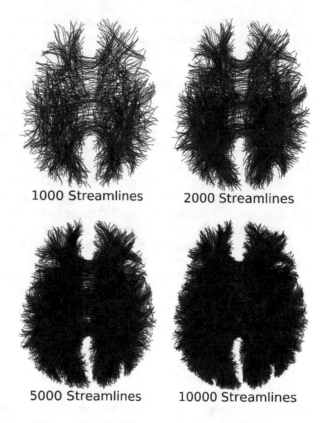

1000 Streamlines   2000 Streamlines

5000 Streamlines   10000 Streamlines

2.3 Finite Element Analysis of ECT Brain Stimulation

To simulate the potential field introduced to the head by ECT, a volumetric FE model of the head was generated from structural MRI data. This included image segmentation, FE mesh generation, electrical property settings and finally running simulations in a numerical solver. Two different ECT electrode placements were simulated in this study, and three different electric conductivity settings were chosen for the white matter compartment. The FE analysis of ECT brain stimulation was based on Bai et al. [25].

2.3.1 Finite Element Model Reconstruction

The structural T1-weighted MRI data was used to reconstruct an FE head model. After contrast and edge enhancements of the structural data, segmentation of the scalp, skull, paranasal sinuses, cerebrospinal fluid, grey matter and white matter were performed using 3D Slicer, an open-source software platform for medical image processing (https://www.slicer.org/). Each tissue compartment was assigned with a label map using a range derived from a histogram, representing the grey levels of the desired tissue type. The selection of desired tissue was performed by setting the threshold on a single slice, and then a paintbrush was used to manually modify and correct the selection. This procedure was repeated at every second or third slice of the image volume, and a full segmentation was created by using a trilinear interpolation to automatically connect the sparse set of contours. A paintbrush was then used again to further modify the tissue map until the desired accuracy was achieved. As the face of the subject was removed from the images to keep the subject identity anonymous, a smooth surface was used to replace the face of the head model. For every tissue compartment, a surface triangular mesh was generated in 3D Slicer. To increase the mesh quality and smoothness, the surface meshes of the head model were imported into Geomagic Wrap (3D Systems, USA). Non-manifold edges were removed, self-intersecting triangles were split, the edge crease was reduced, spikes were smoothed, and existing holes were repaired. The processed surface meshes were then transferred to ANSYS ICEM CFD (ANSYS, USA) to create a volumetric mesh. Edges at the intersections between compartments and at the desired electrode contact locations were defined, and a tetrahedral volumetric mesh with appropriate meshing and coarsening parameters was generated. To calculate the voltage data, the final volumetric mesh, with approximately half a million tetrahedral elements, was exported to COMSOL Multiphysics (COMSOL Inc., Sweden), a cross-platform FE solver.

2.3.2 Tissue Conductivities

To calculate the potential field induced by ECT, electrical properties need to be assigned to the different tissue compartments of the volumetric head model. All parts of the head model, apart from the white matter compartment, were considered electrically homogeneous and isotropic. Conductivity values assigned to the scalp, the skull, the paranasal sinuses, the CSF and grey matter are listed in Table 2. Three different white matter conductivity settings, labelled with G, GW and anisotropic, were simulated in order to compare its influence on the potential distribution and the derived electric field and activating function:

- G: isotropic grey matter conductivity (0.31 S/m)

Table 2 Electrical conductivities of all tissue compartments in the head model [25]

Structure	Conductivity (S/m)
Scalp	0.41
Skull, compact bone	0.06
Skull, spongy bone	0.028
Paranasal sinuses	0
Cerebrospinal fluid	1.79
Grey matter	0.31
White matter (isotropic)	0.14
White matter (anisotropic)	0.065 (transverse to fibre) 0.65 (longitudinal to fibre)

- GW: isotropic conductivity (0.14 S/m)
- Anisotropic: anisotropic conductivity based on conductivity tensor [26, 27]

2.3.3 White Matter Conductivity Anisotropy

The linear relationship between the electrical conductivity tensor and the water diffusion tensor implies that both share the same eigenvectors [11, 12]. The water diffusion tensor was extracted using the diffusion tensor fitting algorithm for DWI data implemented in FSL DTIFIT (https://fsl.fmrib.ox.ac.uk/fsl/fslwiki/FDT). The conductivity tensor σ_C of a white matter finite element was modelled as

$$\sigma_C = E_N diag\left(\sigma_C^{long}, \sigma_C^{trans}, \sigma_C^{trans}\right) E_N^T \tag{2}$$

where E_N is the orthogonal matrix of normalised eigenvectors of the diffusion tensor. σ_C^{long} and σ_C^{trans} are the eigenvalues parallel and perpendicular to the fibre directions with $\sigma_C^{trans} \leq \sigma_C^{long}$ [27]. The eigenvalues of the conductivity tensor were determined using a volume constraint proposed by Wolters et al. [26]. To ensure that only white matter voxels with a high level of anisotropy were considered for the calculation of the conductivity tensor, the fractional anisotropy (FA) derived from the diffusion tensor was used. Regions with a low FA values (typically FA < 0.45), which suggested a low local anisotropy, were removed from the tensor analysis.

2.3.4 ECT Brain Stimulation Settings

In the field of bioelectromagnetism, since the frequency of bioelectric events is low, biological tissues are typically considered as volume conductors, in which the capacitive and inductive components of the electrical impedance is neglected [28]. Thus, all head compartments were formulated as passive volume conductors, and the electric potential resulting from ECT was obtained by solving Laplace's equation

$$\nabla \cdot (-\sigma_C \nabla V) = 0, \tag{3}$$

where V is the electric potential and σ_C is the electrical conductivity tensor. ECT electrodes were defined as circular boundaries with a radius of 2.5 cm on the scalp. A stimulus of 800 mA was delivered through the stimulating electrode. Two bipolar ECT electrode placements were simulated as Fig shown in. 3:

- Bitemporal (A): both electrodes were placed 3 cm superior to the midpoint of a line connecting the external ear canal with the lateral angle of the ipsilateral eye, the stimulating electrode on the right and the return electrode on the left.

Fig. 3 Illustration of
bitemporal (**a**) and right
unilateral (**b**) electrode
placement for ECT. (Image
adapted from [24])

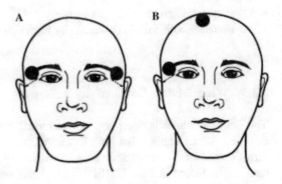

- Right unilateral (B): the stimulating electrode was placed 3 cm superior to the midpoint of a line connecting the right external ear canal with the lateral angle of the right eye and the return electrode placed just right of the vertex of the head.

Furthermore, a constant current density J_n for the stimulating electrode was defined as

$$\int_{S_E} J_n dS = I_S \tag{4}$$

where S_E is the area of the electrode and I_S is the applied stimulus current at a constant DC level of 800 mA. All external boundaries were assigned as electric insulators, and a continuous current density across all interior boundaries was assumed.

2.4 Model Combination

Having both models set up, the FE volume conductor model, solved for the potential distribution within the head for a given electrode placement, can be combined with the white matter tractography model. For every white matter fibre conductivity setting and ECT electrode position, voltage data along the fibres in the tractogram was exported from the COMSOL models. For fibre coordinates not coinciding with a FE mesh point, trilinear interpolation was used. To investigate neural activation, the neural activating function

$$f_A = \frac{1}{r_{ax}c_m} \frac{\partial^2 V}{\partial x^2} \tag{5}$$

was evaluated to link the activation to the second spatial derivative of the potential along the fibre [29]. In Eq. (5), r_{ax} is the axonal resistance determined by the fibre diameter and the axoplasmic resistivity, and c_m is the capacity of the fibre. The activating function was in this chapter approximated by the second spatial derivative of the electrical potential assuming that r_{ax} and c_m were constant.

The electric field (EF) $\boldsymbol{E} = -\nabla V$ is the negative first spatial derivative of the potential along the fibre. The first spatial derivative $\frac{\partial V}{\partial x}$ was approximated by central difference:

$$V'_k = -\boldsymbol{E}_k \approx \frac{V_{k+1} - V_{k-1}}{|r_{k+1,k}| + |r_{k,k-1}|} \tag{6}$$

where k represents the kth node on an individual fibre, V'_k is the first spatial derivative of V at the kth node and $|r_{k+1,k}|$ and $|r_{k,k-1}|$ are the distances between the $(k+1)$th and kth nodes as well as between the kth and $(k-1)$th nodes, respectively. The second spatial derivative along the fibre $\frac{\partial^2 V}{\partial x^2}$ was approximated by the central difference method on the electric field such that

$$V''_k \approx \frac{\frac{V_{k+1}-V_k}{|r_{k+1,k}|} - \frac{V_k-V_{k-1}}{|r_{k,k-1}|}}{\frac{|r_{k+1,k-1}|}{2}} \tag{7}$$

3   Results

3.1   White Matter Fibre Tractography Model

The final neural fibre model – as shown in Fig. 4 – consisted of 2000 individual fibres with the length between 50 and 150 mm. The spatial sampling of each fibre was 0.075 mm and every fibre respected the anatomical boundary condition of starting and ending within the grey matter. Fibres were manually categorised to commissural fibres (red, connecting both hemispheres), projection fibres (blue, connecting the brain to the spinal cord) and association fibres (green, interconnecting brain areas within the same hemisphere). The fibre of the pons, connecting both hemispheres of the cerebellum (red), marked only a small fraction of fibres. The model contained 58% association fibres, 24% commissural fibres (including fibres of the pons) and 18% projection fibres.

3.2   Electric Field and Activating Function for Three White Matter Conductivity Settings

The potential distribution as well as the approximation of the electric field and the activating function along the fibres for a bitemporal and a right unilateral electrode placement is presented in Fig. 5. The comparison of three different conductivity

Fig. 4 Final neural fibre model containing 2000 fibres with a length of between 50 and 150 mm. Red fibres are commissural fibres, interconnecting the left and right hemispheres; blue fibres are projection fibres, projecting down to the spinal cord; green are association fibres that connect brain areas within the same hemisphere

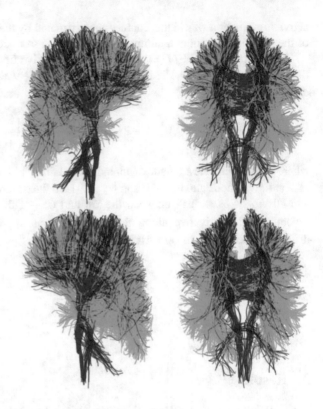

settings for white matter shows that the incorporation of white matter anisotropy resulted in a wider range and a greater non-uniformity of the potential distribution. The non-uniform potential distribution also led to higher amplitudes and a more distinctive distribution of the first and second spatial derivatives when compared to the two isotropic conductivity settings. In addition, the influence on deep brain structures was most pronounced in the models with anisotropic white matter conductivity.

3.3 White Matter Activation

Figure 6 shows the electric potential as well as the first and second spatial derivatives of the potential along the fibre for a single commissural fibre in the anisotropic bitemporal ECT model. High-potential values appeared on the left side of the fibre, i.e. the left hemisphere. The fibre started with a steep descent from the right hemisphere towards the corpus callosum where it ran nearly parallel before it took a turn up and down. The symmetric electrode placement resulted in a strong potential gradient from the right hemisphere to the left. Given the trajectory of the fibre, two peaks appeared in the first spatial derivative resulting in a high second spatial derivative at the beginning and the end of the fibre. While crossing the corpus

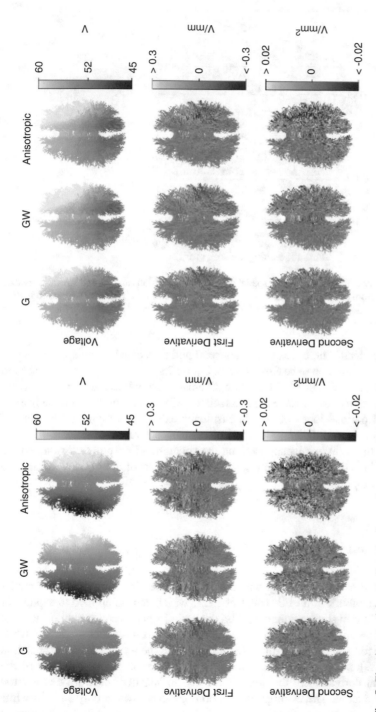

Fig. 5 Electric potential as well as the first and second spatial derivatives of the potential along the fibres for bitemporal (left) and right unilateral (right) electrode placements using three different white matter conductivity settings. In 'G' white matter conductivity is assigned the same value as the grey matter (0.31 S/m); in 'GW', the white matter conductivity is assigned an isotropic value of 0.14 S/m; in 'anisotropic', the white matter fibre conductivity is an anisotropic conductivity tensor derived from the fibre orientation distribution

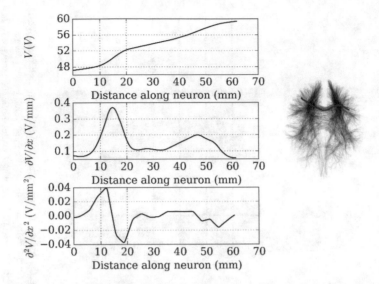

Fig. 6 Electric potential (top left), first spatial derivative (middle left) and second spatial derivative of potential (bottom left) along a single streamline reaching from the right hemisphere to the left hemisphere in the anisotropic bitemporal ECT model. The trajectory of the fibre is indicted in red in the fibre plot (right)

callosum, the electric field was nearly constant and thus causing the approximation of the activating function to be close to zero. Figure 7 shows four views of the second spatial derivative for both ECT models with anisotropic white matter conductivity. For a bitemporal electrode placement, association fibres running through the frontal, temporal and parietal lobes of both hemispheres exhibited a higher probability of activation. For right unilateral electrode placement, activation was more likely to initiate from the right hemisphere, among the splenium of corpus callosum and the association fibres that travelled between the posterior part of the frontal lobe and the parietal and temporal lobes.

4   Discussion

This chapter presented a study that utilised, for the first time, a white matter tractography analysis in an FE model of ECT, which can be applied to models of other types of brain stimulation. In reality, the brain is made up of a tight package of neuronal somas in the grey matter and their extensions — axons — in the white matter connecting to other neurons. According to the theory of external stimulation of neurons [29], neuronal excitation of a homogenous axon can be indicated by the second spatial derivative of the potential V, i.e. the negative derivative of electric field along the axon. The assumption of a homogenous axon is only valid in white matter. In the grey matter, where cell somas are located, changes in the diameter of the axons to the soma violate this condition, and the precise electroanatomy of the

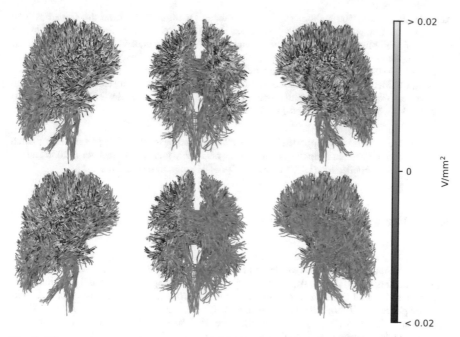

Fig. 7 Three projections of the second spatial derivative along the fibres for bitemporal (top) and right unilateral (bottom) ECT head models with anisotropic white matter conductivity

neuron, together with the variation of the extracellular potential V, has to be evaluated. We have therefore combined a tractography analysis with a volume conductor model and derived brain locations in the white matter, where axons are most likely activated by ECT. It should nevertheless be noted that a more accurate analysis of the activation pattern can only be performed by the inclusion of neuronal cable models. Moreover, the addition of cable models can also be used to simulate the propagation of activation, such as seizure. However, as neurons in varied functional areas of the brain have been found to possess different geometrical and ionic channel properties, the incorporation of cable models across the whole brain may be a challenging and computational expensive task.

To avoid large numerical errors when the second derivative of V along the path of the axon was calculated, it was necessary to smooth the path of the axon and interpolate potential data carefully onto the fibre trajectory. The purpose of the tractography was to study the distribution of electric signals along the fibres on a large scale. Therefore, it is important to have a correct representation of all types of white matter fibres within the brain. For this preliminary study, however, after a downsampling process to select long fibres, the full variety of neural pathways was not well presented, particularly the short projection fibres. In spite of this, a close inspection of the filtered tractogram revealed a good representation of major fibres connecting the two hemispheres, as well as the cerebrum to the brainstem and the cerebellum. Furthermore, it was confirmed that the fibres did not cross liquid-filled areas and that each fibre started and ended within the grey matter.

Acknowledgements The authors would like to thank Dr. Jörg Encke from the Department of Electrical and Computer Engineering, Technical University of Munich, for his support in AF analysis and Dr. Miguel Molina from the Munich School of Bioengineering, Technical University of Munich, for his support in tractography. Image data were provided by the Human Connectome Project, MGH-USC Consortium (principal investigators: Bruce R. Rosen, Arthur W. Toga and Van Wedeen; U01MH093765) funded by the NIH Blueprint Initiative for Neuroscience Research grant, the National Institutes of Health grant P41EB015896 and the Instrumentation Grants S10RR023043, 1S10RR023401 and 1S10RR019307.

Funding The authors were supported by grants from the European Union's Horizon 2020 research and innovation programme under the Marie Skłodowska-Curie grant agreement no. 702030 and the German Research Foundation (DFG) under the D-A-CH programme (HE6713/2-1).

References

1. Bai, S., Loo, C., & Dokos, S. (2013). A review of computational models of transcranial electrical stimulation. *Critical Reviews in Biomedical Engineering, 41*(1), 21–35.
2. Bai, S., Loo, C., Al Abed, A., & Dokos, S. (2012). A computational model of direct brain excitation induced by electroconvulsive therapy: Comparison among three conventional electrode placements. *Brain Stimulation, 5*(3), 408–421.
3. Butson, C. R., & McIntyre, C. C. (2006). Role of electrode design on the volume of tissue activated during deep brain stimulation. *Journal of Neural Engineering, 3*, 1–8.
4. Gunalan, K., et al. (2017). Creating and parameterizing patient-specific deep brain stimulation pathway-activation models using the hyperdirect pathway as an example. *PLoS One, 7*, e0176132.
5. Fan, Q., et al. (2016). MGH-USC Human Connectome Project datasets with ultra-high b-value diffusion MRI. *NeuroImage, 124*, 1108–1114.
6. Fan, Q., et al. (2014). Investigating the capability to resolve complex white matter structures with high b-value diffusion magnetic resonance imaging on the MGH-USC Connectom scanner. *Brain Connectivity, 4*, 718–726.
7. Jenkinson, M., Bannister, P., Brady, M., & Smith, S. (2002, October). Improved optimization for the robust and accurate linear registration and motion correction of brain images. *NeuroImage, 17*(2), 825–841.
8. Jenkinson, M., & Smith, S. (2001, June). A global optimisation method for robust affine registration of brain images. *Medical Image Analysis, 5*(2), 143–156.
9. Greve, D. N., & Fischl, B. (2009). Accurate and robust brain image alignment using boundary-based registration. *NeuroImage, 48*, 63–72.
10. Bai, S., Loo, C., Geng, G., & Dokos, S. (2011). Effect of white matter anisotropy in modeling electroconvulsive therapy. *Conference Proceedings: Annual International Conference of the IEEE Engineering in Medicine and Biology Society, 2011*, 5492–5495.
11. Tuch, D. S., Wedeen, V. J., Dale, A. M., George, J. S., & Belliveau, J. W. (2001). Conductivity tensor mapping of the human brain using diffusion tensor MRI. *Proceedings of the National Academy of Sciences of the United States of America, 98*, 11697–11701.
12. Lee, S., Cho, M., Kim, T., Kim, I., & Oh, S. (2006). Electrical conductivity estimation from diffusion tensor and T2: A silk yarn phantom study. *Proceedings of the 14th Science Meeteeting International Society for Magnetic Resonance in Medicine, 14*, 3034.
13. Ahmad Bakir, A., Bai, S., Lovell, N. H., Martin, D., Loo, C., & Dokos, S. (2019). Finite element modelling framework for electroconvulsive therapy and other transcranial stimulations. In *Brain and human body modeling*. Cham: Springer.
14. Smith, S. M. (2002). Fast robust automated brain extraction. *Human Brain Mapping, 17*, 143–155.

15. Zhang, Y., Brady, M., & Smith, S. (2001). Segmentation of brain MR images through a hidden Markov random field model and the expectation-maximization algorithm. *IEEE Transactions on Medical Imaging, 20*, 45–57.
16. Patenaude, B., Smith, S. M., Kennedy, D. N., & Jenkinson, M. (2011). A Bayesian model of shape and appearance for subcortical brain segmentation. *NeuroImage, 56*, 907–922.
17. Tournier, J. D., Calamante, F., Gadian, D. G., & Connelly, A. (2004). Direct estimation of the fiber orientation density function from diffusion-weighted MRI data using spherical deconvolution. *NeuroImage, 23*, 1176–1185.
18. Tournier, J. D., Calamante, F., & Connelly, A. (2007). Robust determination of the fibre orientation distribution in diffusion MRI: Non-negativity constrained super-resolved spherical deconvolution. *NeuroImage, 35*, 1459–1472.
19. Jeurissen, B., Tournier, J. D., Dhollander, T., Connelly, A., & Sijbers, J. (2014). Multi-tissue constrained spherical deconvolution for improved analysis of multi-shell diffusion MRI data. *NeuroImage, 103*, 411–426.
20. Tournier, J.-D., et al. (2019). MRtrix3: A fast, flexible and open software framework for medical image processing and visualisation. *NeuroImage, 202*, 116137.
21. Tournier, J.-D., Connelly, A., & Calamante, F. (2010). Improved probabilistic streamlines tractography by 2nd order integration over fibre orientation distributions. *Proc. Intl. Soc. Mag. Reson. Med. (ISMRM)*. 1670.
22. Smith, R. E., Tournier, J. D., Calamante, F., & Connelly, A. (2012). Anatomically-constrained tractography: Improved diffusion MRI streamlines tractography through effective use of anatomical information. *NeuroImage, 62*, 1924–1938.
23. Smith, R. E., Tournier, J. D., Calamante, F., & Connelly, A. (2013). SIFT: Spherical-deconvolution informed filtering of tractograms. *NeuroImage, 67*, 298–312.
24. McNally, K. A., & Blumenfeld, H. (2004, February). Focal network involvement in generalized seizures: New insights from electroconvulsive therapy. *Epilepsy & Behavior, 5*(1), 3–12.
25. Bai, S., Dokos, S., Ho, K. A., & Loo, C. (2014). A computational modelling study of transcranial direct current stimulation montages used in depression. *NeuroImage, 87*, 332–344.
26. Wolters, C. H., Anwander, A., Tricoche, X., Weinstein, D., Koch, M. A., & MacLeod, R. S. (2006). Influence of tissue conductivity anisotropy on EEG/MEG field and return current computation in a realistic head model: A simulation and visualization study using high-resolution finite element modeling. *NeuroImage, 30*, 813–826.
27. Shimony, J. S., et al. (1999). Quantitative diffusion-tensor anisotropy brain MR imaging: Normative human data and anatomic analysis. *Radiology, 212*, 770–784.
28. Malmivuo, J., & Plonsey, R. (2012). *Bioelectromagnetism: Principles and applications of bioelectric and biomagnetic fields*. New York: Oxford University Press.
29. Rattay, F. (1986). Analysis of Models for External Stimulation of Axons. *IEEE Transactions on Biomedical Engineering, BME-33*, 974–977.

Personalization of Multi-electrode Setups in tCS/tES: Methods and Advantages

R. Salvador, M. C. Biagi, O. Puonti, M. Splittgerber, V. Moliadze,
M. Siniatchkin, A. Thielscher, and G. Ruffini

1 Introduction

Transcranial current stimulation (tCS), including transcranial direct current stimulation (tDCS), transcranial alternating current stimulation (tACS), and transcranial
random noise stimulation (tRNS), is a family of noninvasive neuromodulatory
techniques that employ weak (1–4 mA) electrical currents applied via electrodes
placed on the scalp for long durations (20–40 min) [1, 2]. Concurrent effects of
stimulation range from changes in cortical excitability [3] to modulation of ongoing
endogenous oscillations [4]. Hebbian-based mechanisms are hypothesized to lead to
long-lasting plastic changes in the brain [5], leading to an increasing interest in
putative therapeutic applications in a range of neurological diseases [6]. One factor
that limits the usefulness of tCS is the widely reported intersubject variability of

R. Salvador (✉) · M. C. Biagi
Neuroelectrics, Barcelona, Spain
e-mail: ricardo.salvador@neuroelectrics.com

O. Puonti · A. Thielscher
Danish Research Centre for Magnetic Resonance, Centre for Functional and Diagnostic
Imaging and Research, Copenhagen University Hospital Hvidovre, Hvidovre, Denmark

Department of Health Technology, Technical University of Denmark, Kongens Lyngby,
Denmark

M. Splittgerber · V. Moliadze
Institute of Medical Psychology and Medical Sociology, University Medical Center Schleswig
Holstein, Kiel University, Kiel, Germany

M. Siniatchkin
Clinic of Child and Adolescent Psychiatry and Psychotherapy, Evangelic Hospital Bethel
(EvKB), Bielefeld, Germany

G. Ruffini
Neuroelectrics, Barcelona, Spain

Starlab, Barcelona, Spain

© The Author(s) 2021
S. N. Makarov et al. (eds.), *Brain and Human Body Modeling 2020*,
https://doi.org/10.1007/978-3-030-45623-8_7

119

responses to stimulation [7]. Several factors can explain variability, but here we will focus on the physical agent of the effects that tCS has on neurons: the electric field (E-field) induced in the tissues.

1.1 Biophysical Aspects of tCS

The distribution of currents in the head can be described mathematically by the electric field (E-field) vector induced by tCS. Depending on the orientation of the latter with respect to neuronal processes, the membrane of pyramidal neurons is polarized (approximately 0.2 mV per 1 V/m of E-field value, [8]), which leads to the observed concurrent effects of stimulation. One common hypothesized mechanism is the polarization of the soma of pyramidal cells due to the component of the E-field perpendicular to the cortical surface (E_n), [9]. However, other mechanisms of interaction are possible, such as the polarization of axon terminals [10].

The E-field distribution depends on factors such as head geometry (thickness and shape of the head tissues), electrical properties of the tissues (electrical conductivity, σ), location and geometry of the electrodes, and the currents that are applied via the electrodes [11]. Since in vivo measurements of the E-field still pose a number of technical challenges [12, 13] and cannot easily be carried out, computational head models based on structural data (usually head MRIs) are usually employed to estimate it [14, 15].

Initial uses of computational head models were limited to a posteriori analysis of the E-field distribution of electrode montages typically applied in experimental protocols [15, 16]. In recent years, several algorithms have been described to leverage these head models with the objective of optimizing some dose parameters (position and currents of the stimulating electrodes) to target a specific brain region and/or cortical network [9, 17–20].

This paradigm shift from "one-model-fits-all" montages to individualized montages leveraging subject-specific head models and dose optimization algorithms has the potential to reduce intersubject variability in the outcomes of tCS and allow for more effective and safe protocols. However, several parameters can affect the outcome of these modeling and optimization pipelines. In this work, we will study how uncertainties in target specification, tissue electrical conductivities, and the threshold for neuromodulatory effects can affect the outcome of the optimization. We will also discuss some of the potential benefits of these pipelines, especially in relation to reduction of intersubject variability of the results of optimization.

2 Methods

2.1 Subjects

We included seven healthy children and adolescents (three males) aged 10–17 years (M 14; SD 2). The study was approved by the Ethics Committee of the Faculty of Medicine, Kiel University, Kiel Germany.[1] All participants and their parents were instructed about the study, and written informed consent according to the Declaration of Helsinki on biomedical research involving human subjects was obtained. The study is part of the STIPED project.[2]

2.2 Head Model Generation

Each subject underwent structural head scanning on a 3 T Philips Achieva scanner, during which the following sequences were acquired: a T1-weighted scan (1 mm^3, TR = 2530 ms, TE = 3.5 ms, TI = 1100 ms, FA = 7°, fast water excitation), a T2-weighted scan (1 mm^3, TR = 3200 ms, TE = 300 ms, no fat suppression), and a diffusion MRI (dMRI) scan (2 mm^3, TR = 6300 ms, TE = 51 ms, 67 directions, b = 1000).

Tissue segmentation was performed using an in-house implementation combining extra-cerebral tissue segmentations from a new segmentation approach, which will be included in a future version of the open-source simulation toolbox SimNIBS,[3] with brain tissue segmentations and cortical gray matter (GM) surface reconstructions from FreeSurfer [21]. Finite element head models were then generated (see Fig. 1), including representations of the scalp, skull, cerebrospinal fluid (CSF), gray matter, and white matter (WM). The head models also contained representations of Pistim electrodes (1 cm radius, cylindrical Ag/AgCl electrodes[4]) placed in 61 positions of the 10-10 EEG system. For the electrodes, only the conductive gel underneath the metal connector was represented in the head model. Unless otherwise stated, the scalp, skull, and CSF were modeled as isotropic with conductivities of 0.33 S/m, 0.008 S/m, and 1.79 S/m, respectively, which are appropriate values for the DC-low frequency values used in tCS [15]. The GM and WM were modeled as anisotropic (volume normalization, [22], with isotropic conductivity values used for diffusion tensor scaling of 0.40 S/m–0.15 S/m, for the GM – WM, [15]). E-field calculations were performed in COMSOL[5] using second-order tetrahedral mesh elements to solve Laplace's equation [11].

---

[1]http://apps.who.int/trialsearch/Trial2.aspx?TrialID=DRKS00008207

[2]http://www.stiped.eu/home/

[3]https://simnibs.github.io/simnibs/build/html/index.html

[4]www.neuroelectrics.com

[5]v5.3, www.comsol.com

Fig. 1 Finite element head model generated for one of the subjects in this study. The model includes representations of the scalp (in yellow), skull (in gray), CSF (in blue), GM (in gray), and WM (in white), as well as gel underneath the electrodes (in green). Air cavities are represented as cavities in the mesh, thus effectively modeling the air as an insulator

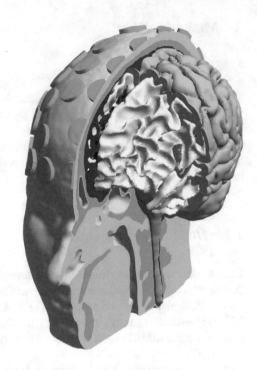

2.3 Montage Optimization Algorithm

The optimization algorithm used in this study is based on the Stimweaver algorithm [9]. We assumed the normal component of the E-field to the cortical (GM-CSF) surface E_n as responsible for the acute effects of stimulation. Positive/negative E_n-values, corresponding to E-fields directed into/out-of-the cortical surface, lead to increased/decreased excitability of the soma of pyramidal cells. Inputs to this algorithm include target E_n-maps, with information about the target E_n-field (E_n^{Target}) in each node of the cortical surface; weight maps, with information about the priority (weight, w) assigned to each node in the optimization; current constraints of the montage (maximum current, in absolute value, per channel $I_{max\ channel} = max_i\{|I_i|\}$ and maximum total injected current $I_{max\ total} = \frac{1}{2}\sum_{i=1}^{N_{Channels}}|I_i|$); and the maximum number of electrodes in the montage. The objective function in the optimization is the error with respect to no intervention (ERNI, with units of mV^2/mm^2):

$$ERNI = \sum_{i=1}^{N} \frac{\left(w_i\sum_{j=1}^{N_{channels}-1}E_{n,i}^{j-Cz}I_j - w_iE_{n,i}^{Target}\right)^2 - \left(w_iE_{n,i}^{Target}\right)^2}{\sum_{j=1}^{N}w_j^2}$$

where N is the number of mesh nodes, $N_{channels}$ is the number of electrodes available for the optimization, $E_{n,i}^{j-Cz}$ is a column vector (lead-field vector) with the normal component of the E-field induced by a bipolar montage that has j as the anode (+1 mA) and Cz as the cathode (-1 mA), and I_j is the current (in mA) of electrode j in the montage that is being evaluated. The term $\sum_{j=1}^{N_{Channels}-1} E_{n,i}{}^{j-Cz} I_j$ yields the normal component of the E-field in the montage being evaluated (for each node i), as follows from the linearity principle [9]. The lead-field terms $E_{n,i}^{j-Cz}$ are calculated on a subject-specific basis using the methods detailed in the previous section. The optimization without the constraint on the number of active electrodes was performed by in-house scripts programmed in Python using the SciPy library.[6] In order to constrain the number of electrodes of the final montage, a genetic algorithm (GA) was implemented following the methods described in [9].

2.4 Studies Performed

In this work, we performed several studies to clarify the impact of several inputs and parameters to the optimization algorithm in the results. The first study (*study a*) aims at determining the impact of target size on the optimization results. This is related to the perceived mechanisms of stimulation underlying the effects of tCS, with some studies focusing on highly localized targets, obtained, for instance, from EEG source localization information [17, 20], where other studies focus on more widespread cortical areas with information extracted from cytoarchitectural information [23] or functional imaging data [24]. Optimization algorithms can tackle both of these cases, but it is unclear how current constraints influence the capability of achieving the desired target E_n-field with increasing target area size.

Another important parameter that affects the E-field distribution, and therefore the results of the optimization, is the electrical conductivity of the head tissues. Measuring these values in vivo still presents several limitations, and data available in the literature has a wide variability, representing different measuring methods and origin of the tissue samples [25]. Furthermore, some reports indicate that this value might change according to the subject's age, at least for the skull [26]. In *study b*, we partially tackled this problem by assessing the influence that different conductivity values for the skull tissue have on the optimization results.

In *study c*, we investigated some of the potential benefits of optimization algorithms on experimental design, namely, less variability on E_n-field distribution.

More details about how each study was performed are presented in the next section. In the studies where surface average E_n values are presented, they were calculated with the following expression:

---

[6]https://www.scipy.org/scipylib/

$$< E_n >= \frac{\int_{Area\,Patch} E_n dA}{A_{Patch}} = \frac{\sum_{i=1}^{N_{Patch}} E_{n,i} A_i}{\sum_{i=1}^{N_{Patch}} A_i}$$

where A_i is the area associated with each node of the mesh (the sum of the areas of all the triangles connected to the node divided by 3) and N_{patch} is the number of nodes in the surface patch where the average is being calculated.

3 Results

3.1 Study A: Effect of Target Size

In this study, we investigated how target size and current constraints affect the actual $<E_n>$ achievable on a target given the current constraints. The target was located on the left precentral gyrus, and its size varied from 9 mm^2 (a tiny spot on the gyrus crown) to ~275 mm^2 (about the entire left frontal lobe). The different areas were first identified on the cortical surface of a template brain model (Colin27[7]), starting from the smallest one and progressively enlarging it up to the maximum size considered. Each area was then remapped separately onto the cortical surface of one of the participants of this study, and single target maps were created: the target area was assigned to excitation, with two values of E_n^{Target} (0.25 V/m and 0.50 V/m), and maximum weight $w_{stim} = 10$; the rest of the cortex was assigned no stimulation ($E_n^{Target} = 0$ V/m) with weight $w_{no\text{-}stim}$ varying for each area, in such a way that both conditions have the same relative importance to the ERNI calculation: $(w_{no\text{-}stim}/w_{stim})^2 = Area_{stim}/Area_{no\text{-}stim}$.

Figure 2 shows $<E_n>$ on the target as a function of target size. The E_n distribution results from optimized montages with an unconstrained number of electrodes, obtained for different combinations of E_n^{Target}, maximum current per electrode, and total injected current ($I_{max\ channel}$, $I_{max\ total}$) including values exceeding the usual safety limit of 4.0 mA.

For all current constraints and both E_n^{Target} values, we observe an overall logarithmic decrease of the $<E_n>$ with the target size. This decrease is more rapid for the higher E_n^{Target}, and it follows a non-monotonic trend that can be correlated to the cortical curvature of the target area and to the distance between the target area and electrodes. In fact, the significant drop at 108 mm^2 w.r.t. the previous size is likely due to the fact that the ROI is now large enough to comprehend both faces of a sulcus, which have surface normals – and consequently normal electric field – pointing in opposite directions: once averaged over the whole ROI, this results in

---

[7]https://www.bic.mni.mcgill.ca/ServicesAtlases/Colin27

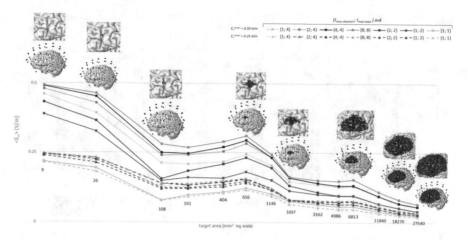

Fig. 2 Average normal component of the E-field on target (red area on the GM surfaces), as a function of the target area (in log scale), for different values of the target normal E-field ($E_n^{Target} = 0.25$ V/m dashed lines, $E_n^{Target} = 0.50$ V/m continuous lines) and combinations of the individual and total current constraints (yellow, $I_{max\ total} = 2$ mA; pink and red, $I_{max\ total} = 2$ mA; gray and black, $I_{max\ total} = 4$ mA; green, $I_{max\ total} = 8$ mA). The pictures also show the available positions of the electrodes on the scalp

a decrease of $\langle E_n \rangle$. With the next area increase, the ROI extends out of the sulcus, over the two adjacent gyri, and approaches the normal projection of the center of the two closest electrodes, which is reflected in the slow increase of $\langle E_n \rangle$, reaching a local peak at 656 mm². Further area increases, in the second half of the plot, repeat this pattern, ultimately created by the compounding effect of the gyrification of the target area and the distance from the covering electrodes.

Concerning the current constraints, in Fig. 3, we look separately at the influence of the $I_{max\ total}$ (3a) and of the $I_{max\ channel}$ (3b). The shaded area in blue in Fig. 3a represents the value of the maximum $\langle E_n \rangle$ achievable in each target, for both E_n^{Target}, normalized with respect to E_n^{Target}. This maximum $\langle E_n \rangle$ is obtained with a montage optimization with unrestricted maximum individual and total injected current and is the same for both values of E_n^{Target}. As we observe, it also decays logarithmically with the target area, from 95% E_n^{Target} for the smallest target to 15% E_n^{Target} for the largest. In this case the decay is to attribute utterly to the effect of head anatomy and electrode positions. The figure also shows the $\langle E_n \rangle$ obtained with different current constraints, normalized with respect to the maximum $\langle E_n \rangle$ achievable, for both E_n^{Target}.

We observe that, for $E_n^{Target} = 0.50$ V/m, only with a $I_{max\ total} = 8.0$ mA it is possible to induce in all areas a $\langle E_n \rangle$ at least over 80% of the maximum $\langle E_n \rangle$ achievable. On the other hand, a total injected current $I_{max\ total}$ of 1.0 mA does not reach even the half of the maximum achievable $\langle E_n \rangle$, in any target, including the smallest. Moreover, we observe that, as a result of the linearity of $I_{max\ total}$ and E_n^{Target}, for the less stringent condition of $E_n^{Target} = 0.25$ V/m, the exact same relative $\langle E_n \rangle$ on each target area can be achieved with half of the total current.

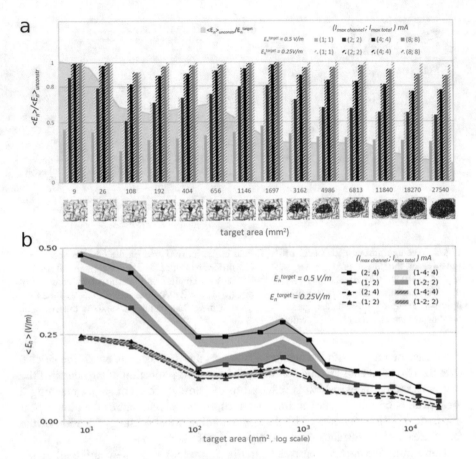

Fig. 3 Effect of the total injected (**a**) and individual (**b**) current constraints on $<E_n>$ on target areas of different sizes, for different target E_n-fields. (**a**) The bars show the relative $<E_n>$ w.r.t. the $<E_n>$ calculated with unconstrained current, per each target area and E_n^{Target}; the area shows the relative unconstrained $<E_n>$ w.r.t. E_n^{Target} (same for both values of E_n^{Target}). (**b**) The shaded areas represent the $<E_n>$ obtained with $I_{max\ channel}$ within 1 mA and $I_{max\ total}$, for $E_n^{Target} = 0.50$ V/m and $I_{max\ total} = 4$ mA (solid gray), $E_n^{Target} = 0.50$ V/m and $I_{max\ total} = 2$ mA (solid red), $E_n^{Target} = 0.25$ V/m and $I_{max\ total} = 4$ mA (dashed gray), $E_n^{Target} = 0.25$ V/m and $I_{max\ total} = 4$ mA (dashed red). The lines represent solutions with $I_{max\ channel} = I_{max\ total}/2$, for $E_n^{Target} = 0.50$ V/m (solid lines) and $E_n^{Target} = 0.25$ V/m (dashed lines)

Consequently, $I_{max\ total} = 4.0$ mA in this case is sufficient to reach at least 80% of the maximum achievable $<E_n>$. In Fig. 3b, we focus on the current constraints considered in studies b and c. This plot indicates that $I_{max\ channel}$ modulates the $<E_n>$ only up to a given target size (which is smaller for the less demanding condition: $E_n^{Target} = 0.25$ V/m). After this threshold area, the only factor influencing $<E_n>$ becomes $I_{max\ total}$.

3.2 Study B: Tissue Conductivity Values

In this study, we assessed how skull conductivity (σ_{skull}) values affected the optimization results. We tried three different conductivity values: 0.008 S/m (our standard conductivity value which corresponds to a ratio of scalp-to-skull conductivity of 41), 0.011 S/m (scalp-to-skull conductivity ratio of 30), and 0.041 S/m (ratio of 8). These values cover a wide range of values reported in the literature [26]. For one of the subjects in this study, we calculated the lead-field matrix and performed optimizations with a common target: the left dorsolateral prefrontal cortex (lDLPFC) as identified by Brodmann area 46 [27]. The cortical surface nodes in this area were set to a target E_n-value of either 0.25 V/m or 0.50 V/m with weight 10. The remaining nodes were set to a target E_n-value of 0 V/m with a weight of 2. Current constraints were set to $(I_{max\ channel}, I_{max\ total}) = (2.0, 4.0)$ mA and $(I_{max\ channel}, I_{max\ total}) = (1.0, 2.0)$ mA.

Figure 4a displays $<E_n>$ on the lDLPFC as a function of σ_{skull} for the different current and target E_n constraints. Average E_n values increase nonlinearly with σ_{skull} for every set of constraints except for the less stringent one: target E_n of 0.25 V/m with $(I_{max\ channel}, I_{max\ total}) = (2.0, 4.0)$ mA. Figure 4b displays the variation of the total injected current (I_{total}) in each montage with σ_{skull}. I_{total} tends to decrease nonlinearly with increasing σ_{skull}, except for the most stringent constraint: target E_n of 0.50 V/m with $(I_{max\ channel}, I_{max\ total}) = (1.0, 2.0)$ mA. In this case, I_{total} stays almost constant at the highest value allowed (2.0 mA).

Figure 5 provides the distribution of E_n in the cortical surface for all the optimizations performed in this study. For all optimizations and for the highest σ_{skull}, the position of the electrodes in the optimized montage is very similar across the different sets of constraints, with only the current values being different. For lower σ_{skull} values, and especially in the more stringent optimization constraints, the montages also differ in the electrode positions, often employing bigger separations between the anodes and cathodes.

We also evaluated the change in average E_n value that would occur when the optimized montages were evaluated in a model with a different σ_{skull} than the one used to derive the montage ($\sigma_{skull}^{Eval} \neq \sigma_{skull}^{Optim}$). This led to changes in average E_n values ranging from -45% to 137% of the values obtained when $\sigma_{skull}^{Eval} = \sigma_{skull}^{Optim}$. The average E_n values decreased when $\sigma_{skull}^{Eval} < \sigma_{skull}^{Optim}$ and they increased otherwise.

3.3 Study C: Intersubject Variability

In study c, we investigated the advantages that montage optimization brings in terms of intersubject variability of the E-field distribution. To do this, we performed subject-specific optimizations for six of the subjects in this study. For each subject, the optimization parameters were the same as the ones employed in study b. Each

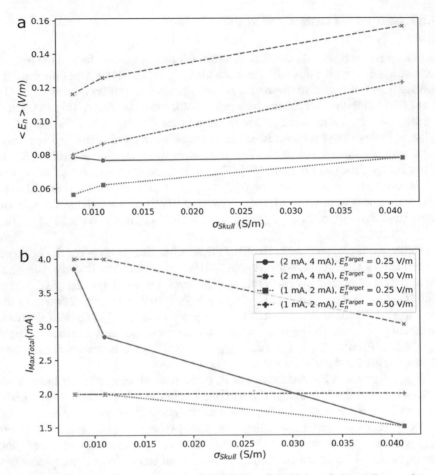

Fig. 4 Average value of E_n (in V/m) in the lDLPFC (**a**) and total injected current ($I_{max\ total}$ in mA, **b**) of the optimized montage as a function of skull conductivity (in S/m). The optimizations were obtained for four different constraints

optimization was then evaluated not only on the subject's head model from which it was derived but also in all the remaining head models. For each evaluation, we calculated the average $<E_n>$ value on the lDLPFC. To test the homoscedasticity of the different distributions, we used Levene's test. To compare the means of the different groups, we used Welch's t-test.

As shown in Fig. 6, the variance of $<E_n>$ across subjects was significantly lower when using a personalized montage as opposed to a non-personalized montage (Levene's test, p-value<0.05), except in the most stringent optimization (lowest current constraints with the highest E_n^{Target}, Levene's test p-value = 0.73). As is also shown in the figure, no statistically significant difference was found between personalized and non-personalized montages when it comes to the group average of $<E_n>$ value across subjects (Welch's t-test p-value>0.651), when the target E_n is

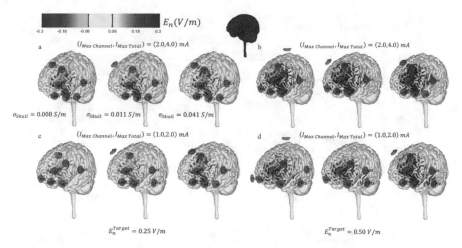

Fig. 5 Distribution of the normal component of the E-field in the cortical surface induced by different montages optimized to increase the excitability of the lDLPFC (shown as an inset in a) as a function of skull conductivity. The current constraints ($I_{max\ channel}$, $I_{max\ total}$) in each optimization as well as the target E_n-field (E_n^{Target}) are shown next to each group of images (a, b, c, and d). The order of the conductivities of the skull within each group of images is the same (see a). The montages were limited to eight channels. The color scale is common to all plots

maintained constant. Increasing the target E_n leads to a statistically significant higher $<E_n>$, regardless of the current constraints (Welch's t-test p-value<7.6×10^{-4}).

4   Discussion

4.1   Interplay of Target Size, Cortical Geometry, and Optimization Constraints

When analyzing the influence of target size and optimization constraints, we found an expected decrease of $<E_n>$ with the target area. As mentioned before, the non-monotonous nature of the decrease could be attributed to the interplay of different parameters: cortical geometry, positions of the electrodes available for the optimization, and optimization parameters (current constraints and target E_n-field). For small targets that do not encompass multiple sulci and/or gyrus, it was possible to achieve even the highest average $<E_n>$ value (0.5 V/m) provided enough (total injected) current was set as a limit. Limiting the currents to the safety values used in most studies [28], ($I_{max\ channel}$, $I_{max\ total}$) = (2.0, 4.0 mA), even $<E_n>$ values of 0.25 V/m cannot be achieved. This depends of course on target position and electrode array. For instance, at the bottom of the sulci under some of the electrode positions available, local maxima have been shown to be created due to the funneling effect of the CSF layer [15]. In these regions, higher $<E_n>$ values might be

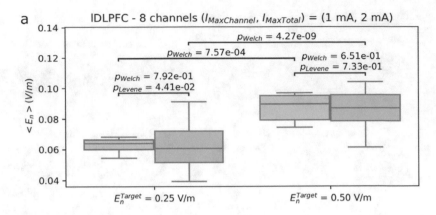

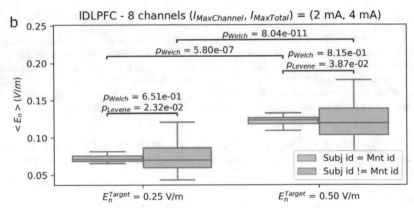

Fig. 6 Effect of current constraints and target E_n value on $<E_n>$ calculated on the lDLPFC for personalized (blue) and non-personalized (pink) optimized montages. The top plots (**a**) show the results for the $(I_{max\ channel},\ I_{max\ total}) = (1.0, 2.0)$ mA current constraints, whereas the bottom plots (**b**) show the results for the $(I_{max\ channel},\ I_{max\ total}) = (2.0, 4.0)$ mA constraints. Welch's t-tests were performed for comparisons between the means of the different groups. Levene's tests were performed to test for homoscedasticity between groups with the same constraints

achievable there with the same current constraints. For these small targets, we also found that the constraint on $I_{max\ channel}$ can limit the achievable $<E_n>$, with higher values being possible when $I_{max\ channel}$ is set to the same value as $I_{max\ total}$. For larger targets, the maximum achievable $<E_n>$ decreases, firstly due to the folded nature of the cortical surface and the electrode distribution (as shown by the rapid decay of $<E_n>$ even with unconstrained currents) and then to the current limitations. In particular, $I_{max\ total}$ is the main limiting factor to $<E_n>$, with the constraint on $I_{max\ channel}$ not mattering as much. This is expected, as larger targets require the distribution of the current on more electrodes to cover the whole area and achieve the same E_n.

Although the effects of having a more dense electrode array available for the optimization were not tested in this study, it is likely that they would be more beneficial for smaller targets (see also [19]), allowing for higher $<E_n>$ to be achieved for the same current constraints.

4.2 Influence of Skull Conductivity

The influence of the conductivity of the skull and other tissues on the E-field distribution in tCS is a well-established fact [29, 30]. Provided enough current was available to the optimization algorithm, all models reached a similar $<E_n>$ value on target. For more stringent constraints (lower currents and/or higher target E_n-fields), it might not be possible to maintain a similar $<E_n>$ across models (this will again depend on target size). For the latter optimizations, we found that the selected montage employs higher currents and a bigger separation between anodes and cathodes to increase $<E_n>$.

In a more realistic scenario, however, the discrepancy between the subject's skull conductivity and the one employed in the model is the main concern. As our results indicate, this can lead to very big discrepancies between the planned and effective $<E_n>$. These results stress the need for assessing subject-specific tissue conductivity values and use them together with subject-specific computational head models.

4.3 Montage Optimization and Intersubject Variability

Consistent with previously published studies [16], we found a large variability when calculating $<E_n>$ induced in six head models by non-personalized montages (on average the standard deviation was 21% of the mean value across all cases). Employing personalized montages significantly reduced the variation (standard deviation of 8% of the mean) in all cases except in the more stringent optimization (top-right boxes in Fig. 6). In that case $((I_{max\ channel},\ I_{max\ total}) = (1.0, 2.0)$ mA and E_n^{Target} of 0.50 V/m), personalization of montages did not reduce variability or result in an increase in average $<E_n>$ at the target. We interpret this as basically showing that the posed optimization problem is very hard to achieve and hence of variable results even with personalization. Increasing the target E_n^{Target} to 0.50 V/m does result in a significant increase in $<E_n>$ for both current constraints. Ultimately, this may prove to be more important than decreasing the variability of the results. Heterogeneity in $<E_n>$ across subjects can always be used as a regressor when analyzing the results of the study (see [31]).

4.4 Study Limitations

As all studies involving computational head models, there are a number of limitations in this study related to the simplifications employed by the models. The biggest simplification is the adoption of a homogeneous compartment for the skull tissue, ignoring the spongy bone region [32]. Although this would certainly influence the $<E_n>$ values reported here, as well as affect the effects of the different current constraints, it is unlikely that the overall qualitative conclusions of the different studies would be influenced.

Another limitation is related to the fact that we focused exclusively on the normal component of the E-field for optimization. The optimization method employed in this study can easily be used with other E-field components, but the optimized montages would employ electrode positions very different from the ones reported here. Again, this is unlikely to affect the overall conclusions of this study, and its general recommendations can be extended when other components of the E-field are of interest.

Regarding the targets, we only considered connected single target regions, despite the fact that interest has arisen lately regarding applications involving multiple distributed targets (as the ones arising from cortical networks, [5, 24]). Again, the stimulation protocol can be readily adapted to these types of targets (with the weights reflecting the statistical significance of the correlation). The conclusions about the influence of current constraints in these types of optimizations are likely to be similar to the ones reported here for the larger areas, but further studies are required.

Finally, we should mention the small number of subjects employed in this study, which limits the generalizability of its results, especially in study c. Future studies are underway which will investigate these findings in a larger population.

4.5 Consequences for Protocol Design

The results presented here clearly demonstrate the advantages of employing optimized montages for determining dose parameters in a tCS protocol. They have the potential of reducing variability in the E-field distribution across subjects in a study by taking into effect idiosyncratic subject properties, such as individual head anatomy, electrical properties, and even target location. Of course, these improvements require availability of appropriate data, such as MRI scans with parameters optimized for tissue segmentation [33], protocols for noninvasive determination of tissue electrical conductivities in a fast and reliable way [34, 35], as well as a combination of functional and structural data to determine the target for optimization. Regarding target size and location, this should guide the determination of the electrical current constraints of the study, as is clearly illustrated by the previous results.

Another important limitation of the usefulness of montage optimization is the lack of information about the mechanisms of tCS. However, several studies have been published illustrating the interaction of the E-field with neurons [10] and the network amplification effects that can be responsible for the ultimate long-term effects of the intervention [4]. The next step in developing montage optimization protocols would be to combine information about biophysical aspects of current propagation and electrophysiological aspects of E-field – neuron interaction and neuron-neuron communication [5, 36, 37].

Acknowledgments This project has received funding from the European Union's Horizon 2020 research and innovation program under grant agreement No 731827 (project STIPED) and from the European FET Open project Luminous (European Union's Horizon 2020 research and innovation program under grant agreement No 686764). The results and conclusions in this article present the authors' own views and do not reflect those of the EU Commission.

References

1. Woods, A. J., Antal, A., Bikson, M., Boggio, P. S., Brunoni, A. R., Celnik, P., Cohen, L. G., Fregni, F., Herrmann, C. S., Kappenman, E. S., Knotkova, H., Liebetanz, D., Miniussi, C., Miranda, P. C., Paulus, W., Priori, A., Reato, D., Stagg, C., Wenderoth, N., & Nitsche, M. A. (2016). A technical guide to tDCS, and related non-invasive brain stimulation tools. *Clinical Neurophysiology, 127*, 1031–1048.
2. Ruffini, G., Wendling, F., Merlet, I., Molaee-Ardekani, B., Mekkonen, A., Salvador, R., Soria-Frisch, A., Grau, C., Dunne, S., & Miranda, P. C. (2013). Transcranial current brain stimulation (tCS): Models and technologies. *IEEE Transactions on Neural Systems and Rehabilitation Engineering, 21*(3), 333–345.
3. Lefaucheur, J.-P., & Wendling, F. (2019). Mechanisms of action of tDCS: A brief and practical overview. *Neurophysiologie Clinique, 49*(4), 269–275.
4. Reato, D., Rahman, A., Bikson, M., & Parra, L. C. (2013, October). Effects of weak transcranial alternating current stimulation on brain activity-a review of known mechanisms from animal studies. *Frontiers in Human Neuroscience, 7*, 1–8.
5. Ruffini, G., Wendling, F., Sanchez-Todo, R., & Santarnecchi, E. (2018). Targeting brain networks with multichannel transcranial current stimulation (tCS). *Current Opinion in Biomedical Engineering, 8*, 70–77.
6. Lefaucheur, J. P., Antal, A., Ayache, S. S., Benninger, D. H., Brunelin, J., Cogiamanian, F., Cotelli, M., De Ridder, D., Ferrucci, R., Langguth, B., Marangolo, P., Mylius, V., Nitsche, M. A., Padberg, F., Palm, U., Poulet, E., Priori, A., Rossi, S., Schecklmann, M., Vanneste, S., Ziemann, U., Garcia-Larrea, L., & Paulus, W. (2017). Evidence-based guidelines on the therapeutic use of transcranial direct current stimulation (tDCS). *Clinical Neurophysiology, 128*(1), 56–92.
7. Polanía, R., Nitsche, M. A., & Ruff, C. C. (2018). Studying and modifying brain function with non-invasive brain stimulation. *Nature Neuroscience, 21*(2), 174–187.
8. Radman, T., Ramos, R. L., Brumberg, J. C., & Bikson, M. (2009). Role of cortical cell type and morphology in subthreshold and suprathreshold uniform electric field stimulation in vitro. *Brain Stimulation, 2*(4), 215–228.
9. Ruffini, G., Fox, M. D., Ripolles, O., Miranda, P. C., & Pascual-Leone, A. (2014, April). Optimization of multifocal transcranial current stimulation for weighted cortical pattern targeting from realistic modeling of electric fields. *NeuroImage, 89*, 216–225.
10. Rahman, A., Reato, D., Arlotti, M., Gasca, F., Datta, A., Parra, L. C., & Bikson, M. (2013). Cellular effects of acute direct current stimulation: Somatic and synaptic terminal effects. *Journal of Physiology (London), 591*(10), 2563–2578.

11. Miranda, P. C., Lomarev, M., & Hallett, M. (2006). Modeling the current distribution during transcranial direct current stimulation. *Clinical Neurophysiology, 117*(7), 1623–1629.

12. Huang, Y., Lafon, B., Bikson, M., Parra, L. C., Liu, A. A., Friedman, D., Wang, X., Doyle, W. K., Devinsky, O., & Dayan, M. (2017). Measurements and models of electric fields in the in vivo human brain during transcranial electric stimulation. *eLife, 6*, 1–26.

13. Opitz, A., Falchier, A., Yan, C. G., Yeagle, E. M., Linn, G. S., Megevand, P., Thielscher, A., Deborah, A. R., Milham, M. P., Mehta, A. D., & Schroeder, C. E. (2016). Spatiotemporal structure of intracranial electric fields induced by transcranial electric stimulation in humans and nonhuman primates. *Scientific Reports, 6*, 31236.

14. Datta, A., Bansal, V., Diaz, J., Patel, J., Reato, D., & Bikson, M. (2009). Gyri-precise head model of transcranial direct current stimulation: Improved spatial focality using a ring electrode versus conventional rectangular pad. *Brain Stimulation, 2*(4), 201–207.

15. Miranda, P. C., Mekonnen, A., Salvador, R., & Ruffini, G. (2013, April). The electric field in the cortex during transcranial current stimulation. *NeuroImage, 70*, 48–58.

16. Laakso, I., Tanaka, S., Koyama, S., De Santis, V., & Hirata, A. (2015). Inter-subject variability in electric fields of motor cortical tDCS. *Brain Stimulation, 8*(5), 906–913.

17. Dmochowski, J. P., Datta, A., Bikson, M., Su, Y. Z., & Parra, L. C. (2011). Optimized multi-electrode stimulation increases focality and intensity at target. *Journal of Neural Engineering, 8*, 4.

18. Guler, S., Dannhauer, M., Erem, B., Macleod, R., Tucker, D., Turovets, S., Luu, P., Erdogmus, D., & Brooks, D. H. (2016). Optimization of focality and direction in dense electrode array transcranial direct current stimulation (tDCS). *Journal of Neural Engineering, 13*(3), 1–31.

19. Saturnino, G. B., Siebner, H. R., Thielscher, A., & Madsen, K. H. (2019). Accessibility of cortical regions to focal TES: Dependence on spatial position, safety, and practical constraints. *NeuroImage, 203*, 116183.

20. Wagner, S., Burger, M., & Wolters, C. H. (2016). An optimization approach for well-targeted transcranial direct current stimulation. *SIAM Journal of Applied Mathematics, 76*(6), 2154–2174.

21. Fischl, B., Salat, D. H., Busa, E., Albert, M., Dieterich, M., Haselgrove, C., Van Der Kouwe, A., Killiany, R., Kennedy, D., Klaveness, S., Montillo, A., Makris, N., Rosen, B., & Dale, A. M. (2002). Whole brain segmentation: Automated labeling of neuroanatomical structures in the human brain. *Neuron, 33*, 341–355.

22. Opitz, A., Windhoff, M., Heidemann, R. M., Turner, R., & Thielscher, A. (2011). How the brain tissue shapes the electric field induced by transcranial magnetic stimulation. *NeuroImage, 58*(3), 849–859.

23. Neri, F., Mencarelli, L., Menardi, A., Giovannelli, F., Rossi, S., Sprugnoli, G., Rossi, A., Pascual-leone, A., Salvador, R., Ruffini, G., & Santarnecchi, E. (2019). A novel tDCS sham approach based on model-driven controlled shunting. *Brain Stimulation, 13*(2), 507–516.

24. Fischer, D. B., Fried, P. J., Ruffini, G., Ripolles, O., Salvador, R., Banus, J., Ketchabaw, W. T., Santarnecchi, E., Pascual-Leone, A., & Fox, M. D. (2017). Multifocal tDCS targeting the resting state motor network increases cortical excitability beyond traditional tDCS targeting unilateral motor cortex. *NeuroImage, 157*, 34–44.

25. Wagner, T., Eden, U., Rushmore, J., Russo, C. J., Dipietro, L., Fregni, F., Simon, S., Rotman, S., Pitskel, N. B., Ramos-Estebanez, C., Pascual-Leone, A., Grodzinsky, A. J., Zahn, M., & Valero-Cabre, A. (2014). Impact of brain tissue filtering on neurostimulation fields: A modeling study. *NeuroImage, 85*(Pt 3), 1048–1057.

26. Wendel, K., Väisänen, J., Seemann, G., Hyttinen, J., & Malmivuo, J. (2010). The influence of age and skull conductivity on surface and subdermal bipolar EEG leads. *Computational Intelligence and Neuroscience, 2010*, 1–7.

27. Petrides, M., & Pandya, D. N. (1999). Dorsolateral prefrontal cortex: Comparative cytoarchitectonic analysis in the human and the macaque brain and corticocortical connection patterns. *European Journal of Neuroscience, 11*, 1–2.

28. Bikson, M., Grossman, P., Thomas, C., Zannou, A. L., Jiang, J., Adnan, T., Mourdoukoutas, A. P., Kronberg, G., Truong, D., Boggio, P., Brunoni, A. R., Charvet, L., Fregni, F., Fritsch, B., Gillick, B., Hamilton, R. H., Hampstead, B. M., Jankord, R., Kirton, A., Knotkova, H., Liebetanz, D., Liu, A., Loo, C., Nitsche, M. A., Reis, J., Richardson, J. D., Rotenberg, A., Turkeltaub, P. E., & Woods, A. J. (2016). Safety of transcranial direct current stimulation: Evidence based update 2016. *Brain Stimulation, 9*, 641–661.

29. Saturnino, G. B., Thielscher, A., Madsen, K. H., Knösche, T. R., & Weise, K. (2019). A principled approach to conductivity uncertainty analysis in electric field calculations. *NeuroImage, 188*, 821–834.

30. Salvador, R., Ramirez, F., V'yacheslavovna, M., & Miranda, P. C. (2012) Effects of tissue dielectric properties on the electric field induced in tDCS: A sensitivity analysis. In: *34th Annual International Conference of the IEEE Engineering in Medicine and Biology Society (EMBC)*, pp. 787–790.

31. Laakso, I., Mikkonen, M., Koyama, S., Hirata, A., & Tanaka, S. (2019). Can electric fields explain inter-individual variability in transcranial direct current stimulation of the motor cortex? *Scientific Reports, 9*, 626.

32. Opitz, A., Paulus, W., Will, S., Antunes, A., & Thielscher, A. (2015). Determinants of the electric field during transcranial direct current stimulation. *NeuroImage, 109*, 140–150.

33. Windhoff, M., Opitz, A., & Thielscher, A. (2013). Electric field calculations in brain stimulation based on finite elements: An optimized processing pipeline for the generation and usage of accurate individual head models. *Human Brain Mapping, 34*, 923–935.

34. Aydin, U., Rampp, S., Wollbrink, A., Kugel, H., Cho, J., Knosche, T. R., Grova, C., Wellmer, J., & Wolters, C. H. (2017). Zoomed MRI guided by combined EEG/MEG source analysis: A multimodal approach for optimizing presurgical epilepsy work-up and its application in a multi-focal epilepsy patient case study. *Brain Topography, 30*, 417–433.

35. Fernandez-Corazza, M., Turovets, S., Luu, P., Price, N., Muravchik, C. H., & Tucker, D. (2018). Skull modeling effects in conductivity estimates using parametric electrical impedance tomography. *IEEE Transactions on Biomedical Engineering, 65*, 1785–1797.

36. Sanchez-Todo, R., Salvador, R. Santarnecchi, E., Wendling, F., Deco, G., & Ruffini, G. (2018). Personalization of hybrid brain models from neuroimaging and electrophysiology data. *bioRxiv* 461350 [Preprint].

37. Ruffini, G., Salvador, R., Tadayon, E., Sanchez-Todo, R., Pascual-Leone, A., & Santarnecchi, E. (2019). Realistic modeling of ephaptic fields in the human brain, *bioRxiv*, 688101 [Preprint].

Part III
Non-invasive Neurostimulation – Spinal Cord and Peripheral Nervous System

Modelling Studies of Non-invasive Electric and Magnetic Stimulation of the Spinal Cord

Sofia Rita Fernandes, Ricardo Salvador, Mamede de Carvalho, and Pedro Cavaleiro Miranda

1 Relevance of Modelling Studies in Non-invasive Spinal Stimulation

The spinal cord (SC) is a complex set of neural pathways and nuclei where essential reflex responses are generated and where transmission of sensory information and motor instructions takes place between peripheral organs and brain centres. Spinal dysfunctions due to various conditions, such as spinal cord injury (SCI), amyotrophic lateral sclerosis (ALS) and stroke, lead to a decrease in motor performance and sensory perception, causing spasticity, pain and muscular weakness [1]. Electric currents have been applied in the spinal cord for the treatment of chronic pain, resulting, in particular, from spinal cord lesion due to trauma or inflammatory diseases. However, the procedure usually involves surgical introduction of electrodes in the epidural space, which is frequently associated with higher risk of infections and medical costs for the patient [2]. Non-invasive electric and magnetic

S. R. Fernandes (✉)
Instituto de Biofísica e Engenharia Biomédica, Faculdade de Ciências da Universidade de Lisboa, Lisbon, Portugal

Instituto de Fisiologia, Instituto de Medicina Molecular João Lobo Antunes, Faculdade de Medicina da Universidade de Lisboa, Lisbon, Portugal
e-mail: srcfernandes@fc.ul.pt

R. Salvador
Neuroelectrics, Barcelona, Spain

M. de Carvalho
Instituto de Fisiologia, Instituto de Medicina Molecular João Lobo Antunes, Faculdade de Medicina da Universidade de Lisboa, Lisbon, Portugal

P. C. Miranda
Instituto de Biofísica e Engenharia Biomédica, Faculdade de Ciências da Universidade de Lisboa, Lisbon, Portugal

© The Author(s) 2021
S. N. Makarov et al. (eds.), *Brain and Human Body Modeling 2020*,
https://doi.org/10.1007/978-3-030-45623-8_8

spinal stimulation can modulate spinal pathway responses in a similar fashion to cortical stimulation techniques, such as transcranial direct current stimulation (tDCS) and transcranial magnetic stimulation (TMS) [3, 4]. Since 2008, exploratory clinical studies in humans using transcutaneous spinal direct current stimulation (tsDCS) have shown evidence of neuromodulation of spinal nociceptive and motor circuitry responses (e.g. [5, 6]). Repetitive spinal magnetic stimulation (repetitive tsMS or r-tsMS) has also been applied in the lumbar region of SCI patients with observed effects in motor spinal function [7, 8].

Computational modelling studies are a powerful tool to understand the biophysics underlying cortical and spinal stimulation and predict possible clinical outcomes. The effects of central nervous system electromagnetic stimulation rely mainly on the electric field (EF) induced in the nervous tissue. These EFs may contribute to inhibit or facilitate neuronal responses and are shaped by the stimulation delivery characteristics. In tDCS and tsDCS, the spatial distribution is influenced by electrode number, location, design (shape and structure), current intensity and polarity (anodal/cathodal); the same applies for coil position, shape, orientation and stimulus parameters in TMS and tsMS [9, 10]. Predictions of the EF and current distribution using realistic human models help to optimize stimulation protocols to address neural targets related with specific clinical dysfunctions [11–17]. The accuracy of these predictions will depend on factors such as the type of volume elements (hexahedral or tetrahedral), the accuracy of the representation of tissue geometry and stimulation source geometry (electrodes, coils) and the accuracy of tissue biophysical properties considered.

Computational studies on the EF spatial distribution during tsDCS are scarce. These studies employ realistic human models, based on high-resolution magnetic resonance imaging (MRI) of healthy volunteers of different ages, and tested different electrode montages in the thoracic and lumbar SC [18–20]. These models assumed mainly simple electrode geometry and isotropic tissues and were mostly based on hexahedral meshes, which do not involve complex tissue interface processing, with a trade-off between computational facility and lack of boundary accuracy. Over the last 5 years, we have developed a tetrahedral-based human trunk model and studied the current density and EF distribution in the spinal cord for tsDCS and tsMS. We have also introduced anisotropic properties for the spinal white matter and muscle tissues [21, 22]. This chapter provides a description of the model design steps and simulation methodology. A summary of the main EF characteristics predicted for tsDCS and tsMS is presented, with a final discussion on the relevance of modelling findings for tsDCS clinical application.

2   Creating a Realistic Human Volume Conductor Model

Similarly to cortical modelling studies, an accurate human realistic body model for spinal stimulation studies requires high-resolution magnetic resonance imaging (MRI) to obtain a correct segmentation of the inner structures of the spinal cord

Fig. 1 Essential steps in the design of a human realistic model

and its surrounding tissues. Resolution should be better than $1 \times 1 \times 1$ mm^3 in each direction to allow distinction between the spinal white matter (spinal-WM) and the spinal grey matter (spinal-GM), since average transverse diameters are only 6–13 mm [23]. Segmentation results in distinct tissue masks that are to be transformed and processed into triangulated surface meshes. All surface meshes have to be assembled into a full model with identification of each tissue boundary through a process named non-manifold assembly, which generates different shells for each tissue and its interfaces. After this operation, each shell can be transformed into a domain filled with tetrahedral volume elements in a process designated by volume mesh creation. Hexahedral meshes are faster to generate than tetrahedral meshes, because these are based directly on the original pixels that comprise each tissue mask. However, these types of meshes may introduce errors in predicted values of current and electric field (EF) at the interfaces between tissues, as was observed in tDCS [24]. Thus, models based on tetrahedral meshes provide a better insight of the stimulation effects on the interfaces between tissues, especially in the spinal CSF/WM and WM/GM interfaces. The basic steps for obtaining a realistic human model based on tetrahedral meshes are summarized in Fig. 1.

Acquisition of a full-body MRI for realistic human spinal modelling is a difficult and time-consuming process. However, there are full-body models available in the web that provide tissue masks ready for the generation of surface and volume meshes, such as the Virtual Population (ViP) Family [25]. ViP is a set of detailed high-resolution anatomical models created from MRI data of human healthy volunteers. These consist of simplified CAD files optimized for modelling using the finite element method (FEM) [25]. These models were used in the aforementioned modelling studies based on hexahedral meshes.

The tetrahedral model used in the studies described ahead was designed based on selected tissue masks from the ViP model Duke, corresponding to a 34-year-old male. The tissue masks considered were the ones proximal to the SC and the stimulation sources. Fourteen tissues were selected: skin, fat (including subcutaneous adipose tissue, SAT), muscle, bone, heart, lungs, viscera (composed by stomach, liver, pancreas, small intestine, large intestine), vertebrae, intervertebral disks, dura mater, cerebrospinal fluid (CSF), cerebellum, brainstem and SC [21, 22]. A spinal-GM tissue was designed and added to the model, considering general knowledge on SC anatomy [26] and relative measurements of GM width and shape at each SC segment using the Visible Human Data Set (VHD) of the National Library of Medicine (NLM) and the Visible Human Project® (www.nlm.nih.gov/research/visible/visible_human.html). The selected masks were converted into surface meshes using the Mimics software (v16) and are represented in Fig. 2. Surface mesh optimization procedures were performed with the 3-matic module to obtain a

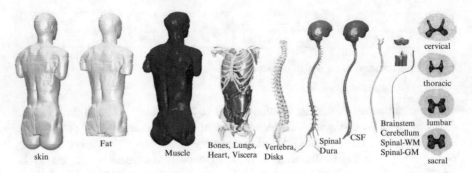

Fig. 2 Top row: surface meshes used in the human model. Bottom row: 2D axial view of WM and GM volume meshes in each spinal region.

Fig. 3 Procedure steps for calculating the EF in NISS

non-manifold assembly of all tissues suitable for successful tetrahedral volume mesh creation, considering a minimum element quality of 0.3. For tsDCS, volume meshing was obtained only after electrode incorporation. The full-body model was truncated at the level of the thighs and above the elbows to shorten computational time. The final model resulted in approximately 20 million tetrahedral volume elements, taking around 6–7 h to be generated in a computer with two quad-core Intel® Xeon® processors clocked at 3.2 GHz and 48 GB of RAM.

3 Electric Field Calculation in Non-invasive Spinal Stimulation (NISS)

The methods for calculating the EF due to non-invasive spinal stimulation (NISS) comprise four steps that are summarized in Fig. 3: design and placement of electrodes and coils in the model, assigning electrical conductivity to tissues and electrode materials (for tsDCS), EF calculation with the finite element method (FEM) and analysis of EF predictions.

3.1   Electrode Model and Stimulation Parameters in tsDCS

Electrode model. The electrode model assumed in the first tsDCS modelling study was a rectangular prism of gel or sponge with an upper rectangular isopotential surface [18–20]. tDCS studies that considered electrode models with higher complexity resulted in slight spatial variations of the cortical EF near electrode connectors [27]. Therefore, the electrodes added to our human model had a more complex design: electrodes were modelled as rectangular prism of conductive rubber in contact with a rectangular layer of gel. The metallic connectors in each pad were represented as a rectangle on the upper surface of the rubber pad, reflecting the dimensions of the electrodes available in our experimental lab (Fig. 4a) [21].

The electrodes were placed over spinous processes (s.p.) of vertebra and in regions not located over the SC: right deltoid (rD), umbilicus (U), right iliac crest (rIC) and cervicomental angle (CMA). A total of 11 different electrode montages were modelled and are shown in Fig. 4b. Montages T10-rD, T10-U, C7-rD, C7-CMA and C4-CMA have already been used in experimental studies [5, 28–32]. Montages T10-rIC, T8-U, T8-rIC, L2-rD and L2-T8 have not been applied yet. Montage C3-T3 was used in an experimental study by our group due to favourable modelling predictions [22].

Determining electrical conductivity of materials. A literature review on the electrical properties of biological tissues was performed. Table 1 presents a list of isotropic electrical conductivity values for DC currents based on this review. A wide range of values is reported in the literature; preference was given to values determined at frequencies lower than 10 kHz (low-frequency range) in human tissue at a body temperature not under 36 °C and less than 24 h post-mortem, using the four-point method for conductivity measurement.

Some isotropic conductivities (σ) were determined as averages:

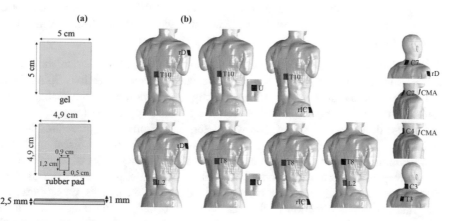

Fig. 4 Electrode geometry considered in the study: (a) gel, rubber pad and connector dimensions; (b) electrode montages simulated with the human model [21, 22]

Table 1 Isotropic electrical conductivities (σ) of tissues in the human trunk model [21, 22]

Tissue	σ (S/m)	References
Skin	0.435	[33]
Fat	0.040	[34]
Muscle	0.355 (av)	[35]
Lungs	0.046 (av)	[35]
Heart	0.535 (av)	[34, 36]
Viscera (liver, pancreas, stomach, small and large intestines, air)	0.123 (av)	[34, 36, 37]
Vertebrae/bone	0.006	[34]
Intervertebral disks	0.200	[34]
Dura mater	0.030	[38]
CSF	1.790	[39]
Brainstem	0.154	[34]
Cerebellum	0.290 (av)	[34]
Spinal-WM	0.143	[34]
Spinal-GM	0.333	[34]

- σ_{muscle} was determined as an arithmetic average of transverse and longitudinal values (0.043 and 0.667 S/m, respectively, from Rush et al. [35]).
- σ_{lungs} was also determined from Rush et al. [35], using DC measurements for dog lungs (human values were not found in the literature search), considering an average value between inflation (0.042 S/m) and deflation (0.051 S/m).
- σ_{heart} is a volume-weighted average of the conductivities of myocardium (0.461 S/m [36]) and heart lumen (considered filled with blood, $\sigma_{blood} = 0.625$ S/m [34]), using the volumes of these tissues in the model.
- $\sigma_{viscera}$ is a volume-weighted average of the conductivities of all visceral tissues present in the model: liver (0.123 S/m [36]), pancreas (0.130 S/m [37]), stomach and large and small intestines (considered as soft tissue – 0.200 S/m [34]).
- $\sigma_{cerebellum}$ is a volume-weighted average of cortical WM and GM conductivities, considering cerebellum WM and GM relative volumes from Damasceno et al. [40].

Spinal-WM is a tissue with considerable anisotropy due to the orientation of its fibres, with a higher conductivity along the caudal-rostral direction. Spinal-WM conductivity was represented by a tensor using information about the spatial orientation of the fibres and the ratio between the conductivity in the longitudinal and transverse directions. A spinal axis was determined from a set of centre-of-mass points of 1-mm-thick cylindrical slices along the SC caudal-rostral direction. The x and y coordinates of this set of points were fitted to a Fourier series of seven and six terms, respectively, as a function of z. These analytical expressions were used to determine the direction of the tangent to the spinal axis. This constitutes the longitudinal (long) direction of the SC, which also defines the transverse (trans) plane, composed of two orthogonal directions, trans1 (right-left, rl) and trans2 (ventral-dorsal, vd). Next, the values of longitudinal and transverse electrical conductivities of the spinal-WM were determined considering the volume constraint, $\frac{4}{3}\pi\sigma_{long}(\sigma_{trans})^2 = \frac{4}{3}\pi\sigma_{isotropic}^3$ [54], with $\sigma_{long} = 10\ \sigma_{trans}$, resulting in $\sigma_{long} = 0.664$ S/m and $\sigma_{trans} = 0.066$ S/m. An initial diagonal conductivity matrix was assigned in a local coordinate system with the diagonal elements equal to the conductivity values σ_{trans1}, σ_{trans2} and σ_{long}, where $\sigma_{trans1} = \sigma_{trans2} = \sigma_{trans}$. This matrix was then rotated in the reference coordinate system using a transformation matrix S according to Eq. (1) for each mesh node.

$$\begin{bmatrix} \sigma_{xx} & \sigma_{xy} & \sigma_{xz} \\ \sigma_{yx} & \sigma_{yy} & \sigma_{yz} \\ \sigma_{zx} & \sigma_{zy} & \sigma_{zz} \end{bmatrix} = S \begin{bmatrix} \sigma_{trans1} & 0 & 0 \\ 0 & \sigma_{trans2} & 0 \\ 0 & 0 & \sigma_{long} \end{bmatrix} S^{-1} \tag{1}$$

The transformation matrix was built from vectors aligned with the longitudinal and transverse directions of the spinal axis determined previously. The conductivity matrix was calculated in each mesh node using a MATLAB script (MATLAB v2015b software). Each component of this conductivity matrix was smoothed to minimize discontinuities. Smoothing was performed with a zero-phase digital filter. Conductivity matrix components were interpolated in COMSOL (COMSOL Multiphysics, version 4.3b) to obtain the conductivity tensor for each volume element.

Anisotropic properties were also considered for muscle: a conductivity tensor was determined for muscle groups close to the stimulation sources and the SC. Different transverse and longitudinal conductivities ($\sigma_{trans} = 0.043$ S/m, $\sigma_{long} = 0.667$ S/m, [35]) were assigned to these muscles, according to the direction of muscle fibres known from anatomy. A diagonal conductivity matrix was assigned for each muscle mesh node, where the diagonal matrix coefficients had values according to fibre orientation: $\sigma_{xx} = \sigma_{yy} = \sigma_{trans}$ and $\sigma_{zz} = \sigma_{long}$ for the neck, deltoid and abdominal muscles and $\sigma_{yy} = \sigma_{zz} = \sigma_{trans}$ and $\sigma_{xx} = \sigma_{long}$ for the pectoral and back muscles. No rotation of the conductivity tensor of muscles was performed.

The gel considered in our model was the one available in our experimental lab – Signa gel (Parker Laboratories, Inc.), which has a conductivity of $\sigma_{gel} = 4 \pm 1$ S/m, according to Minhas et al. [41]. The rubber pad resistance was measured using a four-point probe, and its conductivity estimated to be $\sigma_{rubber} = 44 \pm 1$ S/m, considering the pad as a finite layer of material and following the method described in Smits [42].

EF Calculations Biophysical calculations in the human volume conductor model were performed using the FEM. The current density and EF induced in biological tissues and electrode materials represented in the model were calculated using the AC/DC module in COMSOL Multiphysics 4.3b, which solves Laplace's equation for the electric potential ϕ, $\nabla.(\sigma\nabla\phi) = 0$. Boundary conditions were implemented according to Miranda et al. [15]: continuity of the normal component of the current density in all interior boundaries, electric insulation in the external boundaries and electrode connectors as isopotential surfaces. The potential difference between the anode and cathode was adjusted, using the floating potential boundary condition from COMSOL, so that the current injected through the electrodes was 2.5 mA, following previous published studies (e.g. [5]). The EF ($\vec{E}$) was calculated in all nodes of the mesh elements by taking the gradient of the electric potential ϕ, $\vec{E} = -\nabla\phi$. Current density ($\vec{J}$) was determined using Ohm's law, $\vec{J} = \sigma\vec{E}$, where σ is the electrical conductivity of the corresponding tissue. All tissues were assumed to be purely resistive with unit relative permittivity (ε_r). In lumbar montages, the first electrode was defined as the anode and the second electrode as the cathode (e.g. in L2-T8 montage, L2 is the anode and T8 the cathode), and the reverse was considered in cervical montages (e.g. in C3-T3 montage, C3 is the cathode and T3 the anode). Reversing polarity would invert the direction of the EF but would not affect its magnitude or the magnitude of its components [43].

Simulations were performed for 11 electrode montages considering anisotropy of spinal-WM and muscle. For the seven thoracic and lumbar montages, simulations were also performed considering all tissues isotropic or only with anisotropy of spinal-WM, resulting in a total of $7 \times 3 + 4 = 25$ simulations. Each model had 2.6×10^7 degrees of freedom, and the solution time was about 150 min per simulation on a computer with two quad-core Intel® Xeon® processors clocked at 3.2 GHz and 48 GB of RAM.

EF Analysis Analysis was performed using MATLAB scripts for post-processing the results exported from COMSOL. The following assumption was made regarding neuromodulatory effect predictions: previous tDCS clinical studies reported long-lasting and polarity-dependent changes in neural excitability of the human motor cortex when applying a continuous current of 1 mA to the scalp [44]. Miranda et al. [15] predicted an average EF magnitude in the hand knob of the motor cortex higher than 0.15 V/m, when applying 1 mA to the scalp and reproducing the same conditions mentioned in tDCS clinical trials. The values of the EF calculated in that study are in good agreement with those predicted by other studies [e.g. 13, 27]. Therefore, neuromodulatory effects are considered likely to occur wherever the average EF in the spinal-GM or WM exceeds 0.15 V/m.

The EF was decomposed into three orthogonal vectors according to three relevant directions:

- $\vec{E}_{long}$ – tangent to the longitudinal axis of the SC defined in the previous section and pointing from caudal to rostral
- $\vec{E}_{vd}$ – perpendicular to the first, contained in the yz-plane and pointing from ventral to dorsal
- $\vec{E}_{rl}$ – perpendicular to the first two and pointing from right to left

The spatial distribution of the EF magnitude and of its components along the SC length was studied in the spinal-WM and spinal-GM, considering values averaged over 1-mm-thick axial slices (perpendicular to the z-axis). The spatial variation of the EF in the SC was always reported in terms of SC segments, which have positions that are not always coincident with the position of vertebrae with the same designation, especially in the case of lumbar and sacral spinal segments, which are distant from the vertebra with the same designation.

3.2 Coil Model and Stimulation Parameters in tsMS

Coil Model A Fig. 8 coil, Magstim's double 70 mm coil, was modelled according to Fig. 5, using a MATLAB script [9]. This script requires an input file with all the nodes of the human model and calculates the placement of the centre of the coil through a GUI where the user can set the intended position. The centre (vertex) of the coil was placed over two positions (Fig. 5):

- T12 s.p., with two coil orientations according to the induced EF, at a distance of 24 mm from skin's surface [45]: induced EF pointing in inferior-superior direction (T12-IS); induced EF pointing in the left-right direction (T12-LR)
- C5 s.p., with the coil at a distance of 32 mm from skin's surface and the induced field pointing in the IS direction (C5-IS)

EF Calculation and Analysis The EF induced in tsMS results from the sum of two components. The first term is the primary EF ($d\vec{A}/dt$), and it depends only on the coil's geometry since the quasi-magnetostatic approximation applies in this case [46]. The second term is the secondary EF ($-\vec{\nabla}\phi$), which depends on the geometry of the model, the primary EF and the electrical properties of tissues. The primary EF was determined using a MATLAB script as described in Salvador et al. [9]. Stimulation parameters were introduced considering standard values used in single-pulse

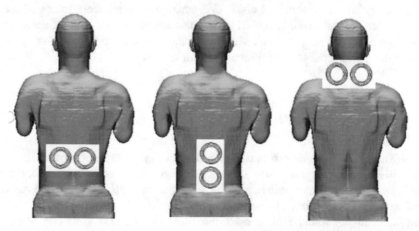

Fig. 5 Coil positions and orientations considered in tsMS simulations. From left to right: T12-IS, T12-LR, C5-IS

TMS for assessment of motor cortex responses: sinusoidal current of frequency f = 3.5 kHz and current amplitude of 2.8 kA, resulting in a maximum dI/dt = 61.54 A/µs, which corresponds to the mean threshold for hand muscle activation [47].

The values of the primary EF were imported into COMSOL to calculate the secondary EF using COMSOL Multiphysics AC/DC module (v5.2a). Electrical properties of tissues were considered the same as in tsDCS simulations, since the stimulus considered is in the low-frequency region (< 10 kHz). The quasi-electrostatic approximation was considered with the same boundary conditions applied for tsDCS. The total EF results from the vector sum of these two components. tsMS simulations comprised 2.7×10^7 degrees of freedom, taking about 60 min to solve on a computer with two quad-core Intel® Xeon® processors clocked at 3.2 GHz and 48 GB of RAM.

EF analysis applied the methodology previously described for tsDCS, for the calculation of E_{long}, E_{vd} and E_{rl} components and their spatial distributions along SC length.

4 Main Characteristics of the Electric Field in NISS

The EF generated during NISS may induce variations in the transmembrane potential of spinal neurons oriented along the EF direction, just as in cortical stimulation techniques [9, 48, 49]. These variations will determine the neuromodulatory potential of this type of stimulation in spinal circuitry. In the context of non-invasive spinal stimulation, modelling studies published so far concern tDCS only and report similar results in terms of EF spatial distribution. Most of the current density spreads along the regions of the skin, fat and muscle located between the electrodes and the target segments of the SC. The shape and morphology of the SC and surrounding tissues (vertebrae, intervertebral disks, CSF and spinal dura) seem to contribute to the presence of local maxima [19–21]. The main features of the EF predicted in our model for NISS and how it varies with tissue characteristics and modelling assumptions will be addressed in the following sections.

4.1 Predictions in tsDCS

The EF magnitude resulting from tsDCS in the SC is predicted to reach its maximum value in the spinal segments that lie in the region comprised between stimulating electrodes in all studies published so far in human models [18–22]. The same is predicted by our tetrahedral model. Figure 6 shows the distribution of the EF magnitude in the spinal-WM considering montages with at least one electrode over the cervical, thoracic or lumbar SC. The EF reaches its maximum value in SC segments located between electrode positions. Values of the EF magnitude and

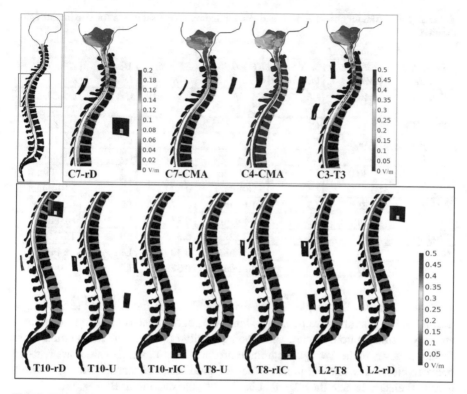

Fig. 6 EF magnitude in the spinal-WM: top row, cervical montages; bottom row, thoracic and lumbar montages. The corresponding colour scale is represented on the right at the end of each row, except for C7-rD, which is represented at the right side of the corresponding distribution. Vertebrae, disks and electrode positions are represented in black and grey in the sagittal plane for each plot

of its components were averaged over 1-mm-thick slices in the spinal-WM and GM as explained in Sect. 3.1. Table 2 indicates the maximum average value and the corresponding spinal segment where it is predicted to occur for each montage. It also presents the range of segments with EF higher than 0.15 V/m. EF maxima are generally located in the same segments in spinal-GM and spinal-WM. EF maximum value is approximately the same in spinal-GM and spinal-WM in thoracic and lumbar montages and higher in spinal-WM for cervical montages. This may be explained by the fact that the highly conductive CSF has a volume that is two to three times smaller in the region between C3 and C6 segments when compared with the rest of the spinal canal, leading to a larger current focusing in those regions that will affect the EF in the adjacent WM segments. L2-T8 and T8-rIC are the montages that maximize the EF in the lumbar regions, where most of the spinal circuitry related with lower limb sensorimotor functions are located. Thus, these montages may result in larger neuromodulatory outcomes in lower limb sensorimotor responses. The same applies for C3-T3 considering upper limb sensorimotor functions.

Table 2 EF maximum absolute value and corresponding spinal segments for maximum region location

Electrode montage	Spinal-GM			Spinal-WM		
	EF max (V/m)	Max location	EF > 0.15 V/m	EF max (V/m)	Max location	EF > 0.15 V/m
C7-rD	0.16; 0.17	C6; T3	C6-C7, T2-T5	0.17; 0.19	C7; T3	C6-C7, T2-T5
C7-CMA	0.24	C7	C5-T1	0.38	C7	C4-T1
C4-CMA	0.14	C4	–	0.29	C5	C2-T1
C3-T3	0.44	C6	C1-T4	0.49	C7	C1-T5
T8-U	0.30	L3	T11-Filum	0.30	L3	T11-Filum
T8-rIC	0.36	L3	T10-Filum	0.36	L3	T10-Filum
T10-U	0.27	L5	L1-Filum	0.27	L5	L1-Filum
T10-rIC	0.34	L5	L1-Filum	0.33	L5	L1-Filum
T10-rD	0.29	T9	T5-T12	0.31	T9	T5-T12
L2-rD	0.29	L1	T6-Filum	0.30	L1	T6-Filum
L2-T8	0.37	L3	T10-Filum	0.37	L3	T10-Filum

Figure 7 shows the profiles of the average EF magnitude along the SC length normalized to maximum value in the spinal-GM and spinal-WM. The distributions are very similar in both tissues, and montages with one electrode in common share similar patterns in the average magnitude profiles. Maximum EF values are reached between electrodes, and minimum values are predicted to occur in montages with a vertebral electrode near the edge that is distal to the maximum EF region.

The EF components along the longitudinal (E_{long}), ventral-dorsal (E_{vd}) and right-left (E_{rl}) directions in the SC have different magnitudes and relative contributions to the total magnitude in the montages studied, especially when comparing thoracic and lumbar montages with CMA montages. In thoracic and lumbar montages, E_{long} is three to six times greater than the other components (Fig. 8, left). In cervical montages, E_{long} can be 10–20 times larger than the other components in the EF magnitude maximum regions, except for dorsal-ventral placements, namely, C4-CMA and C7-CMA (Fig. 8, right, grey lines). In CMA montages, E_{vd} is comparable to E_{long} and even with a higher contribution to the total EF magnitude ($> 0.60\ E_{max}$), however in the spinal-WM segments between the electrodes. Previous studies also predicted an EF with a preferential longitudinal direction [18, 20]. This can be explained by the high electrical conductivity of the CSF, which is one order of magnitude larger that the surrounding tissues, and also by SC cable-shape anatomy.

Electrode position can influence the distribution of the EF and of its components. For montages that have an electrode over the SC, the sign of E_{long} changes near the edge of this electrode that is away from the region between electrodes, leading to a sharp decrease in magnitude (Fig. 8, left). The ventral-dorsal placement characteristic of CMA montages leads to a larger E_{vd} component between electrodes (Fig. 8, right).

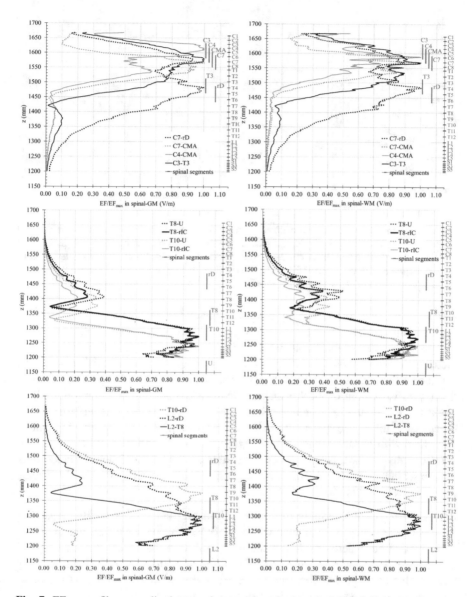

Fig. 7 EF$_{mag}$ profiles normalized to maximum values in the spinal-GM (left) and spinal-WM (right) along the SC length for cervical montages (top row) and thoracolumbar montages (middle and bottom rows). The positions of the electrodes are represented by vertical grey bars, and the positions of spinal segments are represented in grey on the right of each profile. rIC electrode is below the caudal end of the SC and thus not represented

Influence of SC Anatomy on the EF   Anatomy seems to influence the EF induced in the SC due to tsDCS. The EF profiles presented in Fig. 7 have local peaks that appear in the same positions regardless of the montage. This must be due to two main

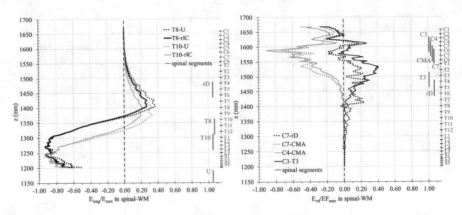

Fig. 8 EF component profiles normalized to maximum values in the spinal-WM in selected montages: left, E_{long}/E_{max} in T8-U, T8-rIC, T10-U and T10-rIC; right, E_{vd}/E_{max} in cervical montages. The positions of the electrodes are represented by vertical grey bars, and the positions of spinal segments are also represented in grey on the right of each profile. rIC electrode is below the caudal end of the SC and thus not represented

reasons: first, the differences in electrical conductivities between neighbouring tissues originate large variations of the EF at interfaces; second, CSF narrowing can occur due to vertebrae bony edges, disk intrusions and, consequently, current focusing in the CSF due to its high conductivity. Figure 9 shows the locations of common EF hotspots for selected thoracic, lumbar and cervical montages. Considering the distribution of the EF magnitude vs. CSF volume in 1-mm-thick slices along the z axis, the EF magnitude increases with decreasing CSF volume for all montages in the regions where the EF is higher than 0.15 V/m. This relation can be quantified with linear fit functions of negative slope for thoracic and lumbar montages and inverse fit functions for cervical montages with coefficient of determination larger than 0.5 [21, 22]. This inverse relation between CSF volume and EF value was also predicted in Fiocchi et al. [19].

Effect of Electrical Conductivity Assumptions on the EF Two additional studies were performed for T10-U montage to quantify what would be the changes induced in the predictions when considering: study (1), full isotropic model vs. anisotropy of the spinal-WM and muscle, and study (2), different isotropic electrical conductivity values. This montage was chosen because it presents higher EF in the LS segments and can be compared with a previous study by Parazzini et al. [18].

Figure 10 summarizes the effects for the EF magnitude when considering anisotropy of spinal-WM and muscle (anisotropic 2), anisotropic of spinal-WM only (anisotropy 1) and isotropy of all tissues (isotropic). The effect of anisotropy on the EF in the spinal-WM is to increase the magnitude in the spinal segments between the electrodes. The differences between the various anisotropy settings are represented by the red lines in Fig. 10 (left panel). Anisotropic 1 and anisotropic 2 have a very small difference, with mean values ≤ 0.024 V/m, indicating that

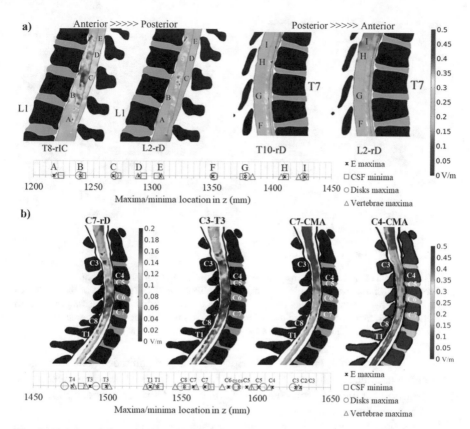

Fig. 9 Volume plots of the EF magnitude showing local maxima in selected regions of the thoracolumbar (a) and cervical (b) spinal-WM. Below each plot, EF maxima and related anatomic features – CSF volume minima, vertebrae and disk volume maxima, are plotted as a function of the z-coordinate. Labels are indicated in each figure marking maxima positions

muscle anisotropy has a small influence on the spinal EF. Muscle tissue is not adjacent to the SC: vertebrae, spinal dura and epidural fat and CSF are located in the current path between muscle and SC, and the combination of these tissue conductivities and shapes may contribute to decrease the effects on the EF due to muscle anisotropy. Anisotropy also introduces local maxima hotspots at the CSF/WM and WM/GM interface. This is easily observed in the EF distributions on transverse slices of selected spinal segments, also shown in Fig. 10 (right panel). The anisotropic 1 and 2 present small hotspot regions in the CSF/WM interface at L2 and L5 segment and near GM horns in S2 segment that do not appear in the isotropic model [21].

The maximum and mean differences between isotropic and both anisotropic models in all montages were determined in the regions where the EF magnitude is higher than 0.15 V/m. Anisotropic 2 presents the highest values for total EF magnitude, just as seen for T10-U. Mean differences between models are on the

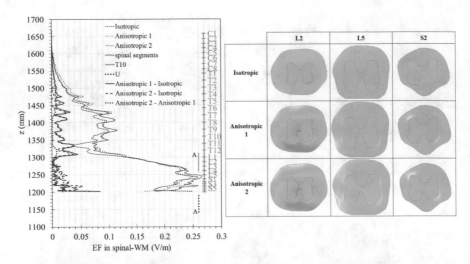

Fig. 10 Left panel, average EF magnitude profiles in T10-U for study 1, considering three conductivity considerations: isotropic, anisotropic 1, anisotropic 2. The positions of the electrodes are represented by vertical grey bars, and the positions of spinal segments are also represented in grey. Right panel: transverse slices of the SC in selected segments in the three different models of conductivity for T10-U. The colour scale is the same as Fig. 6, from 0 to 0.5 V/m

order of 0.01 V/m for the total EF, E_{long} and E_{vd} magnitudes, but maximum differences can reach values of 0.131 V/m for the difference between anisotropic 1 and isotropic for the E_{long} component in T10-rIC and T8-rIC. E_{rl} is negligible in the isotropic model for all montages.

Study 2 considers two sets of conductivity values: isotropic 1, the one used in the tetrahedral model, and isotropic 2, considered in tsDCS modelling studies by Parazzini et al. [18], Fiocchi et al. [19] and Kuck et al. [20]. Isotropic 1 values were taken from Table 1. Isotropic 2 values are as follows: $\sigma_{skin} = 0.100$ S/m; $\sigma_{fat} = 0.078$ S/m; $\sigma_{muscle} = 0.160$ S/m; $\sigma_{bone/vertebrae} = 0.020$ S/m; $\sigma_{lungs} = 0.076$ S/m; $\sigma_{heart} = 0.534$ S/m (av); $\sigma_{viscera} = 0.0254$ S/m; $\sigma_{disks} = 0.161$ S/m; $\sigma_{SC/nerve} = \sigma_{SC/dura} = 0.017$ S/m; $\sigma_{CSF} = 1.59$ S/m; $\sigma_{medula\ oblongata/midbrain/pons} = \sigma_{brainstem} = 0.0276$ S/m. Figure 11 presents the average EF magnitude distribution over the spinal-WM normalized to maximum in each conductivity set. This normalization was considered to evaluate the effect of conductivity on the EF due to electrode positions and anatomical features.

The isotropic 2 set of values resulted in a higher EF in spinal-WM and GM by a factor of 2. The two distributions almost overlap, with the same peak locations, except near the T10 connector, where the difference is larger (0.23 E_{mag}/E_{max}). Similar results were also observed for spinal-GM and for EF components, with isotropic 2 presenting larger E_{long}, E_{vd} and E_{rl} by factors of 2, 4 and 5, respectively, in the LS region for the spinal-GM, and all by a factor of 2 in the spinal-WM. E_{rl} is almost zero in both models. Changing electrical conductivity values will affect the

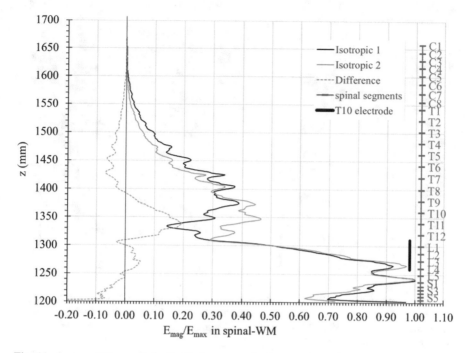

Fig. 11 Average EF magnitude distribution normalized to maximum in spinal-WM along the z direction for isotropic 1 and isotropic 2 sets of conductivity values, for the T10-U montage. The difference between the two distributions is also presented by the dashed line. T10 electrode position is represented in grey. The letter "A" marks the connector position. The vertical grey bar on the right indicates the position of each SC segment

EF magnitude, but not its spatial distribution, preserving the features caused by anatomical morphology.

4.2 Predictions in tsMS

The EF distribution in the SC was calculated for tsMS considering two coil placements, one over the cervical C5 s.p. with an inferior-superior-oriented induced EF (C5-IS) and the other over the lumbar SC, in T12 s.p. In the lumbar placement, two coil alignments were considered, resulting in left-right-oriented induced EF (T12-LR) and inferior-superior-oriented induced EF (T12-IS). The EF magnitude is presented in Fig. 12 for the three tsMS simulations. The EF is higher in the posterior SC regions near the coil in all cases. C5-IS tsMS maximum average EF value is 14.6 V/m and 11.0 V/m in T1 segment for spinal-WM and GM, respectively, 30 times higher than C3-T3, the tsDCS montage with higher EF values. T12-LR tsMS reaches a maximum average value of 14.4 V/m and 5.4 V/m in the spinal-WM

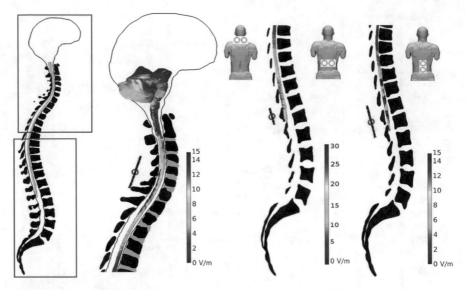

Fig. 12 EF magnitude distributions in the spinal-WM for C5-IS (middle left), T12-IS (middle right) and T12-LR (right) coil placements in tsMS. The left panel illustrates the selected spinal cord region for the cervical tsMS (red line) and lumbar LR and IS tsMS (blue line). The corresponding colour scale is placed on the right of each plot

and GM, respectively, in L3 segment, and T12-IS tsMS reaches maximum average values of 20.5 V/m and 20.4 V/m in the spinal-WM and GM, respectively, in L1 segment. The maximum EFs produced by T12-LR and T12-IS tsMS are 30 and 60 times higher, respectively, than the values reached in tsDCS montages that maximize the EF in the lumbar region, i.e. T8-rIC and L2-T8.

The orientation of the coil will have a strong influence in the EF direction in the SC. When the coil is placed so as to induce an IS-oriented EF, the EF has a larger longitudinal component, and E_{long} has the highest contribution to the total EF. This can be observed for the spinal-GM and spinal-WM profiles, represented in Fig. 13 (top and bottom rows). In C5-IS, there is also a large contribution from the E_{vd} component, which is not present in the spinal-GM, and may be due to spinal-WM anisotropy, since the cervical region presents sections with a large dorsal-ventral orientation, when compared to the lumbar region. LR orientation produces a strong E_{rl} component which contributes for most of the total EF magnitude.

The effects of tsMS decrease faster with distance to source when compared to tsDCS. This is consistent with modelling studies of cortical stimulation, which show a higher EF focality and smaller cortical depth in TMS using Fig. 8 coils [50]. This effect can be observed in the EF magnitude for C3-T3 tsDCS and for C5-IS tsMS in transverse slices of selected spinal segments near the local peaks (Fig. 14). When comparing the EF in C3 to T1 segments, the relative difference between maximum and minimum values is always higher in tsMS, with a pattern of decrease of EF magnitude from dorsal to ventral regions in the SC that is not seen in any of the spinal segments for tsDCS.

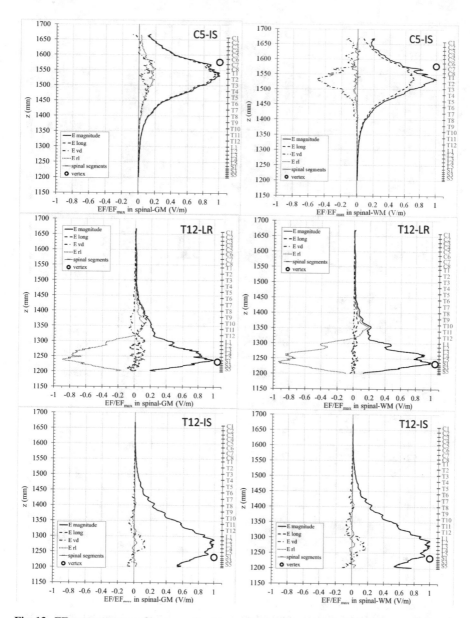

Fig. 13 EF components profiles normalized to maximum values in the spinal-GM and spinal-WM in C5-IS tsMS (top row), T12-LR tsMS (middle row) and T12-IS tsMS (bottom row). The position of the vertex of the coil is represented by a black circle, and the position of spinal segments is also represented in grey on the right of each profile

The anatomy may also play an influence on the EF in tsMS local hotspots, since the profiles in Fig. 13 present local maxima that have the same locations in tsDCS and tsMS. These effects were also observed in other modelling studies on tsDCS,

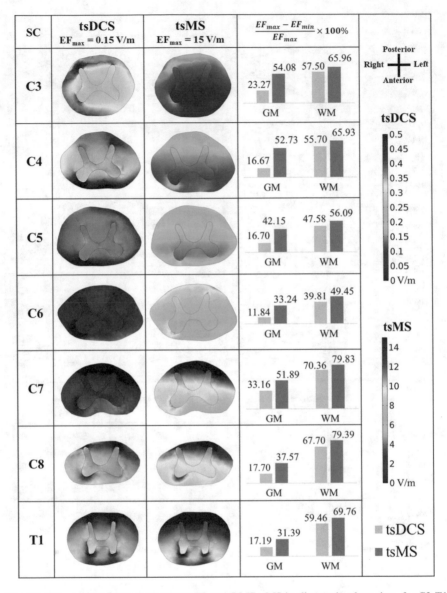

Fig. 14 EF magnitude during C3-T3 tsDCS and C5-IS tsMS in slices at local maxima, for C3-T1 SC segments. On the right, the percentage variation in EF magnitude in each slice is presented as bar plots with the values in the spinal-GM and spinal-WM above each bar. Colour scales and slice orientation are represented in the right column [51]

tDCS and TMS [10, 15, 21]. In the slices represented in Fig. 14, magnitude hotspots appear near dorsal and ventral horns, in the same regions for tsDCS and tsMS in all the segments represented, which is indicative of an anatomical influence due to CSF narrowing and discontinuity of the electrical conductivity at CSF/WM and WM/GM interfaces.

4.3 Implications of Modelling Findings in Clinical Applications of NISS

All modelling studies presented here for tsDCS and tsMS reached values in the SC higher than 0.15 V/m in different spinal regions. This indicates that different

Table 3 Correspondence between tsDCS and tsMS settings and possible clinical targets according to EF predictions

tsDCS		
Electrode montage	*Spinal segments with EF > 0.15 V/ m*	*Related sensorimotor functional area*
C7-rD	C6-C7, T2-T5	Upper extremity, upper thorax
C7-CMA	C4-T1	Scapular girdle, upper limb
C4-CMA	C2-T1	Neck, diaphragm, scapular girdle, upper limb
C3-T3	C1-T5	Neck, diaphragm, scapular girdle, upper limb, upper thorax
T8-U	T11-Filum	Abdominal organs, pelvic girdle, lower limb, pelvic floor
T8-rIC	T10-Filum	Pelvic girdle, lower limb, pelvic floor
T10-U	L1-Filum	Pelvic girdle, lower limb, pelvic floor
T10-rIC	L1-Filum	Pelvic girdle, lower limb, pelvic floor
T10-rD	T5-T12	Medium and lower thorax
L2-rD	T6-Filum	Abdominal organs, pelvic girdle, lower limb, pelvic floor
L2-T8	T10-Filum	Abdominal organs, pelvic girdle, lower limb, pelvic floor
tsMS		
Coil position and orientation	*Spinal segments with EF > 0.15 V/ m and EF > 50% E_{max}*	*Related sensorimotor functional area*
C5-IS	C5-T3	Diaphragm, scapular girdle, upper limb, upper thorax
T12-LR	L2-S3	Pelvic girdle, lower limb
T12-IS	T10-Fillum	Abdominal organs, pelvic girdle, lower limb, pelvic floor

montages and coil positions may enable neuromodulation of specific target regions in the SC. Application of cervical and thoracic tsDCS in healthy human volunteers shows evidence of spinal neuromodulation of upper and lower limb motor and sensory responses, depending on the placement of the vertebral electrode [6, 28, 31]. Recent experimental studies on lumbar repetitive tsMS applied in spinal lesion patients observed reduction of spasticity after stimulation and bladder function improvement [7, 8]. Table 3 presents a summary of the spinal segments with $E > 0.15$ V/m for each tsDCS montage and tsMS coil position, with the corresponding related functional region. These stimulation techniques are considered to be possible coadjutants for motor rehabilitation programs [6, 20, 22]. The EF is known to change neuronal resting potential, depending on the neuron orientation relative to the field, facilitating or inhibiting their firing capability. For instance, in tsMS-LR, the neurons with larger neuromodulation effects will be the ones oriented in the right-left direction. NISS neuromodulation selectivity is not only on the segments comprised between electrodes or near the coil but also on neuron direction relative to the fields obtained, just as observed in previous modelling and experimental studies on cortical stimulation [10].

Although the EF magnitudes predicted in tsMS are ~30 times higher than in tsDCS, comparing magnetic and electric stimulation effectiveness is not straightforward. Unlike DC stimulation, tsMS induces EFs with a more complex temporal profile, which consists of brief stimuli that are repeated at a low frequency. Further experimental studies are required to understand the relative physiological effectiveness of these two techniques and the biophysical mechanisms underlying them. Also, one main advantage of tsDCS is the portability and easier access to stimulating devices, which makes it appealing for home-based therapy, with potentially less side effects [4].

NISS may also induce nerve regeneration. McCaig et al. [52] observed guidance of spinal axonal regeneration in animal models of SCI after epidural stimulation, when applying longitudinal EFs of 0.3–0.4 V/m, which are similar to the EFs predicted for the tsDCS montages modelled. Also, one possible explanation for the functional recovery seen in the tsMS experimental studies referred above is spinal axonal regeneration [7, 8]. Although the values of the EF for human SC regeneration may differ from the ones determined in animals, axonal regeneration should be considered as a possible clinical effect of non-invasive spinal stimulation.

Since different electrode and coil positions result in diverse target spinal regions, due to changes in EF spatial distribution and direction, a combined modelling-experimental approach is recommendable for NISS application, by predicting the appropriate choice of stimulation conditions and parameters according to the intended spinal clinical target.

5 What Lies Ahead in Non-invasive Spinal Stimulation Modelling Studies

Non-invasive spinal stimulation modelling studies can be useful guides for clinical application of these techniques, aiming at the recovery of spinal circuits, frequently damaged by axonal degeneration and neuronal death due to spinal lesions or neurodegenerative diseases. However, little is known about how electromagnetic stimulation changes the way neurons function and regenerate. Understanding the underlying biophysics behind neuronal stimulation will be extremely useful in fine-tuning modelling predictions for the expected outcomes of NISS techniques.

Specifically for tsDCS, experimental studies present variable outcomes in the measurements of spinal motor responses, especially in the cervical SC. Future studies should model the effects of the EF in the transmembrane potential of spinal neurons during and after tsDCS, considering different regions (cervical, thoracic and lumbar) and different spinal reflex circuits, taking into account different electrode geometries and placements, to examine whether differences in the EF distribution can explain the variability of results observed. Spinal neuronal modelling may also be useful to infer on the best electrode settings and montages for stimulating a specific target. After defining the current and EF patterns required for neuromodulation of a specific spinal cellular target, electrode montages can be optimized using an inverse problem approach. This has recently been done for cortical stimulation [53].

Intersubject variability is determinant in the EF distribution for spinal stimulation, as observed in previous studies [18, 20]. Modelling findings indicate that anatomical characteristics, such as shape of the spinal canal and heterogeneity of the electrical conductivity, influence the location of EF hotspots in the SC. To test this, we repeated the C5-IS tsMS and C3-T3 tsDCS calculations with a conductivity of 0.2 S/m (soft tissue conductivity) attributed to all biological tissues in the model (Fig. 15): the peaks in the average EF profiles do not occur in the homogeneous models; thus, those peaks are due to the different electric conductivities of tissues combined with anatomical morphology. As our tetrahedral mesh was based on only one human model, future work should compare the EF distributions in different models to address the influence of anatomical characteristics in the EF spatial profiles. Parazzini et al. [18] presented EF predictions after tsDCS in four different models using hexahedral meshes: in this study, the current density and EF distributions present larger values in children models, which points also to an effect related with age and size.

These observations demonstrate the relevance of personalized modelling. Future biomedical research should be on the development of software that uses pipelines for semi-automatic segmentation of MRI images, which will be useful for the creation of individual models. These models could be applied for NISS computational studies, using neuronal circuitry models and the principle of reciprocity, to optimize tsDCS clinical protocols based on each patient's needs, informing on electrode number, geometry and placement, current and charge delivery. Accurate segmentation of

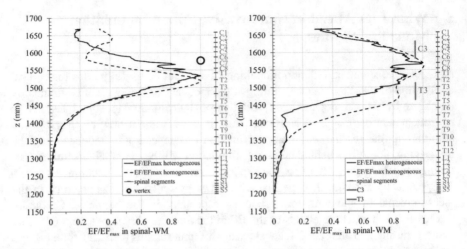

Fig. 15 Average EF magnitude profile along the spinal-WM length normalized to maximum values in C5-IS tsMS (left panel) and C3-T3 tsDCS (right panel). The position of the vertex of the coil is represented by a black circle, the position of the electrodes as vertical grey bars, and the position of spinal segments in grey on the right of each profile

tissues in MRI requires high spatial resolution since the SC structures have dimensions of the order of 1 mm or less, e.g. spinal roots and dorsal root ganglia. The model presented here was based on segmentation of MRI of a healthy volunteer on a 1.5 T scanner, using the sequences that could optimize image contrast and resolution for segmentation for the possible minimum time acquisition. Resolution varied from $0.5 \times 0.5 \times 1.0$ mm^3 to $0.9 \times 0.9 \times 2.0$ mm^3 voxel size, resulting in approximately total scanning time of 6 h [25]. Prolonged and frequent MRI acquisition can be difficult to endure for most patients in the clinical context. MRI scanners of 3–7 T may provide higher resolution in less scan time and help in the optimization of full body models. Thus, the improvement in the model will evolve jointly with strategies to improve MRI resolution and acquisition times.

Modelling studies should be validated by experimental studies to infer on safety limits and adverse effects and to determine the frequency and stimulation time needed for long-lasting effects required for rehabilitation. There has been an increase in experimental studies in human patients and healthy controls in tsDCS, and there are only few studies applying tsMS, but the high variability of results suggest the need for more clinical studies, to increase evidence that could provide gold standards for validation of NISS modelling findings. In vitro and in vivo studies are also relevant to determine the threshold EF values for neuromodulation of spinal circuitry and axonal regeneration of damaged spinal neurons, measured through electrophysiological techniques applied on cell cultures and animal models.

Non-invasive spinal stimulation is an emerging field of research with a multidisciplinary approach. It can be a powerful coadjutant therapy in the treatment of many spinal cord sensorimotor dysfunctions. The combination of modelling and experimental approaches will be essential to optimize NISS application for spinal clinical targets aiming at each patient needs.

Acknowledgements Instituto de Biofísica e Engenharia Biomédica is supported by Fundação para a Ciência e Tecnologia (FCT), Portugal, under grant n° UID/BIO/00645/2013. Instituto de Medicina Molecular João Lobo Antunes is supported by UID/BIM/50005/2019, project funded by Fundação para a Ciência e a Tecnologia (FCT)/ Ministério da Ciência, Tecnologia e Ensino Superior (MCTES) through Fundos do Orçamento de Estado. S.R. Fernandes was supported by a FCT PhD grant, reference SFRH/BD/100254/2014. S.R. Fernandes is currently affiliated as a post-doc researcher at Centre for Rapid and Sustainable Product Development of Polytechnic of Leiria, CDRSP-IPLeiria, 2430-028 Marinha Grande, Portugal.

References

1. Pierrot-Deseilligny, A., & Burke, D. (2012). *The Circuitry of the Human Spinal Cord: Spinal and Corticospinal mechanisms of movement*. Cambridge, UK: Cambridge University Press.
2. Molnar, G., & Barolat, G. (2014). Principles of cord activation during spinal cord stimulation. *Neuromodulation, 17*(Suppl 1), 12–21.
3. Nitsche, M. A., et al. (2008). Transcranial direct current stimulation: State of the art 2008. *Brain Stimulation, 1*(suppl 3), 206–223.
4. Rossi, S., et al. (2009). Safety, ethical considerations, and application guidelines for the use of transcranial magnetic stimulation in clinical practice and research. *Clinical Neurophysiology, 120*(suppl 12), 2008–2039.
5. Cogiamanian, F., et al. (2008). Effect of spinal transcutaneous direct current stimulation on somatosensory evoked potentials in humans. *Clinical Neurophysiology, 119*, 2636–2640.
6. Hubli, M., et al. (2013). Modulation of spinal neuronal excitability by spinal direct currents and locomotion after spinal cord injury. *Clinical Neurophysiology, 124*(suppl 6), 1187–1195.
7. Krause, P., et al. (2004). Lumbar repetitive magnetic stimulation reduces spastic tone increase of the lower limbs. *Spinal Cord, 42*, 67–72.
8. Niu, T., et al. (2018). A Proof-of-Concept Study of Transcutaneous Magnetic Spinal Cord Stimulation for Neurogenic Bladder. *Scientific Reports, 8*(suppl 1), 12549.
9. Salvador, R., et al. (2011). Determining which mechanisms lead to activation in the motor cortex: a modeling study of transcranial magnetic stimulation using realistic stimulus waveforms and sulcal geometry. *Clinical Neurophysiology, 122*, 748–758.
10. Thielscher, A., et al. (2014). Impact of the gyral geometry on the electric field induced by transcranial magnetic stimulation. *NeuroImage, 54*(suppl 1), 234–243.
11. Bikson, M., & Datta, A. (2012). Guidelines for precise and accurate computational models of tDCS. *Brain Stimulation, 5*(3), 430–431.
12. Datta, A., et al. (2009). Gyri-precise head model of transcranial direct current stimulation: Improved spatial focality using a ring electrode versus conventional rectangular pad. *Brain Stimulation, 2*(4), 201–207.
13. Dmochowski, J. P., et al. (2011). Optimized multi-electrode stimulation increases focality and intensity at target. *Journal of Neural Engineering, 8*(4), 046011.
14. Edwards, D., et al. (2013). Physiological and modeling evidence for focal transcranial electrical brain stimulation in humans: a basis for high-definition tDCS. *NeuroImage, 74*, 266–275.
15. Miranda, P. C., et al. (2013). The electric field in the cortex during transcranial current stimulation. *NeuroImage, 70*, 48–58.
16. Opitz, A., et al. (2015). Determinants of the electric field during transcranial direct current stimulation. *NeuroImage, 109*, 140–150.

17. Ruffini, G., et al. (2014). Optimization of multifocal transcranial current stimulation for a weighted cortical pattern targeting from realistic modeling of electric fields. *NeuroImage, 89*, 216–225.

18. Parazzini, M., et al. (2014). Modelling the current density generated by transcutaneous spinal direct current stimulation (tsDCS). *Clinical Neurophysiology, 125*(suppl 11), 2260–2270.

19. Fiocchi, S., et al. (2016). Cerebellar and Spinal Direct Current Stimulation in Children: Computational Modeling of the Induced Electric Field. *Frontiers in Human Neuroscience, 10*, 522.

20. Kuck, A., et al. (2017). Modeling trans-spinal direct current stimulation for the modulation of the lumbar spinal motor pathways. *Journal of Neural Engineering, 4*(suppl 5), 056014.

21. Fernandes, S. R., et al. (2018). Transcutaneous spinal direct current stimulation of the lumbar and sacral spinal cord: a modelling study. *Journal of Neural Engineering, 15*(suppl 3), 036008.

22. Fernandes, S. R., et al. (2019a). Cervical Trans-Spinal Direct Current Stimulation: a modelling-experimental approach. *Journal of Neuroengineering and Rehabilitation, 16*(1), 123.

23. Frostell, A., et al. (2016). A Review of the Segmental Diameter of the Healthy Human Spinal Cord. *Frontiers in Neurology, 7*, 238.

24. Salvador, R., et al. (2017). Tetrahedral vs hexahedral meshes in tCS realistic head modelling. *Brain Stimulation, 10*(2), 436–443.

25. Christ, A., et al. (2010). The Virtual Family - development of surface-based anatomical models of two adults and two children for dosimetric simulations. *Physics in Medicine and Biology, 55* (suppl 2), N23–N38.

26. Standring, S., et al. (2008). *Grey's anatomy: the Anatomical Basis of Clinical Practice 40th edition*. London: Churchill Livingston Elsevier.

27. Saturnino, G. B., et al. (2015). On the importance of electrode parameters for shaping electric field patterns generated by tDCS. *NeuroImage, 120*, 25–35.

28. Cogiamanian, F., et al. (2011). Transcutaneous spinal cord direct current stimulation inhibits the lower limb nociceptive flexion reflex in human beings. *Pain, 152*(2), 370–375.

29. Dongés, S. C., et al. (2017). Concurrent electrical cervicomedullary stimulation and cervical transcutaneous spinal direct current stimulation result in a stimulus interaction. *Experimental Physiology, 102*(10), 1309–1320.

30. Vergari, M., et al. (2012). Additive After-Effects of Spinal and Cortical DC Stimulation on Human Flexion Reflex Pathways. *Journal of the Peripheral Nervous System, 17*, S57–S57.

31. Bocci, T., et al. (2014). Cathodal transcutaneous spinal direct current stimulation (tsDCS) improves motor unit recruitment in healthy subjects. *Neuroscience Letters, 578*, 75–79.

32. Nierát, M., et al. (2014). Does Trans-Spinal Direct Current Stimulation Alter Phrenic Motoneurons and Respiratory Neuromechanical Outputs in Humans? A Double-Blind, Sham-Controlled, Randomized, Crossover Study. *Journal of Neuroscience, 34*(43), 14420–14429.

33. Geddes, L. A., & Baker, L. E. (1967). Specific Resistance of Biological Material-a Compendium of Data for Biomedical Engineer and Physiologist. *Medical & Biological Engineering, 5* (3), 271–292.

34. Haueisen, J., et al. (1997). Influence of tissue resistivities on neuromagnetic fields and electric potentials studied with a finite element model of the head. *IEEE Transactions on Biomedical Engineering, 44*(8), 727–735.

35. Rush, S., et al. (1963). Resistivity of body tissues at low frequencies. *Circulation Research, 12*, 40–50.

36. Surowiec, A., et al. (1987). Invitro Dielectric-Properties of Human-Tissues at Radiofrequencies. *Physics in Medicine and Biology, 32*(5), 615–621.

37. Osswald, K. (1937). Measurement of the conductivity and dielectric constants of biological tissues and liquids by microwave. *Hochfrequentz Tech. Elektroakustik, 49*, 40–49.

38. Struijk, J. J., et al. (1993). Excitation of dorsal root fibers in spinal cord stimulation: a theoretical study. *IEEE Transactions on Biomedical Engineering, 40*(7), 632–639.

39. Baumann, S. B., et al. (1997). The electrical conductivity of human cerebrospinal fluid at body temperature. *IEEE Transactions on Biomedical Engineering, 44*(3), 220–223.

40. Damasceno, A., et al. (2014). The Clinical Impact of Cerebellar Grey Matter Pathology in Multiple Sclerosis. *PLoS One, 9*(5), e96193.
41. Minhas, P., et al. (2010). Electrodes for high-definition transcutaneous DC stimulation for applications in drug-delivery and electrotherapy, including tDCS. *Journal of Neuroscience Methods, 190*(2), 188–197.
42. Smits, F. M. (1958). Measurement of Sheet Resistivities with the Four-Point Probe. *The Bell System Technical Journal, 37*, 711–718.
43. Ruffini, G., et al. (2013). Transcranial Current Brain Stimulation (tCS): Models and Technologies. *IEEE Transactions on Neural Systems and Rehabilitation Engineering, 21*(3), 333–345.
44. Nitsche, M. A., & Paulus, W. (2000). Excitability changes induced in the human motor cortex by weak transcranial direct current stimulation. *The Journal of Physiology, 527*(Pt 3), 633–639.
45. Fernandes, S. R., et al. (2016). Electric Field Distribution in the Lumbar Spinal Cord during Trans-Spinal Magnetic Stimulation. *Clinical Neurophysiology, 128*, e48–e50.
46. Haus, H. A., & Melcher, J. R. (1989). *Electromagnetic fields and energy.* Englewood Cliffs, NJ: Prentice-Hall.
47. Kammer, T., et al. (2001). Motor thresholds in humans: a transcranial magnetic stimulation study comparing different pulse waveforms, current directions and stimulator types. *Clinical Neurophysiology, 112*, 250–258.
48. Silva, S., et al. (2008). Elucidating the mechanisms and loci of neuronal excitation by transcranial magnetic stimulation using a finite element model of a cortical sulcus. *Clinical Neurophysiology, 119*(10), 2405–2413.
49. Rahman, A., et al. (2013). Cellular effects of acute direct current stimulation: somatic and synaptic terminal effects. *The Journal of Physiology, 591*(10), 2563–2578.
50. Deng, Z., et al. (2013). Electric field depth–focality trade-off in transcranial magnetic stimulation: simulation comparison of 50 coil designs *Brain Stimulation, 6*(suppl 1), 1–13.
51. Fernandes, S. R. et al. (2019b). Electric Field Distribution during Non-Invasive Electric and Magnetic Stimulation of the Cervical Spinal Cord. *Conference Proceedings IEEE Engineering in Medicine and Biology Society 2019.*
52. McCaig, C. D., et al. (2009). Electrical dimensions in cell science. *Journal of Cell Science, 122*, 4267–4276.
53. Fernández-Corazza, M., et al. (2016). Transcranial Electrical Neuromodulation Based on the Reciprocity Principle. *Frontiers in Psychiatry, 7*, 87.
54. Wolters, C. (2003). Influence of tissue conductivity inhomogeneity and anisotropy on EEG/MEG based source localization in the human brain (PhD). No. 39 in MPI series in cognitive neuroscience. MPI of Cognitive Neuroscience, Leipzig

A Miniaturized Ultra-Focal Magnetic Stimulator and Its Preliminary Application to the Peripheral Nervous System

Micol Colella, Micaela Liberti, Francesca Apollonio, and Giorgio Bonmassar

1 Introduction

Transcranial magnetic stimulation (TMS) is a noninvasive brain stimulation technique that employs a high-intensity pulsed magnetic field sent through the scalp by a stimulating coil. According to Faraday's law, a time-varying magnetic field induces inside the brain tissue an electric field, which may elicit a neuronal response. Due to the lack of physical contact, TMS results in almost a painless stimulation compared to electric noninvasive techniques, and for this reason, it has been deeply investigated over the past decades. Many studies have demonstrated the effectiveness of TMS as a therapeutic solution for the treatment of different neuropsychiatric conditions, among which are major depression [1], chronic pain [2–5], epilepsy [6], and obsessive-compulsive disorder (OCD) [7]. Furthermore, TMS is extensively adopted in neuroscience research to investigate intracortical, cortico-cortical, and cortico-subcortical interactions [6, 8, 9] and to assess causal relations between brain activity and behavior, as during speech [10–12] and motor mapping [13–19]. Despite the success, there are critical barriers to employing TMS in neuroscience research. One

M. Colella
Athinoula A. Martinos Center for Biomedical Imaging, Department of Radiology, Massachusetts General Hospital, Harvard Medical School, Boston, MA, USA

Department of Information Engineering, Electronics and Telecommunications (DIET), University of Rome "La Sapienza", Rome, Italy

M. Liberti · F. Apollonio
Department of Information Engineering, Electronics and Telecommunications (DIET), University of Rome "La Sapienza", Rome, Italy

G. Bonmassar (✉)
Athinoula A. Martinos Center for Biomedical Imaging, Department of Radiology, Massachusetts General Hospital, Harvard Medical School, Boston, MA, USA
e-mail: giorgio.bonmassar@mgh.harvard.edu

© The Author(s) 2021
S. N. Makarov et al. (eds.), *Brain and Human Body Modeling 2020*,
https://doi.org/10.1007/978-3-030-45623-8_9

major restriction of TMS in its present form is its limited spatial resolution, which makes the task of focusing the stimulation exclusively to the targeted cortical region very challenging. The spatial resolution of TMS depends on the stimulating coil's geometry and dimensions. A deep investigation of the focality of the electric field induced by 50 TMS coils has been conducted by Deng et al. [20]. In this work, the authors showed that figure-of-eight coils are more focal than circular coils. For example, the Magstim 70 mm figure-8 coil (P/N 9925, 3190) has a 15 cm^2 focality [20, 21], while the Magstim 90 mm circular coil (P/N 3192) has a 70 cm^2 focality [20, 21]. Also, smaller coils are, in general, more focal than larger coils, with focality that can be as small as 5 cm^2 [20]. For this reason, several attempts have been made to increase the spatial resolution of TMS by reducing the dimension of the stimulating coil [22–24]. To date, however, all the coils with reduced dimensions were designed for either invasive [24] or noninvasive animal applications [23]. In fact, as the dimension of the coil shrinks (diameter less than 2 cm), larger electric currents (few hundreds of kiloamperes) are required to produce enough induced electric field in the human cortex (~60–100 V/m). Given the conventional engineering approach which uses a few turns (<30) of wire winding inside an insulation housing, passing such large currents would produce excessive amounts of heat and large magnetic forces that may raise safety concerns or exceed specifications of conventional wires. To overcome such technological barriers, we have introduced flex circuit technology to design a new generation of miniaturized coils, which allows for a high number of turns (100 or greater) in a multilayer structure characterized by a reduced diameter (15–20 mm). Such an approach has been at first used to develop microscopic coils for invasive magnetic stimulation (μMS) capable of activating neuronal circuitry both in vitro [25] and in vivo [26]. Based on these results, a flex circuit technology was adapted to develop the first generation of miniaturized coils for noninvasive magnetic stimulation (μCoil). The concept behind the structure of the μCoil consists of a certain number of parallel copper traces, named layers, deposited on a long flexible substrate wound in N turns around a core. The prototype of μCoil was built using a four-trace Kapton sheet, wound in 123 turns around a copper pin with a 1 mm diameter. Since each layer was 1 mm high, with a distance of 0.5 mm between two adjacent layers, we fabricated a single 8-mm-tall circular solenoid (μCoil), with an outer diameter of 15 mm. A figure-of-eight stimulator was obtained by pairing two single μCoils, thus obtaining a structure with a maximum dimension of 30 mm. To date, these coils are the only figure-of-eight coils capable of performing stimulation of the peripheral nervous system on healthy volunteers. In this chapter, we studied numerically the behavior of six different μCoil geometries to show how each coil geometry affects the peripheral nerve stimulation threshold.

2 Models and Methods

2.1 μCoil Modeling

The electromagnetic simulation software Sim4Life (v.5, Zurich MedTech, Zurich) was adopted in all of the simulations. A planar loop coil geometry represented the actual copper trace configuration. The space between two consecutive loops of 1.5 mm was selected to model the trace thickness and spacing. Six different geometries were considered, as shown in Fig. 1. The first geometry studied was a two-layered single circular coil, with each layer wound in 123 turns around a 1 mm air core. Second, two other layers, with the same characteristics as above, were added on top to make a four-layered single circular coil. The third geometry herein studied was the figure-of-eight configuration, which consisted of two four-layered single circular coils paired together and with currents circulating in the opposite phase. Three additional figure-of-eight coil models were built to study how the presence and length of an ideal iron core (i.e., linearity assumption: $\mathbf{B} = \mu_r\mathbf{H}$ with $\mu_r = 60$) would modify the magnetic and electric field generated by the μCoil. Each solenoid was wound in 100 turns and had an inner and outer diameter of 10 mm and 21 mm, respectively. Three lengths of the iron core were considered: 9 mm, 23 mm, and 41 mm. Of all the geometries, the circular four-layered μCoil and the figure-of-eight μCoil were manufactured and tested. The iron-core μCoils were studied to

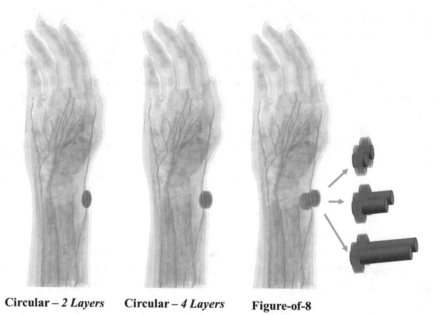

Circular – 2 Layers   **Circular – 4 Layers**   **Figure-of-8**

Fig. 1 Four figure-of-eight μCoil models. μCoil models and placement over the right arm of the neurofunctionalized model Yoon-Sun and the point where the two figure-of-eight μCoils touch was precisely above the nerve trajectory in the wrist area

enhance the magnetic field strength of the newly developed μTMS coil for future cortical applications. The electromagnetic simulations were conducted using the Magneto Quasi-Static solver of the software Sim4life (v.5.9, Zurich MedTech, Zurich). This solver implements the quasi-static approximation that allowed decoupling the electric field from the magnetic field. Considering the source current density **J**, magnetic vector potential **A** was calculated from Ampere's law and used to compute the **B** and **E** field. EM simulations were performed by feeding a sinusoidal current with a 1.9 kHz frequency in each μCoil and a phase of 0° in the first μCoil and of 180° in the second μCoil when considering the figure-of-eight coil. The two coils were placed in a figure-of-eight coil over the arm of the anatomical, neurofunctionalized body model Yoon-Sun (ViP., v4 [27]), in correspondence of the superficial branch of the radial nerve, as shown in Fig. 1. The Yoon-Sun model is characterized by a detailed representation of the peripheral nervous system and the various nerve fibers. This simulation setup was validated by experiments conducted on healthy volunteers [28]. The μCoil driving sinusoidal currents were set to 1 A for comparison among different configurations.

2.2 Modeling Peripheral Nerve Stimulation: Titration Analysis

To study the interaction between the induced electric field and the peripheral nervous system, the results of the EM simulations were fed into the dynamic Neuron solver version (v. 5.9) embedded in Sim4Life in a Multiphysics approach. The excitable behavior of the fibers was simulated using the McIntyre-Richardson-Grill model, which is based on a double-cable representation of the axon that allows separating electrical representations for the myelin and underlying internodal axolemma [29, 30].

Nerves can be excited by an external potential, V_{ext}, defined as in Eq. 1, where C is the linear path connecting each data point to an arbitrary zero voltage reference [31].

$$V_{ext} = -\int_C \mathbf{E} \cdot d\mathbf{l} \tag{1}$$

In this work, a titration analysis was conducted for all the simulated μCoil geometries. Titration is the iterative process that allows determining the stimulation threshold of a nerve by stimulating it with pulses of increasing intensity.

3 Results

3.1 Magnetic Field Generated by the μCoils

Figure 2 shows the vector view of the **B** field generated by each μCoil on the central cross-sectional plane. Inside the coil, the presence of more layers directs the **B** field along the axis of the coil. The magnetic field lines of forces diverge outside the coil and form closed loops. Doubling the number of layers increased the maximum intensity of **B** induced on the skin from 30 to 40 mT, thus by a factor of 33% without changing its direction. When forming the eight-shaped coil by pairing two circular μCoils fed in counter phase, the **B** field lines formed a third loop at the center of the figure-of-eight coil, due to the coupling between the two coils. The use of the iron core modified the direction of the **B** field that was along the coil axis for the whole length of the core and parallel to the surface of the coil in the space between the two cores. At the interface between iron and air, vector **B** was deviated to respect the continuity of the normal component to the surface. Moreover, at the level of the nerve (i.e., 2.5 mm from the surface of the skin) **B** field intensity for the coil with the 41 mm, long core is higher than that generated with the 23 mm long core and with the 9 mm long core, by 15% and 50%, respectively.

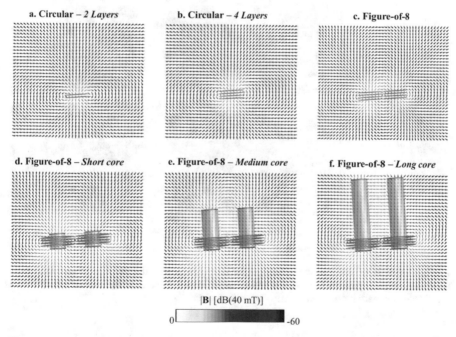

Fig. 2 Vector view of the **B** field generated by each μCoil model. Arrows show direction, orientation, and intensity of **B**. Intensities of **B** are represented by a color bar in a logarithmic scale with 0 dB referenced to 40 mT, which is the maximum |B| induced on the surface of the skin by the air core figure-of-eight μCoil (Panel c) when fed with 1 A

3.2   Electric Field Induced by the μCoils

The single circular coil that was composed of four layers induced inside the tissues an electric field stronger than the field induced by the two-layered coil: the maximum E field induced at the level of the nerve with 1 A was 0.27 V/m for the two-layered circular coil and 0.40 V/m for the four-layered one (see Fig. 3, Panel a and b). The figure-of-eight coil configuration increased the intensity of the induced electric field on the nerve up to 0.6 V/m (Fig. 3, panel C). Moreover, the direction of the **B** field (Fig. 2) induced an **E** field-oriented mostly along the longitudinal axis of the arm and (therefore) along the radial nerve. The presence of an iron core further oriented E along the nerve, as shown by the vector arrows in Fig. 3 d–f. Moreover, the longer the core, the higher was the maximum intensity of the induced **E**: 2.5 V/m for 41 mm, 2.3 V/m for 23 mm, and 1.6 V/m for 9 mm. Figure 4 shows electric field components tangential (E_{tan}) and normal (E_{norm}) to the nerve fibers, for the two circular μCoils and the air core figure-of-eight μCoil. The figure-of-eight μCoil induced a tangential component that was five and seven times higher than that induced by a four-layered and a two-layered single circular μCoil, respectively, and a normal component that was two times lower.

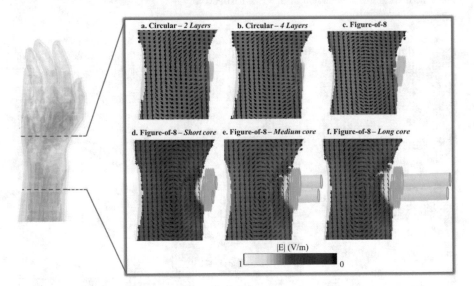

Fig. 3 Direction, orientation, and intensity of the **E** field induced in the arm of the Yoon-Sun model on the longitudinal plane of the arm that crosses the center of the stimulating coil. A vector view of the **E** field is superimposed on the intensities distribution map

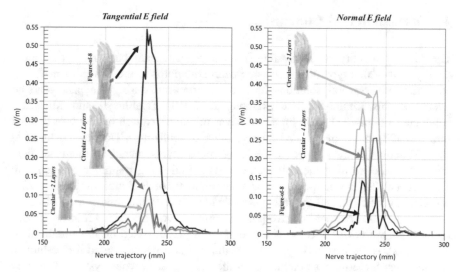

Fig. 4 The tangential (left) and normal (right) components of the **E** field-induced along the nerve by the two-layered circular μCoil, the four-layered circular μCoil, and figure-of-eight μCoil

3.3 Variation of the Peripheral Nerve Stimulation Threshold

When simulating exposure of the arm to the two-layered single circular coil, the computed stimulation threshold for the radial nerve fibers varied between 259 A and 269 A. Nerve action potentials were generated near the extremities of the coil. Increasing the number of layers from two to four reduced the coil current threshold in each layer/coil to values between 177.56 A and 184.39 A, while the activation site of the neuronal response was not affected. The stimulation threshold was further reduced when simulating with a figure-of-eight coil: the five times higher induced tangential electric field of the figure-of-eight-coil reduced the threshold to values between 58 A and 58.5 A; such a small range of variability may be due to the augmented focality. For all fibers, the activation site of the neuronal response was at the center of the figure-of-eight coil. Onset timing was not affected by the number of layers or by the change of coil shape and remained around an average time of 0.8 ms. A reduction of the onset timing from 0.8 to 0.5 ms, meaning an earlier spike, was obtained when changing from air to iron core. The presence of the core also reduced the stimulation threshold from 58 A to 14 A, 10.1 A, and 9.4 A for the short, medium, and long core, respectively, and further oriented **E** field along the direction of the nerve.

4   Discussion and Conclusion

The lack of focality of TMS stimulators is a significant barrier for the future deployment of this type of noninvasive neurostimulator in the clinics or neuroscience research. Several solutions have been proposed during the past years to overcome this limitation. For example, in 2010, Talebinejal and Musallam conducted a numerical study on a multilayer miniaturized coil made up of braided *Litz* wire [23] and demonstrated the feasibility of realizing a TMS coil reduced in size. Moreover, they showed that adding more layers increased magnetic field intensity while keeping the coil area constant. Following a similar geometry, in Tischler et al. 2011, a mini-coil with an outer diameter of 26.5 mm was developed. It is crucial to notice that all the attempts made to reduce dimensions of the TMS coils were conceived for either invasive [24] or noninvasive [23] animal models. To our knowledge, the newly developed figure-of-eight μCoil, based on flex circuit technology [32], is the first prototype of miniaturized coil manufactured for on-human applications. Experiments conducted on the peripheral nervous system of healthy volunteers showed that the μCoil could elicit somatosensory nerve action potentials (SNAPs) when the coil is placed over the superficial branch of the radial nerve. In this chapter, we conducted a numerical study that aimed to show how different possible μCoil geometries can affect the radial nerve stimulation threshold. Six different topologies of μCoils were modeled, based on the three different μCoils actually manufactured: the single circular four-layered and the figure-of-eight coils were originally designed to be tested on the PNS, and the iron core figure-of-eight coil developed to target the cerebral cortex. Each coil was placed over the radial nerve superficial branch of the neurofunctionalized model Yoon-Sun [27]. A multiphysics approach was herein considered, and electromagnetic simulations were coupled with the dynamic neuronal solution. To study the effect of adding more layers, we first studied a two-layered single circular μCoil and compared it with the four-layered one. The results showed that increasing the number of layers from two to four increased the intensity of the **B** field by 50%, thus inducing a stronger **E** field. As a consequence, the stimulation threshold was lower for the four-layered coil. A figure-of-eight coil was obtained by pairing together two four-layered circular coils. Such μCoil was able to generate a stronger magnetic field, thus inducing a stronger electric field on the nerve. Moreover, the figure-of-eight coil oriented the **E** field along the trajectory of the nerve fibers, leading to a better coupling between the coil and the nerve and a lower stimulation threshold. This coupling is further enhanced by addition of an iron core. Additionally, the results showed that the presence of the core not only increased intensity of the induced electric field, as expected, but also reduced the onset timing of the first spike. Nevertheless, it should be noted that saturation of the iron core was not considered in this study. Thus, computed stimulation threshold was underestimated, due to the overestimation of both **B** and **E** field intensities. However, we are planning to model next the nonlinearities of a real core. We confirmed that the figure-of-eight coil was more efficient than the circular coil as it enhanced the focality and the coupling between nerve fibers and **E** field. Moreover, the results showed that the presence of an iron core further reduced stimulation threshold while increasing depth, making this geometry suitable for investigating the μCoil stimulation of the brain cortex.

References

1. Pascual-Leone, A., Walsh, V., & Rothwell, J. (2000). Transcranial magnetic stimulation in cognitive neuroscience - virtual lesion, chronometry, and functional connectivity. *Current Opinion in Neurobiology, 10*(2), 232–237.
2. Kobayashi, M., & Pascual-Leone, A. (2003). Basic principles of magnetic stimulation. *Lancet, 2*, 145–156.
3. Fitzgerald, P. B., Fountain, S., & Daskalakis, Z. J. (2006). A comprehensive review of the effects of rTMS on motor cortical excitability and inhibition. *Clinical Neurophysiology, 117* (12), 2584–2596.
4. Najib, U., Bashir, S., Edwards, D., Rotenberg, A., & Pascual-Leone, A. (2011). Transcranial brain stimulation: Clinical applications and future directions. *Neurosurgery Clinics of North America, 22*(2), 233–251.
5. Valero-Cabré, A., Pascual-Leone, A., & Coubard, O. A. (2011). Transcranial magnetic stimulation (TMS) in basic and clinical neuroscience research. *Revue Neurologique (Paris), 167*(4), 291–316.
6. Rotenberg, A. (2010). Prospects for clinical applications of transcranial magnetic stimulation and real-time EEG in epilepsy. *Brain Topography, 22*(4), 257–266.
7. Carmi, L., Alyagon, U., Barnea-Ygael, N., Zohar, J., Dar, R., & Zangen, A. (2018). *Clinical and electrophysiological outcomes of deep TMS over the medial prefrontal and anterior cingulate cortices in OCD patients* (Vol. 11, p. 158). Brain Stimulation.
8. Horvath, J. C., Perez, J. M., Forrow, L., Fregni, F., & Pascual-Leone, A. (2011). Transcranial magnetic stimulation: A historical evaluation and future prognosis of therapeutically relevant ethical concerns. *Journal of Medical Ethics, 37*(3), 137–143.
9. Rotenberg, A., Horvath, J., & Pascual-Leone, A.. (2014). Transcranial magnetic stimulation series editor.
10. Jennum, P., Friberg, L., Fuglsang-Frederiksen, A., & Dam, M. (1994). Speech localization using repetitive transcranial magnetic stimulation. *Neurology, 44*(2), 269.
11. Wassermann, E. M., et al. (1999). Repetitive transcranial magnetic stimulation of the dominant hemisphere can disrupt visual naming in temporal lobe epilepsy patients. *Neuropsychologia, 37*, 537–544.
12. Epstein, C. M., et al. (1999). Localization and characterization of speech arrest during transcranial magnetic stimulation. *Clinical Neurophysiology, 110*, 1073.
13. Wassermann, E. M., McShane, L. M., Hallett, M., & Cohen, L. G. (1992). Noninvasive mapping of muscle representations in human motor cortex. *Electroencephalography and Clinical Neurophysiology/Evoked Potentials, 85*(1), 1–8.
14. Brasil-Neto, J. P., Cohen, L. G., Panizza, M., Nilsson, J., Roth, B. J., & Hallett, M. (1992). Optimal focal transcranial magnetic activation of the human motor cortex: Effects of coil orientation, shape of the induced current pulse, and stimulus intensity. *Journal of Clinical Neurophysiology, 9*(1), 132–136.
15. Cincotta, M., et al. (2006). Mechanisms underlying mirror movements in Parkinson's disease: A transcranial magnetic stimulation study. *Movement Disorders, 21*(7), 1019–1025.
16. Cincotta, M., & Ziemann, U. (2008). Neurophysiology of unimanual motor control and mirror movements. *Clinical Neurophysiology, 119*, 744.
17. Civardi, C., Vicentini, R., Collini, A., Boccagni, C., Cantello, R., & Monaco, F. (2009). *Motor cortical organization in an adult with hemimegalencephaly and late onset epilepsy* (Vol. 460, p. 126). Neuroscience Letters.

18. Martinez, M., Brezun, J. M., Zennou-Azogui, Y., Baril, N., & Xerri, C. (2009). Sensorimotor training promotes functional recovery and somatosensory cortical map reactivation following cervical spinal cord injury. *The European Journal of Neuroscience, 30*(12), 2356–2367.

19. Karl, A., Birbaumer, N., Lutzenberger, W., Cohen, L. G., & Flor, H. (2001). Reorganization of motor and somatosensory cortex in upper extremity amputees with phantom limb pain. *The Journal of Neuroscience, 21*(10), 3609–3618.

20. De Deng, Z., Lisanby, S. H., & Peterchev, A. V. (2013). Electric field depth-focality tradeoff in transcranial magnetic stimulation: Simulation comparison of 50 coil designs. *Brain Stimulation, 6*(1), 1–13.

21. Hovey, C., & Jalinous, R. (2006). The guide to magnetic stimulation. *Magstim, 20*(4), 284–287.

22. Salinas, F. S., Lancaster, J. L., & Fox, P. T. (2009). 3D modeling of the total electric field induced by transcranial magnetic stimulation using the boundary element method. *Physics in Medicine and Biology, 54*, 3631.

23. Talebinejad, M., & Musallam, S. (2010). Effects of TMS coil geometry on stimulation specificity. In *2010 Annual international conference of the IEEE Engineering in Medicine and Biology Society EMBC'10*, pp. 1507–1510.

24. Tischler, H., et al. (2011). Mini-coil for magnetic stimulation in the behaving primate. *Journal of Neuroscience Methods, 194*(2), 242–251.

25. Bonmassar, G., Lee, S. W., Freeman, D. K., Polasek, M., Fried, S. I., & Gale, J. T. (2012). Microscopic magnetic stimulation of neural tissue. *Nature Communications, 3*, 910–921.

26. Park, H. J., Bonmassar, G., Kaltenbach, J. A., Machado, A. G., Manzoor, N. F., & Gale, J. T. (2013). Activation of the central nervous system induced by micro-magnetic stimulation. *Nature Communications, 4*, 1–9.

27. Gosselin, M.-C., et al. (2014). Development of a new generation of high-resolution anatomical models for medical device evaluation: The virtual population 3.0. *Physics in Medicine and Biology, 59*(18), 5287–5303.

28. Colella, M., et al. (2019). A microTMS system for peripheral nerve stimulation. *Brain Stimulation, 12*(2), 521.

29. McIntyre, C. C., & Grill, W. M. (2002). Extracellular stimulation of central neurons: Influence of stimulus waveform and frequency on neuronal output. *Journal of Neurophysiology, 88*(4), 1592–1604.

30. McIntyre, C. C., Richardson, A. G., & Grill, W. M. (2002). Modeling the excitability of mammalian nerve fibers: Influence of afterpotentials on the recovery cycle. *Journal of Neurophysiology, 87*(2), 995–1006.

31. *Sim4life Documentation*. ZMT Zurich MedTech AG.

32. Colella, M., et al. (2019). Ultra-focal magnetic stimulation using a μTMS coil: A computational study, 2019. In *41st Annual international conference of the IEEE Engineering in Medicine and Biology Society*, pp. 3987–3990

Part IV
Modeling of Neurophysiological Recordings

Combining Noninvasive Electromagnetic and Hemodynamic Measures of Human Brain Activity

Fa-Hsuan Lin, Thomas Witzel, Matti S. Hämäläinen,
and Aapo Nummenmaa

1 Introduction

Each of the presently available technologies for noninvasive electromagnetic or hemodynamic measurements of brain activity offers different spatiotemporal resolution and physiological sensitivity. Human functional MRI [6, 30] is temporally limited by the slow hemodynamic response (~ seconds) due to relative cerebral blood flow (CBF), cerebral blood volume (rCBV), and metabolism changes, which are indirect markers of neuronal signaling. Using the echo-planar imaging [36] technique, fMRI can typically provide a spatial sampling on a millimeter scale with homogeneous volumetric sensitivity and whole-brain coverage.

Magnetoencephalography (MEG) and electroencephalography (EEG) in turn detect extracranial magnetic fields and electric potential differences on scalp, which are both elicited by spatially clustered and temporally coherent postsynaptic neuronal currents [40]. Consequently, MEG/EEG can be used to study neuronal dynamics with millisecond resolution. Different from fMRI, where tomographic images are usually obtained, the spatial resolution of MEG/EEG is related to the capability to resolve intracranial current sources from extracranial measurements. To characterize the distribution of postsynaptic neuronal currents responsible for the

F.-H. Lin (✉)
Department of Medical Biophysics, University of Toronto, Toronto, ON, Canada

Physical Sciences Platform, Sunnybrook Research Institute, Toronto, ON, Canada

Athinoula A. Martinos Center for Biomedical Imaging, Massachusetts General Hospital,
Charlestown, MA, USA
e-mail: fhlin@sri.utoronto.ca

T. Witzel · M. S. Hämäläinen · A. Nummenmaa
Athinoula A. Martinos Center for Biomedical Imaging, Massachusetts General Hospital,
Charlestown, MA, USA

© The Author(s) 2021
S. N. Makarov et al. (eds.), *Brain and Human Body Modeling 2020*,
https://doi.org/10.1007/978-3-030-45623-8_10

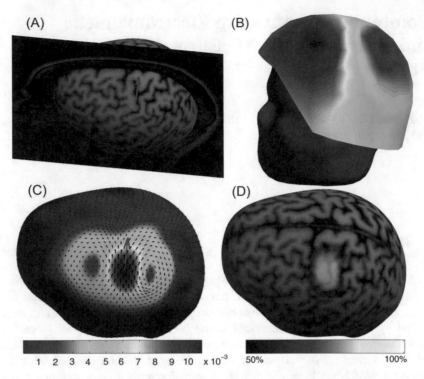

Fig. 1 (**a**) A simulated "point-like" current dipole shown on a simulated cortical surface obtained by shrinking the inner skull segmented from anatomical MRI by about 1 cm. The orientation of the dipole is tangential to the cortical surface. (**b**) The simulated MEG measurement shown with a realistic outer skin surface and sensor locations from an actual measurement. (**c**) The Minimum-Norm Estimate (MNE) vector field is displayed with black arrows, and the MNE amplitude is shown in color scale. (**d**) The full width at half maximum (FWHM) of the MNE point-spread function gives an estimate for the spatial resolution of the source localization method

macroscopically measured voltage and magnetic field, we have to solve an electromagnetic inverse problem, which admits no unique solution [26, 28]. With auxiliary mathematical and physiological assumptions, we can get reasonable estimates of the spatial distribution of the neuronal currents that generate the MEG (and EEG) measurements (see, Fig. 1).

There are two main lines of approach to the MEG/EEG inverse problem (for a review, see [22]). The most traditional one assumes that the measurements are generated by a small number of equivalent current dipoles (ECDs). The main challenges with this approach are that i) the optimization problem for finding the best matching parameters for the ECDs is nonlinear and ii) the optimization becomes more and more difficult with increasing number of dipoles. The problem of handling the *a priori* unknown number of dipoles is also a nontrivial one, because it requires rather involved numerical methods [3, 29]. The Minimum-Norm Estimate (MNE) approach circumvents the nonlinearity of estimating current source location and orientation by assuming a discrete grid of source locations, each of which has three orthogonal dipole orientations. With respect to estimating dipole *amplitudes*, the inverse problem is linear. A least squares estimator (i.e., the MNE) can be calculated by standard numerical methods. The benefit is that the number of sources need not be known *a priori*, and a closed form solution is available in a computationally robust form. The natural drawback of this method is that the identity of each "localized source" becomes less clear-cut. In the case of Fig. 1, both the MNE and the single ECD explain the data equally well, which is a manifestation of the non-uniqueness of the inverse solution.

In principle, fMRI and EEG/MEG data can be integrated in order to achieve high spatiotemporal brain imaging (for a review, see, e.g., [44]). The basic rationale for such data integration is based on neurophysiological evidence: invasive studies in primates suggest that BOLD fMRI signal increases are closely related to the same postsynaptic neuronal activity [34] that generates MEG responses [22, 40]. Tight coupling between neuronal and vascular events has also been reported in the somatosensory system of rodents [12, 13]. Tentatively, these observations support the computational strategy of using fMRI, a vascular marker of neuronal events, as a physiological constraint for reducing the spatial ambiguity in the source localization of MEG/EEG. For example, the ECD fitting method in MEG/EEG source localization can be informed by fMRI [2, 20, 48]. The statistical maps derived from fMRI data can also be used as a spatial prior for the distributed source reconstruction [10, 32, 33]. A further study using simulations demonstrated the advantage of combined fMRI and EEG for a higher efficiency of cortical current density estimation at different signal-to-noise ratios (SNRs) with the presence of both fMRI-visible and fMRI-invisible sources [5]. MEG has a millisecond temporal resolution, ideal for studying cortical oscillations. It has been shown that integrating fMRI and MEG can also improve the localization of cortical sources of oscillatory activity [31].

In this article, we use the cortically constrained distributed source modeling framework to illustrate how fMRI information can be used to assist MEG/EEG localization and what are the potential benefits and pitfalls of this approach. We then briefly discuss the further modeling efforts and extensions of the fMRI-weighted MNE that have been presented in the literature. We also elaborate on the practical aspects of designing a successful MEG/EEG/fMRI experiment, data analysis, and interpretation of the results. The neurovascular coupling, technical challenges, and opportunities for further optimizing the integration are also described. As the MEG and EEG signals have a similar physiological origin, but their sensors have different sensitivity profiles, the combination of the two yields theoretically the best localization results. However, the EEG is substantially more sensitive to the volume conductor model: the poorly conducting skull distorts and smears the electric scalp potentials, whereas the currents in the skull and scalp make only a minor contribution to MEG [23]. In what follows, all models and methods could be formulated in terms of both MEG and EEG measurements, but we use MEG as our main example due to the less error-prone forward model. Ultimately, we expect to develop multimodal MEG/EEG/fMRI neuroimaging methodology for characterizing spatiotemporal functional connectivity in large-scale neural networks of the human brain with high sensitivity and accuracy.

2 Methods

2.1 Minimum-Norm Estimates

Under the quasi-static approximation of Maxwell's equations [22], the measured MEG signals and the underlying current source are related by a linear transformation:

$$\mathbf{Y}(t) = \mathbf{AX}(t) + \mathbf{N}(t), \tag{1}$$

where $\mathbf{Y}(t)$ is an m-dimensional vector containing measurements from m sensors at time instant t; $\mathbf{X}(t)$ is a $3n$-dimensional vector denoting the unknown amplitudes of the three components of n current sources; $\mathbf{A}$ is the gain matrix representing the mapping from the unit dipole components to MEG sensors, i.e., the solution of the forward problem; and $\mathbf{N}(t)$ denotes noise in the measured data. For typical analysis of evoked responses, the measurement noise $\mathbf{N}(t)$ can be assumed to be Gaussian with zero mean and a time-independent spatial covariance matrix $\mathbf{C}$, which can be estimated from the data. The number of sensors is some hundreds, and any realistically spaced grid covering the cortex requires thousands of source points. Thus, the inverse problem is severely underdetermined as the number of equations m (sensors) is an order of magnitude smaller than the number of unknowns $3n$ (source amplitudes). With the presently available accurate reconstructions of cortical surfaces [8, 15, 16], the locations of the sources can be constrained according to the individual anatomy. If we assume that apical dendrites of pyramidal cells, which are mainly oriented perpendicular to the cortical mantle, are the principal generators of the MEG signals [40], we can also fix the orientation of the sources and reduce the number of unknowns from $3n$ to n (see, Fig. 2).

If we further assume that the source amplitudes have a Gaussian *a priori* distribution with a time-independent covariance matrix $\mathbf{R}$, we obtain the *maximum a posteriori* (MAP) estimate or the ℓ_2 minimum-norm solution, which is linearly related to the measurements [9]:

$$\begin{aligned}
\mathbf{X}^{\mathrm{MNE}}(t) &= \mathbf{RA}^T \left(\mathbf{ARA}^T + \lambda^2 \mathbf{C}\right)^{-1} \mathbf{Y}(t) = \lambda^{-2}\mathbf{RA}^T \left(\lambda^{-2}\mathbf{ARA}^T + \mathbf{C}\right)^{-1} \mathbf{Y}(t) \\
&= \mathbf{WY}(t),
\end{aligned} \tag{2}$$

where λ^2 is a regularization parameter, which is introduced to avoid noise amplification in the matrix inversion, and the superscript T indicates matrix transpose. The parameter λ^2 can be estimated from the amplitude signal-to-noise ratio (SNR) of the whitened data: $\lambda^2 = \mathrm{tr}\left(\widetilde{\mathbf{A}}\mathbf{R}\widetilde{\mathbf{A}}^T\right)/\mathrm{tr}(\mathbf{I}_{m \times m})/\overline{SNR}^2$. Here $\mathrm{tr}(\cdot)$ denotes the trace of a

Fig. 2 (**a**) A surface model of the left cortical hemisphere gray-white matter boundary reconstructed by FreeSurfer. (**b**) A close-up view of the sensorimotor cortex, showing the dipole sources as red arrows, oriented perpendicular to the cortical surface

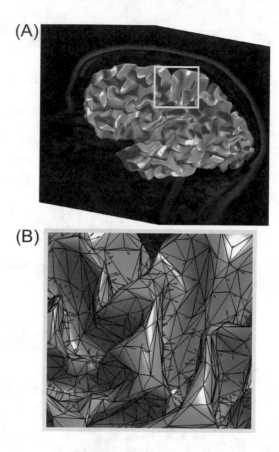

matrix. The whitened forward operator is $\widetilde{\mathbf{A}} = \mathbf{C}^{-1/2}\mathbf{A}$, and $\mathbf{I}_{m \times m}$ is the whitened (unit) noise covariance. $\overline{SNR}$ is an estimate for the SNR of the data.

The solution $\mathbf{X}^{\mathrm{MNE}}(t)$ in Eq. (2) provides the values of the current amplitudes that best fit the MEG measurements in the least squares sense, with the additional constraint of having the minimal (Euclidean) ℓ_2 norm. It may be desirable to further transform the resulting current distribution estimate into a statistical map that takes into account the spatial distribution of fluctuations in the source estimate caused by noise [10]. To this end, we need to consider the variance of the linear inverse estimates, when the data consists of noise only:

$$w_k^2 = \left(\mathbf{W}\mathbf{C}\mathbf{W}^T\right)_{kk} = \left(\widetilde{\mathbf{W}}\widetilde{\mathbf{W}}^T\right)_{kk}. \tag{3}$$

For fixed-orientation sources, we now obtain the noise-normalized activity estimate for the kth dipole and t^{th} time point as the ratio

$$X_k^{dSPM}(t) = \frac{X_k^{MNE}(t)}{w_k}.$$ (4)

The dSPM thus normalizes the actual MNE by the standard deviation of the fluctuation of the MNE that results from inverting data, which consists of noise only.

To incorporate the spatial information from fMRI, it has been suggested that MEG source locations coinciding with significant fMRI activity were given a higher variance in the *a priori* source covariance matrix **R** [10, 32]. Specifically, the source covariance matrix was assumed diagonal, and the fMRI weighting for source location k was encoded as:

$$R_{kk} = \begin{cases} \sigma_1^2 & \text{if } k \text{ active in fMRI} \\ \sigma_0^2 & \text{otherwise} \end{cases}$$ (5)

A weighting ratio of 10:1 between active (σ_1^2) and inactive cortical locations (σ_0^2) has been suggested by a simulation study [32].

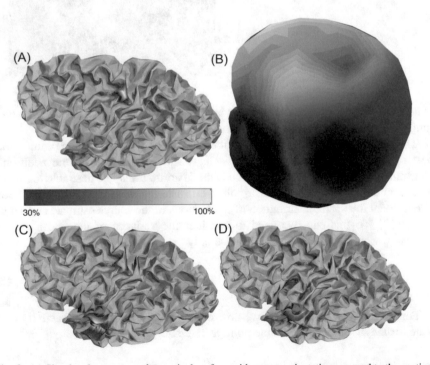

Fig. 3 (**a**) Simulated sources on the cortical surface with source orientations normal to the cortical sheet. (**b**) Simulated MEG measurements corresponding to the source configuration of (**a**). (**c**) The resulting cortically constrained MNE. (**d**) The noise sensitivity normalized MNE (dSPM)

2.2   Example: MNE Analysis and the Effect of fMRI Weighting

Figure 3a shows a simulated source consisting of two patches of activated cortex, located approximately at motor and auditory areas. As dictated by physics of quasi-static magnetic fields generated by dipolar current sources in a nearly spherical volume conductor, MEG is mostly sensitive to sources in the sulcal walls. The auditory cortical source is located entirely in the wall of the Sylvian fissure, whereas the motor cortical source extends over the precentral gyrus, thus giving less optimal summation of the MEG fields. This is readily visible in Fig. 3b: the dipolar field pattern from the auditory cortex is much more prominent. This translates directly to the MNE of Fig. 3c: even though originally of similar amplitude, the motor cortical source estimate is weaker, and the gyral part of the source is missing. As smaller more superficial sources can produce similar MEG field as larger deep sources, the minimum-norm constraint has a tendency to push the estimates toward the more superficial parts of the brain surface. However, as the superficial parts are also more prone to noise fluctuations in the inverse estimates, the effect of the noise normalization of dSPM counteracts this and pushes the source maxima deeper (Fig. 3d).

Continuing with the same simulated source, Fig. 4 demonstrates the effects of incorporating an fMRI weighting. The first row shows the case where the fMRI weighting matches closely to the true source. Consequently, both sources are recovered, and the extra ripples are suppressed. The second row corresponds to the case where the motor cortical activity is not visible in the fMRI. Then, the fMRI weighting also abolishes this source from the MEG inverse solution with the selected MNE threshold. For the last row, we demonstrate a case where an extra activation cluster is present in the fMRI, leading to some false-positive sources at the corresponding fMRI-weighted MNE.

Note that for all the cortical images, the threshold was set to be 30% of the maximum amplitude, and the full color scale was used to display the sources or estimates on the cortical surface. This selection is rather arbitrary and has an obvious effect on the visual appearance of the estimates to be, for example, apparently more focal.

3   Discussion

3.1   Developments of the fMRI-Weighted MNE

As shown by simulation examples, the main problem with the simple "fMRI-weighted MEG" is the relatively strong bias toward the fMRI data. Biophysically, we have reasons to expect that the "active" areas detected by MEG and fMRI may be only partially overlapping. On one hand, the temporal synchronization and summation of the neuronal activity on a millisecond scale is crucial for elicitation of a

fMRI prior   fMRI weighted MNE

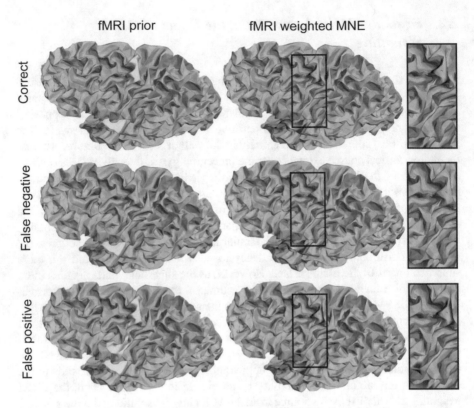

Fig. 4 First row: the fMRI-weighted MNE in case where the fMRI prior information is concordant with the MEG sources. Second row: the motor cortical source is missing from the fMRI map. Third row: a superfluous activation cluster is present in the fMRI map near the auditory cortical source

measurable MEG response. On the other hand, fMRI hemodynamic response allows the summation of activity over seconds. For instance, a sharp transient stimulus may produce a response clearly visible in MEG but may not be strong enough to push the vascular system to produce a robust hemodynamic response. On the contrary, weak but asynchronously sustained activity may temporally integrate to a measurable fMRI response but remain undetected by MEG. In addition to temporal summation, MEG and fMRI have different spatial sensitivities: fMRI has no spatial cancellation due to neuronal currents having incoherent orientations as MEG [1], and it has equal sensitivity for detecting activity in gyri and sulci.

Different from using fMRI as a spatial prior, Daunizeau and colleagues propose a symmetric approach for multimodal integration of fMRI and MEG/EEG data by constructing a model where the spatial activation profile in each anatomically defined parcel of the cortex is assumed to be similar in both modalities [11]. Activations, which are not present in both modalities, are modeled as Gaussian residuals, allowing for natural discrepancies between the different types of data. The model becomes computationally rather complex as variables between modalities become spatiotemporally entangled. Although the explicit modeling of coupled and uncoupled sources is appealing from a theoretical viewpoint, it is somewhat questionable how desirable it is to let MEG/EEG data substantially influence the estimated waveform of the fMRI hemodynamic response. Henson and colleagues maintain the asymmetry between modalities but consider more flexible fMRI priors while preserving the basic Gaussian model structure, which renders computations highly tractable [27]. The work expresses the prior covariance as multiple variance/covariance components $\mathbf{R}_i$: $\mathbf{R} = \Sigma_i \lambda_i \mathbf{R}_i$, each with an adaptive weight parameter λ_i. The covariance components $\mathbf{R}_i$ are generated based on fMRI analysis, whereas appropriate values for the covariance component weights $\widehat{\lambda}_i$ are estimated from the MEG/EEG data with the parametric empirical Bayesian (PEB) method. Once the prior covariance $\mathbf{R}$ is fixed to the PEB estimate $\widehat{\mathbf{R}} = \Sigma_i \widehat{\lambda}_i \mathbf{R}_i$, the solution for the dipole amplitudes reduces to the fMRI-weighted MNE. The obvious question is then how to partition the fMRI activation map into different variance/covariance components $\mathbf{R}_i$ – the limiting cases are that all locations determined active form one diagonal variance component $\mathbf{R}_1$, and the inactive form another $\mathbf{R}_0$, in the spirit of the original fMRI-weighted MNE (see, Eq. (5)) or that each fMRI-activated location/cluster is assigned to an individual variance component. Moreover, off-diagonal covariance terms in the components can be also introduced: we may have an *a priori* reason to believe that, for instance, left and right primary auditory cortices should be activated in a similar fashion if identical stimulation is delivered to both auditory pathways. As the number of ways in which the fMRI data can be split into covariance components is rather large, the practical question of how to generate a reasonable prior structure of appropriate complexity remains an important challenge. The fMRI-Informed Regional Estimation (FIRE) method [41] combines elements of the symmetric approach and the automatic relevance determination (ARD) approaches [19, 39, 45, 50]. FIRE also utilizes the anatomical parcellation of the cortex [17] and assumes that both electromagnetic and hemodynamic activity have a common spatial profile at each parcel but independent temporal waveforms. The overall source variance is adaptively estimated for each region, the ARD structure allowing adequate variability of source strengths for each parcel and letting thus non-active regions to be suppressed. Different from the symmetric approach, the FIRE approach assumes that the hemodynamic responses are directly observed. Thus, FIRE is computationally tractable and conceptually simple. The result is that for those regions where clear fMRI responses are detected, the MEG source localization leans on the available fMRI spatial information, and if fMRI information is missing but MEG signals detected, the localization results are similar to the basic MNE.

3.2   Experimental Design, Model Comparison and Validation, and Neurovascular Coupling Models

As mentioned above, different physiological origins of the signals introduce a natural challenge in how to design the experiment such that multimodal data fusion is meaningful. In MEG/EEG, stimuli eliciting transient responses may be optimal for high temporal resolution MEG/EEG measurements, while delivering "continuous" trains of stimuli may be more appropriate to drive the vascular system into a steady state for high contrast-to-noise ratio (CNR) fMRI data. Accordingly, the integrative analysis of the data requires some special attention. The fMRI weighting prior for the MNE, by definition, should encode information that we have about the phenomenon *before (*a priori*) we see the MEG/EEG data*. Thus, determining the fMRI weighting for the MNE should be based solely on the fMRI data. Strictly speaking, it is incorrect to tweak the fMRI prior *after* seeing the results of the fMRI-weighted MNE. However, saying that the (fMRI) prior should not depend on the (MEG/EEG) data to be modeled does not mean that the prior cannot have unknown parameters that are estimated from the data, such as the prior covariance weights λ_i in the PEB approach [27]. The symmetrical approach [11, 27] and the FIRE approach [41] partially avoid this problem as the prior is mainly fixed by the cortical parcellation and other explicit modeling assumptions about the spatial concordance of the electromagnetic and hemodynamic responses.

If multiple fMRI-MEG/EEG integration methods are applied to a given dataset, what can be said about the validity or accuracy of these models? Can we compare the models and select the most likely one in some sense? There are criteria for Bayesian model comparison and selection such as the model evidence, as utilized in [27], which can be useful for evaluating the model complexity and guiding the model selection process. In principle, due to the MEG/EEG inverse problem, there is no "true" solution that can be singled out by any statistical test – no matter how much MEG/EEG data we collect, or how realistic we make the volume conductor (head) model, there will be multiple source configurations that fit equally well to a given set of MEG/EEG data. Even in the presence of almighty Bayes, the silent sources remain silent. If the hypothetical silent MEG/EEG source configuration is detected by fMRI and incorporated into the spatial prior, we do not gain further information of its electromagnetic characteristics – the product between an inverse operator and zero measurements (due to silent MEG/EEG source) still yields zero source estimates. Obviously poor models can be detected by comparing, e.g., the data fit of a given model against a *standard* model, such as the MNE, which operates on the minimal assumptions. However, it is unlikely that any model will yield a substantially better data fit for the MEG/EEG data, since the MNE is also an optimal estimator in least squares sense. Some forms of cross-validation might also be applied to test and compare prediction errors and to detect over-fitting.

3.3 Neurovascular Coupling: The Physiological Bases of Integrating fMRI and MEG Source Modeling

A better understanding of neurovascular coupling is of fundamental importance in integrating hemodynamic and neuronal activity data (for review, see [25, 35]). BOLD contrast fMRI has been suggested to be closely related to the input synaptic activity [34] and to neuronal output spiking [24, 37]. When postsynaptic neuronal signals are highly synchronous, they constitute the magnetic fields measured by MEG in guinea pigs [40, 43]. Invasive recordings can directly measure the neurovascular coupling and therefore offer more detailed information for further testing and validating the noninvasive models. Animal models and intracranial measurements in humans will be needed to provide a backbone for the development of noninvasive imaging approach, which will always rest on some weighty modeling assumptions due to the indirect nature of observations. Population-level models of the neurovascular coupling can also provide insights and predictions about the noninvasive data in various circumstances [4, 47].

A linear relationship between the strength of neuronal signal and hemodynamic signal has been suggested by studies in the human visual system [24, 38, 43, 46, 49] and the motor system [42]. However, using a rodent model, a nonlinear relationship between the strength of the local hemodynamic response and neuronal activity has also been reported [12]. Such hemodynamic output may be explained as the spatio-temporal convolution of local electrophysiological responses [13]. A more complicated correlation structure between MEG and BOLD fMRI responses was also found in the human auditory system: the same auditory clicks can elicit transient and sustained MEG responses with the transient response more closely related to the BOLD fMRI signal [21]. Taken together, the mechanism through which the MEG and fMRI signals become coupled, as well as the ensuing degree of observed correlation in the macroscopic responses, remains only partially elucidated.

From the MEG/EEG source modeling perspective, it is highly motivated to explore and exploit the fMRI data as a spatial prior to complement the non-unique nature of estimating neuronal current source distributions using extracranial recordings [26]. Such a data fusion technique has been supported by studies showing reliable correlations between hemodynamic responses and neuronal activity as discussed above: brain areas showing significant hemodynamic responses measured by fMRI are expected to be engaged in corresponding neuronal activity, the synchronous synaptic components of which are measured by MEG. This rationale is also supported by studies showing that MEG and fMRI can co-localize to the same cortical areas in the visual system [7] and motor system [18]. However, it should be noted that the spatial distribution between electrophysiological activity and hemodynamic responses is not in a complete agreement. In the somatosensory area, the distance between the center of fMRI map and the center of the electrophysiological maps can be separated by approximately 1 cm [14]. In summary, further investigation of the neurovascular coupling should provide support for developing the mathematical models of integrating fMRI and MEG data for spatiotemporally sensitive and functionally specific detection of human brain activation.

In conclusion, the field of multimodal integration of noninvasive imaging technologies such as MEG/EEG and fMRI is still in a rather early stage, and the methodology will continue to evolve as more data about the physiological origins of the signals are accumulated.

Acknowledgments This work was supported by the National Institutes of Health Grants by R01HD040712, R01NS037462, R01NS048279, P41RR014075, R01MH083744, R21DC010060, R21EB007298, R00EB015445, National Center for Research Resources, Natural Sciences and Engineering Research Council, Canada (RGPIN-2020-05927), Academy of Finland (127624 and 298131), Finnish Cultural Foundation, and Finnish Foundation for Technology Promotion.

References

1. Ahlfors, S. P., Han, J., Lin, F. H., Witzel, T., Belliveau, J. W., Hamalainen, M. S., & Halgren, E. (2010). Cancellation of EEG and MEG signals generated by extended and distributed sources. *Human Brain Mapping, 31*, 140–149.
2. Ahlfors, S. P., Simpson, G. V., Dale, A. M., Belliveau, J. W., Liu, A. K., Korvenoja, A., Virtanen, J., Huotilainen, M., Tootell, R. B., Aronen, H. J., & Ilmoniemi, R. J. (1999). Spatiotemporal activity of a cortical network for processing visual motion revealed by MEG and fMRI. *Journal of Neurophysiology, 82*, 2545–2555.
3. Auranen, T., Nummenmaa, A., Hamalainen, M. S., Jaaskelainen, I. P., Lampinen, J., Vehtari, A., & Sams, M. (2007). Bayesian inverse analysis of neuromagnetic data using cortically constrained multiple dipoles. *Human Brain Mapping, 28*, 979–994.
4. Babajani, A., & Soltanian-Zadeh, H. (2006). Integrated MEG/EEG and fMRI model based on neural masses. *IEEE Transactions on Biomedical Engineering, 53*, 1794–1801.
5. Babiloni, F., Babiloni, C., Carducci, F., Romani, G. L., Rossini, P. M., Angelone, L. M., & Cincotti, F. (2003). Multimodal integration of high-resolution EEG and functional magnetic resonance imaging data: A simulation study. *NeuroImage, 19*, 1–15.
6. Belliveau, J. W., Kennedy, D. N., Jr., McKinstry, R. C., Buchbinder, B. R., Weisskoff, R. M., Cohen, M. S., Vevea, J. M., Brady, T. J., & Rosen, B. R. (1991). Functional mapping of the human visual cortex by magnetic resonance imaging. *Science, 254*, 716–719.
7. Brookes, M. J., Gibson, A. M., Hall, S. D., Furlong, P. L., Barnes, G. R., Hillebrand, A., Singh, K. D., Holliday, I. E., Francis, S. T., & Morris, P. G. (2005). GLM-beamformer method demonstrates stationary field, alpha ERD and gamma ERS co-localisation with fMRI BOLD response in visual cortex. *NeuroImage, 26*, 302–308.
8. Dale, A., Fischl, B., & Sereno, M. (1999). Cortical surface-based analysis. I. Segmentation and surface reconstruction. *Neuroimage, 9*, 179–194.
9. Dale, A., & Sereno, M. (1993). Improved localization of cortical activity by combining EEG and MEG with MRI cortical surface reconstruction: A linear approach. *Journal of Cognitive Neuroscience, 5*, 162–176.

10. Dale, A. M., Liu, A. K., Fischl, B. R., Buckner, R. L., Belliveau, J. W., Lewine, J. D., & Halgren, E. (2000). Dynamic statistical parametric mapping: Combining fMRI and MEG for high-resolution imaging of cortical activity. *Neuron, 26*, 55–67.

11. Daunizeau, J., Grova, C., Marrelec, G., Mattout, J., Jbabdi, S., Pélégrini-Issac, M., Lina, J.-M., & Benali, H. (2007). Symmetrical event-related EEG/fMRI information fusion in a variational Bayesian framework. *NeuroImage, 36*, 69–87.

12. Devor, A., Dunn, A. K., Andermann, M. L., Ulbert, I., Boas, D. A., & Dale, A. M. (2003). Coupling of total hemoglobin concentration, oxygenation, and neural activity in rat somatosensory cortex. *Neuron, 39*, 353–359.

13. Devor, A., Ulbert, I., Dunn, A. K., Narayanan, S. N., Jones, S. R., Andermann, M. L., Boas, D. A., & Dale, A. M. (2005). Coupling of the cortical hemodynamic response to cortical and thalamic neuronal activity. *Proceedings of the National Academy of Sciences of the United States of America, 102*, 3822–3827.

14. Disbrow, E. A., Slutsky, D. A., Roberts, T. P., & Krubitzer, L. A. (2000). Functional MRI at 1.5 tesla: A comparison of the blood oxygenation level-dependent signal and electrophysiology. *Proceedings of the National Academy of Sciences of the United States of America, 97*, 9718–9723.

15. Fischl, B., Liu, A., & Dale, A. M. (2001). Automated manifold surgery: Constructing geometrically accurate and topologically correct models of the human cerebral cortex. *IEEE Transactions on Medical Imaging, 20*, 70–80.

16. Fischl, B., Sereno, M., & Dale, A. (1999). Cortical surface-based analysis. II: Inflation, flattening, and a surface-based coordinate system. *NeuroImage, 9*, 195–207.

17. Fischl, B., van der Kouwe, A., Destrieux, C., Halgren, E., Ségonne, F., Salat, D. H., Busa, E., Seidman, L. J., Goldstein, J., Kennedy, D., Caviness, V., Makris, N., Rosen, B., & Dale, A. M. (2004). Automatically parcellating the human cerebral cortex. *Cerebral Cortex, 14*, 11–22.

18. Formaggio, E., Storti, S. F., Avesani, M., Cerini, R., Milanese, F., Gasparini, A., Acler, M., Pozzi Mucelli, R., Fiaschi, A., & Manganotti, P. (2008). EEG and FMRI coregistration to investigate the cortical oscillatory activities during finger movement. *Brain Topography, 21*, 100–111.

19. Friston, K., Harrison, L., Daunizeau, J., Kiebel, S., Phillips, C., Trujillo-Barreto, N., Henson, R., Flandin, G., & Mattout, J. (2008). Multiple sparse priors for the M/EEG inverse problem. *NeuroImage, 39*, 1104–1120.

20. George, J. S., Aine, C. J., Mosher, J. C., Schmidt, D. M., Ranken, D. M., Schlitt, H. A., Wood, C. C., Lewine, J. D., Sanders, J. A., & Belliveau, J. W. (1995). Mapping function in the human brain with magnetoencephalography, anatomical magnetic resonance imaging, and functional magnetic resonance imaging. *Journal of Clinical Neurophysiology, 12*, 406–431.

21. Gutschalk, A., Hamalainen, M. S., & Melcher, J. R. (2010). BOLD responses in human auditory cortex are more closely related to transient MEG responses than to sustained ones. *Journal of Neurophysiology, 103*, 2015–2026.

22. Hamalainen, M., Hari, R., Ilmoniemi, R. J., Knuutila, J., & Lounasmaa, O. V. (1993). Magnetoencephalography – Theory, instrumentation, and application to noninvasive studies of the working human brain. *Review of Modern Physics, 65*, 413–497.

23. Hamalainen, M. S., & Sarvas, J. (1989). Realistic conductivity geometry model of the human head for interpretation of neuromagnetic data. *IEEE Transactions on Biomedical Engineering, 36*, 165–171.

24. Heeger, D. J., Huk, A. C., Geisler, W. S., & Albrecht, D. G. (2000). Spikes versus BOLD: What does neuroimaging tell us about neuronal activity? *Nature Neuroscience, 3*, 631–633.

25. Heeger, D. J., & Ress, D. (2002). What does fMRI tell us about neuronal activity? *Nature Reviews. Neuroscience, 3*, 142–151.
26. Helmholtz, H. (1853). Ueber einige Gesetze der Vertheilung elektrischer Strome in korperlichen Leitern, mit Anwendung auf die thierisch-elektrischen Versuche. *Annals of Physical Chemistry, 89*(211–233), 353–377.
27. Henson, R. N., Flandin, G., Friston, K. J., & Mattout, J. (2010). A parametric empirical Bayesian framework for fMRI-constrained MEG/EEG source reconstruction. *Human Brain Mapping, 31*, 1512–1531.
28. Ilmoniemi, R. J. (1995). Magnetoencephalography-a tool for studies of information processing in the human brain. In H. Lubbig (Ed.), *The inverse problem* (pp. 89–106). Berlin: Akademie Verlag.
29. Jun, S. C., George, J. S., Pare-Blagoev, J., Plis, S. M., Ranken, D. M., Schmidt, D. M., & Wood, C. C. (2005). Spatiotemporal Bayesian inference dipole analysis for MEG neuroimaging data. *NeuroImage, 28*, 84–98.
30. Kwong, K. K., Belliveau, J. W., Chesler, D. A., Goldberg, I. E., Weisskoff, R. M., Poncelet, B. P., Kennedy, D. N., Hoppel, B. E., Cohen, M. S., Turner, R., Cheng, H., Brady, T. J., & Rosen, B. R. (1992). Dynamic magnetic resonance imaging of human brain activity during primary sensory stimulation. *Proceedings of the National Academy of Sciences of the United States of America, 89*, 5675–5679.
31. Lin, F. H., Witzel, T., Hamalainen, M. S., Dale, A. M., Belliveau, J. W., & Stufflebeam, S. M. (2004). Spectral spatiotemporal imaging of cortical oscillations and interactions in the human brain. *NeuroImage, 23*, 582–595.
32. Liu, A. K., Belliveau, J. W., & Dale, A. M. (1998). Spatiotemporal imaging of human brain activity using functional MRI constrained magnetoencephalography data: Monte Carlo simulations. *Proceedings of the National Academy of Sciences of the United States of America, 95*, 8945–8950.
33. Liu, A. K., Dale, A. M., & Belliveau, J. W. (2002). Monte Carlo simulation studies of EEG and MEG localization accuracy. *Human Brain Mapping, 16*, 47–62.
34. Logothetis, N., Pauls, J., Augath, M., Trinath, T., & Oeltermann, A. (2001). Neurophysiological investigation of the basis of the fMRI signal. *Nature, 412*, 150–157.
35. Logothetis, N. K. (2008). What we can do and what we cannot do with fMRI. *Nature, 453*, 869–878.
36. Mansfield, P. (1977). Multi-planar image formation using NMR spin echos. *Journal of Physics, C10*, L55–L58.
37. Mukamel, R., Gelbard, H., Arieli, A., Hasson, U., Fried, I., & Malach, R. (2005). Coupling between neuronal firing, field potentials, and FMRI in human auditory cortex. *Science, 309*, 951–954.
38. Niessing, J., Ebisch, B., Schmidt, K. E., Niessing, M., Singer, W., & Galuske, R. A. (2005). Hemodynamic signals correlate tightly with synchronized gamma oscillations. *Science, 309*, 948–951.
39. Nummenmaa, A., Auranen, T., Hämäläinen, M. S., Jääskeläinen, I. P., Sams, M., Vehtari, A., & Lampinen, J. (2007). Automatic relevance determination based hierarchical Bayesian MEG inversion in practice. *NeuroImage, 37*, 876–889.
40. Okada, Y., Wu, J., & Kyuhou, S. (1997). Genesis of MEG signals in a mammalian CNS structure. *Electroencephalography and Clinical Neurophysiology, 103*, 474–485.
41. Ou, W., Nummenmaa, A., Ahveninen, J., Belliveau, J. W., Hamalainen, M. S., & Golland, P. (2010). Multimodal functional imaging using fMRI-informed regional EEG/MEG source estimation. *NeuroImage, 52*, 97–108.
42. Parkes, L. M., Bastiaansen, M. C., & Norris, D. G. (2006). Combining EEG and fMRI to investigate the post-movement beta rebound. *NeuroImage, 29*, 685–696.
43. Rees, G., Friston, K., & Koch, C. (2000). A direct quantitative relationship between the functional properties of human and macaque V5. *Nature Neuroscience, 3*, 716–723.

44. Rosa, M. J., Daunizeau, J., & Friston, K. J. (2010). EEG-fMRI integration: A critical review of biophysical modeling and data analysis approaches. *Journal of Integrative Neuroscience, 9*, 453–476.
45. Sato, M.-a., Yoshioka, T., Kajihara, S., Toyama, K., Goda, N., Doya, K., & Kawato, M. (2004). Hierarchical Bayesian estimation for MEG inverse problem. *NeuroImage, 23*, 806–826.
46. Singh, M., Kim, S., & Kim, T. S. (2003). Correlation between BOLD-fMRI and EEG signal changes in response to visual stimulus frequency in humans. *Magnetic Resonance in Medicine, 49*, 108–114.
47. Sotero, R. C., & Trujillo-Barreto, N. J. (2008). Biophysical model for integrating neuronal activity, EEG, fMRI and metabolism. *NeuroImage, 39*, 290–309.
48. Vanni, S., Warnking, J., Dojat, M., Delon-Martin, C., Bullier, J., & Segebarth, C. (2004). Sequence of pattern onset responses in the human visual areas: An fMRI constrained VEP source analysis. *NeuroImage, 21*, 801–817.
49. Wan, X., Riera, J., Iwata, K., Takahashi, M., Wakabayashi, T., & Kawashima, R. (2006). The neural basis of the hemodynamic response nonlinearity in human primary visual cortex: Implications for neurovascular coupling mechanism. *NeuroImage, 32*, 616–625.
50. Wipf, D., & Nagarajan, S. (2009). A unified Bayesian framework for MEG/EEG source imaging. *NeuroImage, 44*, 947–966.

Multiscale Modeling of EEG/MEG Response of a Compact Cluster of Tightly Spaced Pyramidal Neocortical Neurons

Sergey N. Makarov, Jyrki Ahveninen, Matti Hämäläinen, Yoshio Okada, Gregory M. Noetscher, and Aapo Nummenmaa

1 Introduction

Electroencephalography (EEG) [19, 25] and magnetoencephalography (MEG) [7] noninvasively record electric potentials and magnetic fields, respectively, due to neural currents. These methods are used as tools in clinical research, in basic neuroscience, and as diagnostic and monitoring measures in clinical practice. In addition, EEG, as well as invasive neurophysiological recordings, may be applied to

S. N. Makarov (✉)
Electrical and Computer Engineering Department, Worcester Polytechnic Institute, Worcester, MA, USA

Athinoula A. Martinos Center for Biomedical Imaging, Massachusetts General Hospital, Charlestown, MA, USA
e-mail: makarov@wpi.edu

J. Ahveninen · M. Hämäläinen
Athinoula A. Martinos Center for Biomedical Imaging, Massachusetts General Hospital, Charlestown, MA, USA

Harvard Medical School, Boston, MA, USA

Y. Okada
Harvard Medical School, Boston, MA, USA

Division of Newborn Medicine, Department of Medicine, Boston Children's Hospital, Boston, MA, USA

G. M. Noetscher
Electrical and Computer Engineering Department, Worcester Polytechnic Institute, Worcester, MA, USA

A. Nummenmaa
Athinoula A. Martinos Center for Biomedical Imaging, Massachusetts General Hospital, Charlestown, MA, USA

© The Author(s) 2021
S. N. Makarov et al. (eds.), *Brain and Human Body Modeling 2020*,
https://doi.org/10.1007/978-3-030-45623-8_11

195

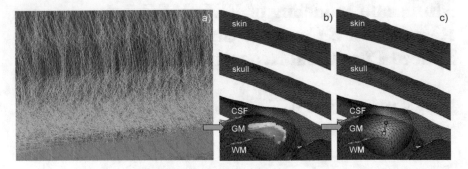

Fig. 1 (**a**) Computer reconstruction of neuronal arbor made of clones of a realistic pyramidal neuron ID NMO_86955 from the NeuroMorpho.Org inventory Version 7.5 in layers II and III with a density of approximately 150 neurons per mm^2 [2]. For this work, we did not study the field originating in layer V even though the large PNs in this layer can also produce strong electric and magnetic fields locally inside the brain. (**b**) A neuronal cluster with an area of approximately 16 mm^2 and 2450 individual neurons reconstructed in layers II and III of the anterior central gyrus for subject #101309 of the Human Connectome Project [31]; see also the Population Head Model Repository [11, 30]. (**c**) Equivalent electric-current dipole model located at the "electric" gravity center of the cluster. GM stands for gray matter, WM for white matter, and CF for cerebrospinal fluid conductivity boundaries

enable brain-computer interfaces or BCIs (see, e.g., [1, 12, 26]) with the goal of mitigating various neurological disabilities [34].

In the most demanding clinical evaluations, EEG and/or MEG is followed by direct recordings with subdural or intraparenchymal depth electrodes. A modern high-resolution intracranial recording technique – intracranial electroencephalography or iEEG – is blossoming in various fields of human neuroscience [21]. At present, intracortical arrays with electrodes as small as 20 μm in size and with 25–100 μm electrode spacings are designed and tested (see, e.g., [37]). Local field potential (LFP) electrodes are, for example, 50-μm-diameter tungsten microwires [36]. A similar tendency toward fine resolution is observed for more accurate MEG measurement techniques [9, 24].

The ultimate goal of neurophysiological recordings of any type is an estimation of the sources generating the measured signal patterns. These sources are electric currents flowing in the micrometer-size sparse neuronal arbor; consider, for example, the arbor of pyramidal neurons in layers II and III of the neocortex shown in Fig. 1a. A large group or a cluster of such synchronously activated cortical neurons shown in Fig. 1b is the basic block in the analysis of EEG and MEG. At present, direct modeling of such extremely complicated current distributions is not possible with commonly used numerical methods, i.e., the finite element method, the boundary element method, and the finite difference method.

Therefore, a lumped macroscopic electric-current dipole model shown in Fig. 1c, which consists of a closely spaced or coinciding source and sink of electric current in a conducting medium, has traditionally been used as a source substitute for the cluster of synchronously activated cortical neurons in the analysis of EEG and MEG [7, 15, 19, 25]. Several excellent open-source software packages for the dipole-based EEG/MEG analysis are available, including Brainstorm [29], FieldTrip [20], and MNE [5].

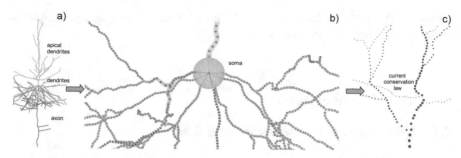

Fig. 2 (**a**) Morphology of neocortical pyramidal neuron ID NMO_86955 from the NeuroMorpho. Org inventory Version 7.5. (**b**) Realistic current paths within the microscopic arbor are schematically indicated by small circles. The circles are simultaneously the poles (9318 in the present case) or the sources and sinks of elementary current sources – microscopic electric dipoles – situated within the firing arbor. (**c**) Current conservation law illustrated by different sizes of the poles – microscopic current dipole strengths. We assume the same current inflow/outflow at all synaptic connections (arbor terminations) of apical dendrites with the total current accumulating toward the soma

This model is indeed a physically valid substitute for any ensemble of microscopic dipolar current sources in a *homogeneous* conducting medium and in the far field, i.e., at distances significantly exceeding the cluster size. However, when irregular conductivity boundaries are present in the immediate vicinity of the cluster, they may disturb the current distribution. As a result, even the integral far-field response might be different from that of the equivalent dipole.

The fast multipole method or FMM [6, 23] enables computing the response of many millions of microscopic electric sources for a comparable or even larger number of observation or target points in a short amount of time. A proper coupling of the fast multipole method and the boundary element method – the BEM-FMM approach suggested in [8, 13, 14] – further enables computing the corresponding induced charge distribution at tissue conductivity boundaries which, in turn, results in obtaining precise current, voltage, and magnetic field distributions at the boundaries and everywhere in space.

Using the BEM-FMM, one may be in position to depart from the simplified dipole model in Fig. 1c toward a more realistic computational model which follows actual microscopic electric current flow in every dendritic or axonal branch of a neuron as shown in Fig. 2a, b. Along with this, current splitting and combining according to Kirchhoff's current law or KCL is enforced as illustrated in Fig. 2c.

Moreover, one may be in position to model a large group of such tightly spaced neurons firing simultaneously, i.e., directly model the entire compact cluster of cortical neurons. Such a cluster may be located anywhere in the cerebral cortex. The realistic cluster size may be as large as 10,000 individual neurons, while the overall computation times do not exceed several minutes on a standard server.

This study is aimed to apply the developed method to answer the following question: how well does the conventional dipole model approximate a cluster of neurons with an area of 16–25 mm^2 (1/7500 to 1/4800 of the total cerebral cortex area) when approaching the cortical boundaries?

2 Materials and Methods

2.1 Gyrus Cluster Construction and Analysis

Due to the geometry and electrophysiological characteristics of cortical neurons, a gyrus cluster, which is essentially parallel to the skull surface, is expected to generate a strong EEG response but a weak MEG response. Figure 3a–c shows one reconstructed gyrus cluster with an area of approximately 16 mm^2 and 2450 individual pyramidal neurons (ID NMO_86955 from the NeuroMorpho.Org) located in layers II and III of the anterior central gyrus for subject #101309 of the Human Connectome Project [31]. A realistic neuronal density of approximately 150 neurons per mm^2 [2] has been implemented. To do so, we cloned the individual neuron

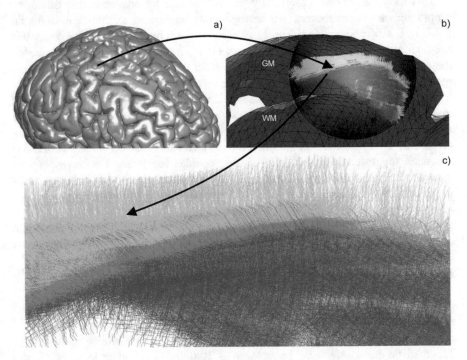

Fig. 3 (**a**) Position of the gyrus cluster beneath the gray matter shell along with the coronal and sagittal observation planes. (**b**) Zoomed in position of the cluster between gray matter and white matter shells – the nearest macroscopic conductivity boundaries. (**c**) Zoomed in display of the cluster topology with a length of approximately 4 mm. Pale ivory color corresponds to apical dendrites

model shown in Fig. 2a for a large number of locations between the white matter (WM) and gray matter (GM) surfaces, aligned the neurons with the surface normal vectors, then moved them toward a position that was approximately 1 mm away from the WM triangular surface in the direction of its outer normal vector.

The microscopic neural origin of the primary currents in EEG/MEG is thought to be the aggregate of postsynaptic longitudinal currents flowing inside the apical dendrites of the large, spatially aligned neocortical pyramidal neurons or PNs [18, 27]. We therefore assumed that the synchronized electric currents flow only in the apical dendrites of the neurons, which are shown pale ivory in Fig. 3c. The longest apical dendrite branch has a length of approximately 500 μm. We also assumed equal outflowing currents at all available synaptic connections with the total current accumulated (via accurately traversing the dendritic tree) and then terminated at the soma.

The apical dendrite branches of a single neuron have been divided into $M = 2387$ individual straight segments – microscopic current dipoles – each with an average length of 1.2 μm. The corresponding poles are seen in Fig. 2c. This detailed model is somewhat superfluous for EEG or iEEG purposes since all intermediate current sources along a branch will cancel out and only the end synapse sources and the soma source of opposite polarity will remain significant. However, it is meaningful for MEG purposes since the entire current path along the neuronal arbor will be reflected in the measurements. The total number of microscopic dipole sources in the present cluster is approximately 6 M.

To choose a realistic value of current dipole moment density q_0 (current dipole moment per unit cross-sectional area of the active cortex) in the source region, we used the value $q_0 = 1\,nA \cdot m/mm^2$ found by Murakami and Okada [17]. This value is invariant across the cerebral cortex, hippocampus, and cerebellum over a wide phylogenetic scale from reptiles to humans. This value also agrees with the dipole moment density estimated from a neural current magnetic resonance imaging (MRI) study [28]. When the microscopic dipole vector length is d_m and its relative weight (equal to one at synapses and equal to 18 at the soma in the present case) is w_m, an expression for the resulting current constant I_0 follows from

$$\frac{Neurons}{mm^2} \times I_0 \sum_{m=1}^{M} w_m d_m = q_0 \tag{1}$$

which yields $I_0 = 1.8\,nA$.

The moment of an equivalent lumped dipole shown in Fig. 1c was found as a vector sum of all individual dipole moments in the cluster. The center of an equivalent lumped dipole was found as the weighted average of all individual dipole centers in the cluster. The weights are the magnitudes of individual dipole moments.

Finally, the underlying macroscopic head model used surface meshes for seven brain compartments of the Population Head Model Repository [11, 30]. Further, the surface mesh was refined (oversampled) using a 1 × 4 barycentric triangle subdivision, and then surface-preserving Laplacian smoothing [32, 33] was applied. This resulted in the surface mesh resolution (edge length) of 0.75 mm in the cortex.

Two measurable output quantities obtained via numerical computations are the electric potential for EEG/iEEG and the magnetic field for MEG. We used a linear scale for all surface plots including inner skull or pia mater surface, skin surface, and a surface at the distance of 18 mm from the skin (a "magnetometer" surface used for MEG purposes only). For the surface plots, only the normal component of the magnetic field recorded by the flat MEG magnetometers was plotted and analyzed.

For volumetric plots corresponding to intracranial recordings close to the cluster, larger potential/field variations may be observed. In the last case, we used the log-modulus transformation [10]

$$\varphi_{dB} = sign(\varphi) \cdot 20 \log_{10}\left(\frac{\varphi}{\varphi_0} + 1\right), \varphi_0 = 0.4 \, \mu V \tag{2}$$

A similar logarithmic transformation but without the additive constant equal to one was applied to the magnetic field magnitude with $B_0 = 0.4$ pT.

To analyze the surface/interface data, we used two error measures to distinguish between topography and magnitude errors, respectively. These are the relative difference measure or RDM defined here as [3, 16, 22, 35]:

$$E = \left\| \frac{\varphi_1}{\|\varphi_1\|} - \frac{\varphi_2}{\|\varphi_2\|} \right\| \tag{3}$$

and the magnitude (MAG) error defined as [16]:

$$MAG = \frac{\|\varphi_1\|}{\|\varphi_2\|} \tag{4}$$

with φ_1 being the cluster potential.

Along with this, we computed the ratio of maximum potential differences. The identical definitions were applied for the normal component of the magnetic field at the interfaces.

2.2   Sulcus Cluster Construction and Analysis

A sulcus cluster, which is essentially perpendicular to the skull surface, is expected to generate a weak EEG response but a strong MEG response. Figure 4a–c shows one reconstructed sulcus cluster with an area of approximately 25 mm^2 and 3175 individual pyramidal neurons (ID NMO_86955 from the NeuroMorpho.Org) with approximately 8 M microscopic dipole sources located in layers II and III of the superior frontal sulcus for the same subject #101309 of the Human Connectome

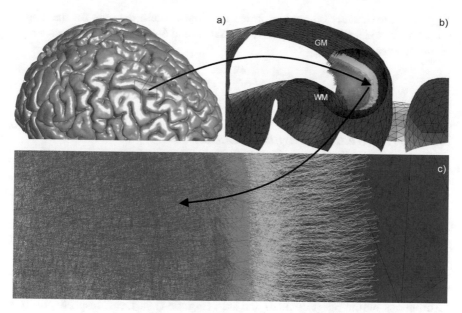

Fig. 4 (**a**) Position of the sulcus cluster beneath the gray matter shell along with the coronal and sagittal observation planes. (**b**) Zoomed in position of the cluster between gray matter and white matter shells – the nearest macroscopic conductivity boundaries. (**c**) Zoomed in display of the cluster topology with a length of approximately 2 mm. Pale ivory color corresponds to apical dendrites

Project [31]. A slightly lower neuronal density of approximately 112 neurons per mm^2 was used. However, for the current dipole moment density, we again used the value $q_0 = 1 \, nA \cdot m/mm^2$ found by Murakami and Okada [17] so that the total dipole moment of the cluster appears approximately 25/16 times greater than in the previous case. This was done to compensate for a larger distance from the skull surface, which is larger in the present case by a factor of approximately 5/4. Otherwise, all other parameters and the method of analysis remain the same.

2.3 Modeling Algorithm

The complete mathematical algorithm of the boundary element fast multipole method, along with justification examples, will be described elsewhere. The present computations use the most recent version of the FMM library originally developed by Gimbutas and Greengard [4] and run on an Intel Xeon E5-2683 v4 CPU (2.1 GHz) server with 256 GB RAM, Windows Server 2008 R2 Enterprise, implemented on the MATLAB 2018a platform. Apart from the computations of static model-specific parameters – potential surface integrals for macroscopic boundaries – the corresponding iterative solution reaches a relative residual of

10^{-3} in approximately 80 sec and in ten iterations. These data are for the macroscopic head model with approximately 3 M facets and 6–8 M individual microscopic dipole sources.

3   Results

3.1   Gyrus (Nearly Horizontal) Cluster

Figure 5 shows the data for the electric potential. The left column corresponds to the cluster model, while the right column corresponds to the equivalent macroscopic current dipole. Figure 5a,b displays the volumetric potential distribution in the immediate vicinity of the cluster and the dipole, respectively. A logarithmic scale in decibels given by Eq. (2) is used. As expected, the dipole response is "sharper," i.e., more localized in space, especially close to the pia or inner skull surface. This circumstance potentially leads to a larger RDM error given by Eq. (3), although "centers of gravity" of both responses nearly coincide, as shown by the potential distributions on the pia (Fig. 5c,d) and skin (Fig. 5e,f) surfaces, respectively. It is worth noting that the maximum values of the surface potential differ by a factor of approximately two for the pia mater.

Figure 6 shows the corresponding data for the magnetic field. The left column corresponds to the cluster model, while the right column corresponds to the equivalent macroscopic current dipole. Figure 6a,b displays the volumetric distribution of the magnitude of the total magnetic field in the immediate vicinity of the cluster and the dipole, respectively. The cluster response is more inhomogeneous. A logarithmic scale in decibels given by Eq. (2) is used. Figure 6c–f shows the normal surface component (in the direction of the outer normal vector) of the magnetic field recorded by a magnetometer for the pia and skin surfaces, respectively. We observe a modest change in the field distribution pattern.

3.2   Sulcus (Predominantly Vertical) Cluster

Figure 7 shows the data for the electric potential. The left column corresponds to the cluster model, while the right column corresponds to the equivalent macroscopic current dipole. Figure 7a,b displays the volumetric potential distribution in the immediate vicinity of the cluster and the dipole, respectively. A logarithmic scale in decibels given by Eq. (2) is used. Again, the dipole response is somewhat "sharper," i.e., more localized in space. A case in point is an isocurve corresponding to 38 dB in Fig. 7a,b. Potential distributions on the pia (Fig. 7c,d) and skin (Fig. 7e,f) surfaces visually look similar, at least at first sight.

It is worth noting that the maximum surface potential differences appear to be higher *for the cluster*, which is exactly the opposite of the previous case.

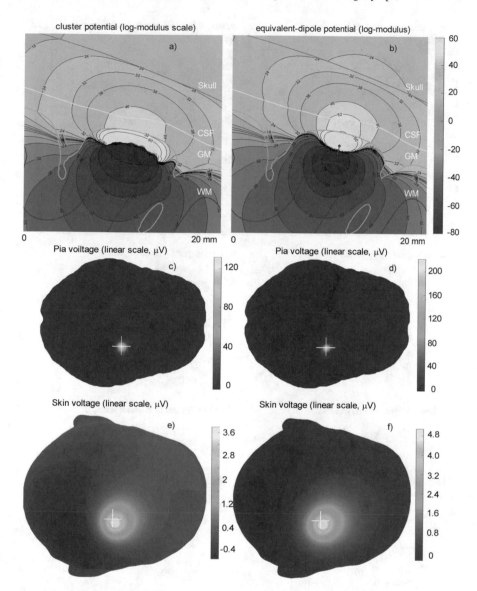

Fig. 5 Electric potential data for the gyrus cluster (left column) versus the equivalent-dipole data (right column). (**a, b**) Volumetric potential distribution in the immediate vicinity of the cluster and the dipole, respectively, using a logarithmic scale. (**c, d**) Surface potential distribution on the inner skull (pia) surface using a linear scale. (**e, f**) Surface potential distribution on the skin surface using a linear scale

Figure 8 shows the corresponding data for the magnetic field. The left column corresponds to the cluster model, while the right column corresponds to the equivalent macroscopic current dipole. Figure 8a,b displays the volumetric distribution of the magnitude of the magnetic field in the immediate vicinity of the cluster and the

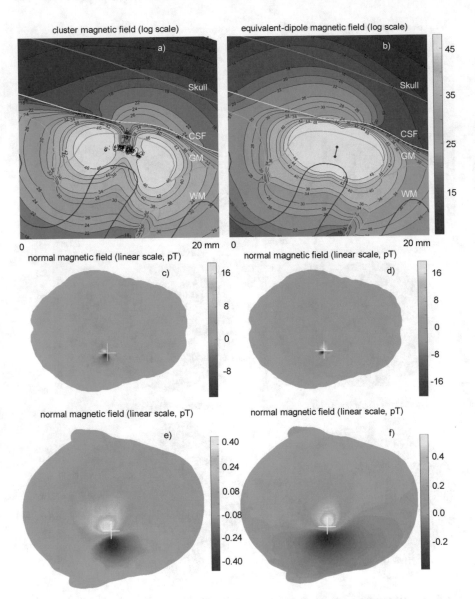

Fig. 6 Magnetic field data for the gyrus cluster (left column) versus the equivalent-dipole data (right column). (**a, b**) Volumetric field magnitude distribution in the immediate vicinity of the cluster and the dipole, respectively, using a logarithmic scale. (**c, d**) Normal magnetic field distribution on the inner skull (pia) surface using a linear scale. (**e, f**) Normal magnetic field on the skin surface using a linear scale

dipole, respectively. The cluster response is more inhomogeneous. A logarithmic scale in decibels given by Eq. (2) is used. Figure 8c–f shows the normal surface component (in the direction of the outer normal vector) of the magnetic field

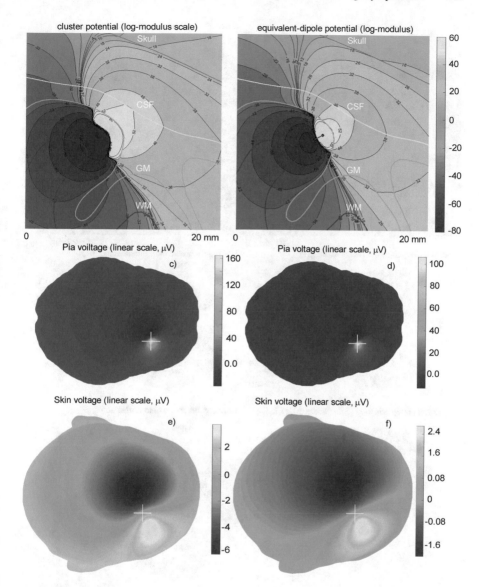

Fig. 7 Electric potential data for the sulcus cluster (left column) versus the equivalent-dipole data (right column). (**a, b**) Volumetric potential distribution in the immediate vicinity of the cluster and the dipole, respectively, using a logarithmic scale. (**c, d**) Surface potential distribution on the inner skull surface using a linear scale. (**e, f**) Surface potential distribution on the skin surface using a linear scale

recorded by a magnetometer for the pia and skin surfaces, respectively. We observe a visual similarity in the field distribution patterns.

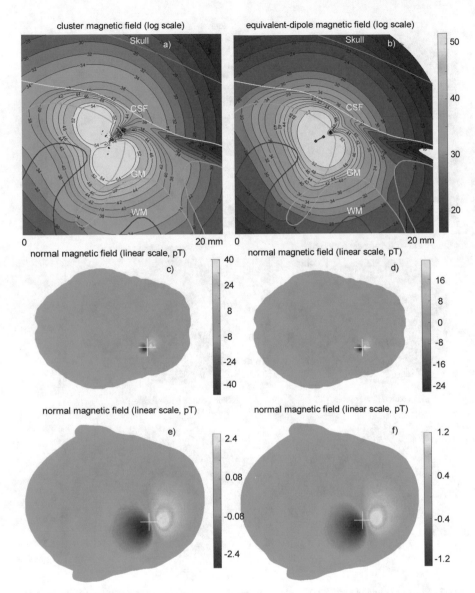

Fig. 8 Magnetic field data for the sulcus cluster (left column) versus the equivalent-dipole data (right column). (**a, b**) Volumetric field magnitude distribution in the immediate vicinity of the cluster and the dipole, respectively, using a logarithmic scale. (**c, d**) Normal magnetic field distribution on the inner skull surface using a linear scale. (**e, f**) Normal magnetic field on the skin surface using a linear scale

Table 1 Quantitative potential/normal magnetic field deviations: gyrus cluster of 16 mm^2

Surface	Potential (from Fig. 5)			Normal magnetic field (From Fig. 6)		
	RDM (shape) error, E, %	Ratio of max.potential diff. (cluster/dipole)	MAG metric	RDM (shape) error, E, %	Ratio of max.field diff. (cluster/dipole)	MAG metric
Pia matter	27	0.6	0.8	74	0.8	0.8
Skin	58	0.8	0.8	36	1.0	1.0
18 mm from skin surface				31	1.0	1.1

Table 2 Quantitative potential/normal magnetic deviations: sulcus cluster of 25 mm^2

Surface	Potential (from Fig. 7)			Normal magnetic field (from Fig. 8)		
	RDM (shape) error, E, %	Ratio of max.potential diff. (cluster/dipole)	MAG metric	RDM (shape) error, E, %	Ratio of max.field diff. (cluster/dipole)	MAG metric
Pia matter	27	1.7	2.1	18	1.8	2.1
Skin	46	2.1	2.3	5	2.3	2.3
18 mm from skin surface				9	2.3	2.3

3.3 Quantitative Error Measures

Tables 1 and 2 summarize data for the RDM error given by Eq. (3), logarithmic magnitude (lnMAG) error given by Eq. (4), and the error in the maximum swing of the electric potential or the normal surface magnetic field, respectively, for both cases. For the normal magnetic field, we additionally include data for a magnetometer surface that was chosen to be located at a distance of 18 mm from the skin surface. Quite surprisingly, a relatively large topographic error is generated for the electric potential, despite a good visual agreement observed in Figs. 5 and 7, respectively.

4 Conclusions

When the absolute response values are ignored and only the response topology or distribution in space is concerned, the representative error measure is the RDM error marked blue in Tables 1 and 2, respectively. It follows from these tables that, quantitatively, the MEG data generally indicate a *better* agreement between the distributed multiscale neuronal cluster model and the equivalent macroscopic lumped-dipole model. This is especially true for the magnetometer surface separated from the skin and for the most important case of the MEG sulcus cluster. The MEG

RDM error also generally decreases when the distance from the cluster increases. This is in contrast to the EEG/iEEG data where the RDM error might even increase (!) when moving from the pia surface to the skin surface (Table 1).

As to the absolute response values, we observe from Figs. 5, 6, 7, and 8, from Tables 1 and 2, and from the relevant modeling data that the EEG/iEEG lumped-dipole model

(i) Slightly overestimates the maximum iEEG/EEG response for the gyrus cluster
(ii) Significantly underestimates the maximum iEEG/EEG response for the sulcus cluster

On the other hand, the MEG dipole model

(i) Is in good agreement with the cluster model on the skin surface and the magnetometer surface (18 mm away from skin) for the gyrus cluster
(ii) Significantly underestimates the maximum MEG pia/skin/magnetometer surface response for the sulcus cluster.

These observations were confirmed by running several additional relevant cases.

Since the developed BEM-FMM algorithm is quite fast, it might be possible in future to replace the entire macroscopic dipole approach by the distributed neuronal arbor modeling.

Acknowledgments This work has been partially supported by the National Institutes of Health under award numbers R00EB015445, R44NS090894, R01MH111829, R01NS104585, R01EB022889, and R01DC016915.

References

1. Al-Qaysi, Z. T., Zaidan, B. B., Zaidan, A. A., & Suzani, M. S. (2018). A review of disability EEG based wheelchair control system: Coherent taxonomy, open challenges and recommendations. *Computer Methods and Programs in Biomedicine,* pii: S0169–2607(18)30462–0. https://doi.org/10.1016/j.cmpb.2018.06.012.
2. Benes, F. M., Vincent, S. L., & Todtenkopf, M. (2001). The density of pyramidal and nonpyramidal neurons in anterior cingulate cortex of schizophrenic and bipolar subjects. *Biological Psychiatry, 50*(6), 395–406.
3. Engwer, C., Vorwerk, J., Ludewig, J., & Wolters, C. H. (2017). A discontinuous Galerkin method to solve the EEG forward problem using the subtraction approach. *SIAM Journal on Scientific Computing, 39*(1), B138–B164. https://doi.org/10.1137/15M1048392.
4. Gimbutas, Z., & Greengard, L. (2015). Simple FMM libraries for electrostatics, slow viscous flow, and frequency-domain wave propagation. *Communications in Computational Physics, 18* (2), 516–528. https://doi.org/10.4208/cicp.150215.260615sw.
5. Gramfort, A., Luessi, M., Larson, E., Engemann, D. A., Strohmeier, D., Brodbeck, C., Parkkonen, L., & Hämäläinen, M. S. (2014). MNE software for processing MEG and EEG data. *NeuroImage, 86,* 446–460. https://doi.org/10.1016/j.neuroimage.2013.10.027.
6. Greengard, L., & Rokhlin, V. (1987). A fast algorithm for particle simulations. *Journal of Computational Physics, 73*(2), 325–348. https://doi.org/10.1016/0021-9991(87)90140-9.

7. Hämäläinen, M. S., Hari, R., Ilmoniemi, R. J., Knuutila, J., & Lounasmaa, O. V. (1993). Magnetoencephalography—Theory, instrumentation, and applications to noninvasive studies of the working human brain. *Reviews of Modern Physics, 65*(2), 413–449. https://doi.org/10. 1103/RevModPhys.65.413.

8. Htet, A. T., Saturnino, G. B., Burnham, E. H., Noetscher, G., Nummenmaa, A., & Makarov, S. N. (2019). Comparative performance of the finite element method and the boundary element fast multipole method for problems mimicking transcranial magnetic stimulation (TMS). *Journal of Neural Engineering, 16*, 1–13. https://doi.org/10.1088/1741-2552/aafbb9.

9. Iivanainen, J., Stenroos, M., & Parkkonen, L. (2017). Measuring MEG closer to the brain: Performance of on-scalp sensor arrays. *NeuroImage, 147*, 542–553. https://doi.org/10.1016/j. neuroimage.2016.12.048.

10. John, J. A., & Draper, N. R. (1980). An alternative family of transformations. *Journal of the Royal Statistical Society. Series C (Applied Statistics), 29*(2), 190–197. https://www.jstor.org/ stable/2986305.

11. Lee, E., Duffy, W., Hadimani, R., Waris, M., Siddiqui, W., Islam, F., Rajamani, M., Nathan, R., & Jiles, D. (2016). Investigational effect of brain-scalp distance on the efficacy of Transcranial magnetic stimulation treatment in depression. *IEEE Transactions on Magnetics, 52*(7), 1–4. https://doi.org/10.1109/TMAG.2015.2514158.

12. Leuthardt, E. C., Schalk, G., Wolpaw, J. R., Ojemann, J. G., & Moran, D. W. (2004). A brain-computer interface using electrocorticographic signals in humans. *Journal of Neural Engineering, 1*(2), 63–71. https://doi.org/10.1088/1741-2560/1/2/001.

13. Makarov, S. N., Noetscher, G. M., Raij, T., & Nummenmaa, A. (2018a). A quasi-static boundary element approach with fast multipole acceleration for high-resolution bioelectromagnetic models. *IEEE Transactions on Biomedical Engineering, 65*(12), 2675–2683. https://doi.org/10.1109/TBME.2018.2813261.

14. Makarov, S. N., Noetscher, G. M., & Sundaram, P. (2018b). Microscopic and macroscopic response of a cortical neuron to an external electric field computed with the boundary element fast multipole method. *bioRxiv* Preprint Aug. 13, 2018; https://doi.org/10.1101/391060.

15. Malmivuo, J., & Plonsey, R. (1995). *Bioelectromagnetism. Principles and applications of bioelectric and biomagnetic fields.* Oxford: Oxford University Press. Ch. 8. ISBN-10: 0195058232.

16. Meijs, J. W., Weier, O. W., Peters, M. J., & van Oosterom, A. (1989). On the numerical accuracy of the boundary element method. *IEEE Transactions on Biomedical Engineering, 36* (10), 1038–1049. https://doi.org/10.1109/10.40805.

17. Murakami, S., & Okada, Y. (2015). Invariance in current dipole moment density across brain structures and species: Physiological constraint for neuroimaging. *NeuroImage, 111*, 49–58. https://doi.org/10.1016/j.neuroimage.2015.02.003.

18. Murakami, S., & Okada, Y. (2006). Contributions of principal neocortical neurons to magnetoencephalography and electroencephalography signals. *The Journal of Physiology, 575*, 925–936. https://doi.org/10.1113/jphysiol.2006.105379.

19. Nunes, P. L., & Srinivasan, R. (2006). *Electric fields of the brain. The neurophysics of EEG* (2nd ed.). Oxford: Oxford University Press. ISBN-10:019505038X.

20. Oostenveld, R., Fries, P., Maris, E., & Schoffelen, J. M. (2011). FieldTrip: Open source software for advanced analysis of MEG, EEG, and invasive electrophysiological data. *Computational Intelligence and Neuroscience, 2011*, 156869. https://doi.org/10.1155/2011/156869.

21. Parvizi, J., & Kastner, S. (2018). Promises and limitations of human intracranial electroencephalography. *Nature Neuroscience, 21*(4), 474–483. https://doi.org/10.1038/s41593-018-0108-2.

22. Piastra, M. C., Nüßing, A., Vorwerk, J., Bornfleth, H., Oostenveld, R., Engwer, C., & Wolters, C. H. (2018). The discontinuous galerkin finite element method for solving the MEG and the combined MEG/EEG forward problem. *Frontiers in Neuroscience, 12*(Article 30), 1–18. https://doi.org/10.3389/fnins.2018.00030.

23. Rokhlin, V. (1985). Rapid solution of integral equations of classical potential theory. *Journal of Computational Physics, 60*(2), 187–207. https://doi.org/10.1016/0021-9991(85)90002-6.

24. Sander, T. H., Preusser, J., Mhaskar, R., Kitching, J., Trahms, L., & Knappe, S. (2012). Magnetoencephalography with a chip-scale atomic magnetometer. *Biomedical Optics Express, 3*(5), 981–990. https://doi.org/10.1364/BOE.3.000981.

25. Schomer, D. L., & Lopes da Silva, F. H. (Eds.). (2017). *Niedermeyer's electroencephalography: Basic principles, clinical applications, and related fields* (7th ed.). Oxford: Oxford University Press. ISBN-10: 0190228482.

26. Semprini, M., Laffranchi, M., Sanguineti, V., Avanzino, L., De Icco, R., De Michieli, L., & Chiappalone, M. (2018). Technological approaches for neurorehabilitation: From robotic devices to brain stimulation and beyond. *Frontiers in Neurology, 9*, 212. https://doi.org/10.3389/fneur.2018.00212.

27. Sherman, M. A., Lee, S., Law, R., Haegens, S., Thorn, C. A., Hämäläinen, M. S., Moore, C. I., & Jones, S. R. (2016). Neural mechanisms of transient neocortical beta rhythms: Converging evidence from humans, computational modeling, monkeys, and mice. *Proceedings of the National Academy of Sciences of the United States of America, 113*(33), E4885–E4894. https://doi.org/10.1073/pnas.1604135113.

28. Sundaram, P., Nummenmaa, A., Wells, W., Orbach, D., Orringer, D., Mulkern, R., & Okada, Y. (2016). Direct neural current imaging in an intact cerebellum with magnetic resonance imaging. *NeuroImage, 132*, 477–490. https://doi.org/10.1016/j.neuroimage.2016.01.059.

29. Tadel, F., Baillet, S., Mosher, J. C., Pantazis, D., & Leahy, R. M. (2011). Brainstorm: A user-friendly application for MEG/EEG analysis. *Computational Intelligence and Neuroscience, 2011*, 879716. https://doi.org/10.1155/2011/879716.

30. The Population Head Model Repository. (2017). IT'IS Foundation website. doi: https://doi.org/10.13099/VIP-PHM-V1.0. Retrieved from: https://www.itis.ethz.ch/virtual-population/regional-human-models/phm-repository/

31. Van Essen, D. C., Ugurbil, K., Auerbach, E., Barch, D., Behrens, T. E., Bucholz, R., Chang, A., Chen, L., Corbetta, M., Curtiss, S. W., Della Penna, S., Feinberg, D., Glasser, M. F., Harel, N., Heath, A. C., Larson-Prior, L., Marcus, D., Michalareas, G., Moeller, S., Oostenveld, R., Petersen, S. E., Prior, F., Schlaggar, B. L., Smith, S. M., Snyder, A. Z., Xu, J., & Yacoub, E. (2012). The Human Connectome Project: A data acquisition perspective. *NeuroImage, 62*(4), 2222–2231. https://doi.org/10.1016/j.neuroimage.2012.02.018.

32. Vollmer, J., Mencl, R., & Müller, H. (1999a). Improved Laplacian smoothing of noisy surface meshes. *Computer Graphics Forum, 18*(3), 131–138. https://doi.org/10.1111/1467-8659.00334.

33. Vollmer, J., Mencl, R., & Müller, H. (1999b). Improved Laplacian smoothing of noisy surface meshes. *EUROGRAPHICS* '99. Brunet P., & Scopigno R. (Guest Editors), vol. 18. https://doi.org/10.1111/1467-8659.00334.

34. Wolpaw, J. R., Birbaumer, N., McFarland, D. J., Pfurtscheller, G., & Vaughan, T. M. (2002). Brain-computer interfaces for communication and control. *Clinical Neurophysiology, 113*(6), 767–791.

35. Wolters, C., Koestler, H., Moeller, C., Haerdtlein, J., Grasedyck, L., & Hackbusch, W. (2007). Numerical mathematics of the subtraction method for the modeling of a current dipole in EEG source reconstruction using finite element head models. *SIAM Journal on Scientific Computing, 30*(1), 24–45. https://doi.org/10.1137/060659053.

36. Xu, W., de Carvalho, F., & Jackson, A. (2019). Sequential neural activity in primary motor cortex during sleep. *Journal of Neuroscience* pii: 1408–18. https://doi.org/10.1523/JNEUROSCI.1408-18.2019.

37. Zátonyi, A., Fedor, F., Borhegyi, Z., & Fekete, Z. (2018). In vitro and in vivo stability of black-platinum coatings on flexible, polymer microECoG arrays. *Journal of Neural Engineering, 15*(5), 054003. https://doi.org/10.1088/1741-2552/aacf71.

Part V
Neural Circuits. Connectome

Robustness in Neural Circuits

Jeffrey E. Arle, Longzhi Mei, and Kristen W. Carlson

1 Introduction: Stability and Resilience – "Robustness"

Complex systems are found everywhere – from scheduling to traffic, food to climate, economics to ecology, the brain, and the universe. Complex systems typically have many elements, many modes of interconnectedness of those elements, and often exhibit sensitivity to initial conditions. Complex systems by their nature are generally unpredictable and can be highly unstable.

However, most *highly connected* complex systems are actually quite stable and resistant to disruption from minor changes in parameters [1]. This is a concept originating from Bernard and Cannon (homeostasis) and now is a central hypothesis of "robustness" in theoretical biology [2–5]. A more contemporary review by Demongeot and Demetrius [6] captures the relationship between the concepts of robustness, entropy, and complexity:

> The hypothesis that a positive correlation exists between the complexity of a biological system, as described by its connectance, and its stability, as measured by its ability to recover from disturbance, derives from the investigations of the physiologists, Bernard and Cannon, and the ecologist Elton. Studies based on the ergodic theory of dynamical systems and the theory of large deviations have furnished an analytic support for this hypothesis. Complexity in this context is described by the mathematical object evolutionary entropy, stability is characterized by the rate at which the system returns to its stable conditions (steady state or periodic attractor) after a random perturbation of its robustness. This article reviews the

Kristen W. Carlson: *Member, IEEE*

J. E. Arle
Department of Neurosurgery, Beth Israel Deaconess Medical Center, Boston, MA, USA

Department of Neurosurgery, Harvard Medical School, Boston, MA, USA

Department of Neurosurgery, Mount Auburn Hospital, Cambridge, MA, USA

L. Mei · K. W. Carlson (✉)
Department of Neurosurgery, Beth Israel Deaconess Medical Center, Boston, MA, USA
e-mail: kwcarlso@bidmc.harvard.edu

analytical basis of the entropy — robustness theorem — and invokes studies of genetic regulatory networks to provide empirical support for the correlation between complexity and stability. Earlier investigations based on numerical studies of random matrix models and the notion of local stability have led to the claim that complex ecosystems tend to be more dynamically fragile. This article elucidates the basis for this claim which is largely inconsistent with the empirical observations of Bernard, Cannon and Elton. Our analysis thus resolves a long-standing controversy regarding the relation between complex biological systems and their capacity to recover from perturbations. The entropy-robustness principle is a mathematical proposition with implications for understanding the basis for the large variances in stability observed in biological systems having evolved under different environmental conditions.

Stability is characterized by measures of how perturbing initial conditions or default parameters result in permanent deviation from baseline behavior vs a tendency to return toward initial conditions (e.g., a basin of attraction) [7–9]. Biological robustness is the ability for a system to recover from disturbing its natural healthy equilibrium into deleterious regions of its parameter space [7, 10–13]. Mathematical analyses in nonlinear dynamical systems theory provide a theoretical foundation for robustness theory [14–16]. Since the advent of control theory, engineers have emphasized negative feedback as a restoring mechanism [17–20]. Much more empirical work at many biological systems levels is needed to marry theory and practice [13, 21–26].

On the practical side, by way of example, the medical and neuroscience literature has many examples where thresholds, numbers of neurons, synapses, locations, branching, conductivities, capacitances, impedances, time-varying dynamics, and so forth are not accurate, known, or are even different across publications. Often, sensitivity analysis is critical to understand the underlying dynamics, and the system must compensate for sensitivity to maintain stability and robustness [27]. In this regard, the complexity of neural circuitry systems may compensate for large error bars in parameter accuracy [28–31].

For practical and theoretical interest, we sought to examine stability and robustness vs complexity in neural circuitry parameter spaces using Monte Carlo simulation.

2   Methods

2.1   Node Parameters at Several Systems Levels Granularity

Using the universal neural circuitry simulation (UNCuS) software [32–34], we specified a subset of ten neuron parameters for sample spaces composed of different numbers of populations, groups, and numbers of cells (neurons). To globally sample the combined parameter space, we used Monte Carlo methods (Table 1) [27]. Out of a space of $\sim 5 \times 10^{11}$ parameter configurations we sampled 5×10^3 combinations. UNCuS and the Monte Carlo sampling program were written by us in C++ (Microsoft, Redmond, WA, USA) and Java (Oracle Corp., Redwood City, CA,

Table 1 Neural circuit topology that varied total # (number of) nodes at several systems levels of granularity in the model

Circuit parameter	Range
#Populations	Set 1: 2 or 9. Set 2: 3, 4, 5, 6, 7, or 8
#Cells/group	Randomly selected from range of 5 – 150 in steps of 5: {5, 10, …, 145, 150}
#Intergroup connections	Constant at 40% – each source neuron connected to 40% of the target group's neurons
Node parameter	Definition
#Cells/configuration	(#Populations-1) * 2 * #Cells
#Projections/cell	40% * #Cells/Group
#Projections/ configuration	#Cells * #Projections/Cell
#Synapses/ configuration	#Dendritic Compartments * #Projections
#Synapses	#Synapses * #Cells * #Projections

USA). UNCuS was designed specifically to perform computationally efficient neural circuitry simulations for large as well as small circuits. The program integrates essential behavior at the underlying systems level of ion channel conductance and dendritic tree emission current and scales well in projects of the current size [32, 35].

The number of populations ranged from two to nine, and each population always had just two groups of neurons, one excitatory and one inhibitory. Groups within a population were never connected to each other but were always connected to all other groups in the other populations, with the same density. What was varied in the configurations was the number of (1) cells per group and (2) neuron parameters of the cells. Figure 1 shows screenshots from UNCuS of example configurations.

We created 50 separate project configurations with identical numbers of cells per group in each case. Initially we looked at the extremes of the parameter space, 7 configurations with 2 populations and 7 configurations with 9 populations, totaling 14 configurations. Then we looked at the intermediate numbers of populations and groups using 6 configurations each with 3, 4, 5, 6, 7, and 8 populations, totaling 36 configurations.

To set the varying number of cells in each population, the program randomly selected a number of cells from the set {5,10, …, 145, 150}.

The number of connections from source to target group was fixed at 40%; for example, from a source group of any number of neurons to a target group of, e.g., 100 neurons, each source neuron connected to 40 of the target group's neurons. Groups never recurred back to themselves or the other group in its population, always to exogenous target groups in other populations (Fig. 1); hence in calculating the total number of connections, we subtract one from the total number of populations in the case.

At each target cell, the axon synapsed on two of ten electrotonic compartments in the middle of the dendrite, specifically compartments four and five. See Limitations

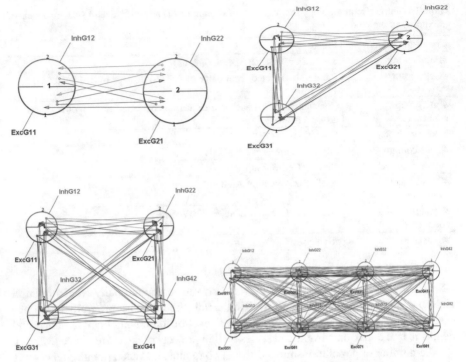

Fig. 1 Four sample project configurations from universal neural circuitry simulation software UNCuS. Neuron populations (circles) contain numbered groups and connections between groups identified as excitatory (blue) or inhibitory (green). Each neuron in each group connected to 40% of the neurons it targeted. The total number of synapses can be thought of abstractly as the total nodes in the complex system

below for the reasoning behind the choice of fixing connection strength at 40% and the dendritic compartments at two in the middle of the proximal-distal range.

2.2   Neuron Cell Parameters

Just as for each project case there were a random number of cells in each group, the sampling program also randomly selected a set of neuron cell parameters. In UNCuS, there are 12 parameters to characterize a neuron type. We excluded refractory time and spike width, leaving 10 cell parameters to select from. The parameters, their default baseline values, and the value ranges subjected to random sampling are shown in Table 2. All cells in all groups in each case used the same random selection of cell parameters. For the equations governing cell behavior, see Arle et al. [32, 33].

In previous projects, we usually drove a circuit with specific stimuli on specific cell groups such that their initial firing rate corresponded to what was reported in the

Table 2 Neuron parameters sampled via Monte Carlo simulation

Symbol	Definition	(Min, max, interval), (baseline values)
Ek	Reversal potential of the rectifying potassium conductance	(−105, −25, 1),−65 mV
Eb	Reversal potential of basic channel	−65 mV
Gk	Rectifying potassium conductance	(5, 20, 1), 10 nS
Gb	"Lumped" ion channel conductance	(5, 20, 1), 10 nS
Tm	Membrane time constant	(1, 11, 1), 5 ms
Tgk	Time constant of rectifying potassium conductance	(0.5, 5, 0.5), 2 ms
tTh	Time constant of membrane threshold	(0.5, 5, 0.5), 5 ms
C	Threshold accommodation term	(0.1, 0.9, 0.1), 0.2 (dimensionless)
B	Delayed rectifier potassium conductance strength	(100, 1000, 1), 500 nS
Th0	Threshold transmembrane potential	(2.0, 20, 1), 12 mV

Fig. 2 Typical membrane potential over time of a neuron in the simulations. The transient period of the first 100 ms is discarded in our calculations. To more clearly show the spiking activity, we plot the middle 600 ms of activity between 200 ms and 800 ms

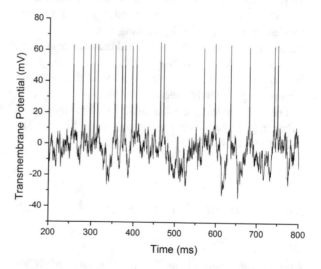

literature. Here we raised background noise to a level that activated all circuits, specifically 1.7 nA, which we found to be sufficient.

A plot of a typical spiking neuron's membrane voltage over time is shown in Fig. 2.

2.2.1 Dynamic Adjustment of Input Amplitude

The program dynamically condensed the input amplitude to each cell from all incoming excitatory or inhibitory post-synaptic potentials (EPSP, IPSP) to a normalized range between −5 and +5 nA. The reasoning here that there is some range in actual neurons beyond stimulating or inhibiting the cell has no greater effect.

However, in modeling and simulations, there is often a range of a parameter for which one has validated the behavior of the simulated entity, while, beyond that validation value, behavior is unknown and may be unpredictable. Essentially, the method used was to calculate the maximum possible input current for a given project case based on UNCuS' dendritic emission current look-up table, the known dendritic compartments (#4 and #5), the known number of inputs (40%, e.g., 40 per 100 cells), and then, at each time step, multiply each cell's input by the factor required to condense the scale to −5 to +5 nA.

2.3   Simulation Duration, Time Step, and Calculation of Firing Rates

All simulations were run for 1000 ms in 0.25 ms time steps, which we have found to give stable cell behavior, often after an initial transient period. We defined a "momentary firing rate" by moving a window of 100 ms, time step by time step, through the 1000 ms, taking the total number of spikes in each interval, and dividing that number by the interval length, 100 ms, to give spikes/second. This method discards the first 100 ms, which, as said above, often involved a transient period transitioning to steady state behavior.

The program recorded the spikes for each neuron, took the momentary firing rate, and averaged over all cells in a group to give a group firing rate, which underlies the data shown in Results.

2.4   Definition of "Robustness" via Coefficient of Variance (CV)

We used coefficient of variance (aka "variation") (CV) as a metric for "robustness," where the standard deviation was divided by the spike rate of baseline neuron cell instead of the mean of the Monte Carlo sampled data:

$$cv = \frac{1}{FR_{base}} \sqrt{\sum_{i=1}^{n} \frac{(FR_i - FR_{base})}{n - 1}} \qquad (1)$$

where $n = 50$, the number of population-group configurations (each with a random set of 10-cell parameters), FR_i is the firing rate of a given configuration, and FR_{base} is the firing rate with the baseline neural parameters. Thus, the baseline parameters, shown in Table 2, established a firing rate against which the deviation of all Monte Carlo sampled configurations would be checked.

2.5 Definition of "Robustness" via an Adapted Lyapunov Exponent

To give a second perspective on how the level of complexity affects the deviation of neural system from its baseline performance, we adapted a standard formula for the Lyapunov exponent as was done with the definition of *CV*. Thus, the baseline neuron cell firing rate was used as the initial condition against which deviations over time were compared in calculating the Lyapunov exponent (*LE*) [35, 36]:

$$\lambda = \lim_{s \to \infty} \frac{1}{s} \sum_{k=0}^{s-1} ln \left| \frac{FR_k}{FR_{base,k}} \right| \tag{2}$$

where s is the time step, FR_k is the momentary firing rate of the sampled parameter point at time step k, $FR_{base,k}$ is the baseline firing rate at time step k, and s is the total number of time steps. When the sampled configuration firing rate is equal to that of the baseline, there is no divergence and $LE = 0$.

2.6 Cumulative Firing Rate vs Momentary Firing Rate

We calculated *cumulative firing rate* (*CFR*) as a global reflection of the circuitry in time, as distinct from the momentary firing rate. The *CFR* is defined at time t as the total number of spikes up to t divided by t to give spike rate/second.

2.7 Limitations

There are many other ways to vary neural circuit parameters than those we chose to explore. Notably the many different topologies of connecting circuits were programmed to perform a specific behavior. Of those, we did not, for instance, explore circuits that must be stable to perform their function, such as clocks and rhythm generators, or that evolved to become emergently stable, such as the various networks revealed by functional connectivity analyses [37]. The space of connection strengths could also be explored more fully. Our past experience building small- and large-scale connectome models [34, 38, 39] and preliminary results in this study showed that, at the connection strength extremes, very low or high connection density and/or few distal dendritic compartment connections or many proximal dendritic compartment connections gave either no activity or hyperactivity, respectively. On the other hand, one can build circuits with, e.g., high connection density in some parts of the circuit if they are regulated by other parts of the circuit; an effectively infinite number of such specific circuits are possible.

Other neural circuitry simulation software exists, and the variety is evolving rapidly, driven by rapidly increasing interest in building connectome models (see Carlson et al. elsewhere in this book [24]).

3   Results

In general, in non-quiescent circuits, the momentary firing rate, which characterizes the system's short-term behavior in the time domain, was never constant but dynamic. For the robustness measures, the behavior of excitatory and inhibitory groups was similar, so only results for excitatory groups are shown below.

3.1   Sample Time Course of Firing Rate of Two Population-Group Configurations

Figure 3 allows us to visualize the variance of relatively simple vs relative complex neural circuits over time before we apply metrics to get a more precise understanding. The figure plots the time course of firing rates in the two extreme population-group configurations; in the left, 2 populations with 5 cells per group are shown, and in the right, 9 populations with 150 cells per group are shown. In those configurations, each curve is a unique set of ten neuron parameter values. Note that the two

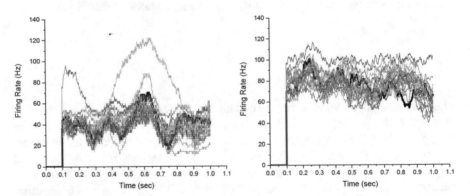

Fig. 3 Momentary firing rate over 1 second in a simple vs complex neural circuit. Left: 2 interconnected populations with just 5 cells in each group (there are always 2 groups/populations in all configurations). Right: 9 interconnected populations with 150 cells/group. In those configurations, the firing rates of 50 different neural parameters are plotted over 1 second (after a 100 ms transient period), with a baseline parameter case shown in black. Firing rates in the more complex system are higher than in the simple one due to recurrent input from a higher number of cells. The outmost curve in the left case implies that a particular set of ten-parameter value is more sensitive than others, i.e., one of the ten parameters may be more sensitive than the rest of the parameters. CV in the simple system is 0.754 and in the more complex system 0.236

outlier parameter configurations easily seen in the simpler model are also present in the complex model but are to some extent "tamed" by the higher connectedness and number of circuit elements. One effect is that in more complex systems, errant random inputs can cancel each other.

3.1.1 Plots of Firing Rate of All Sample Points vs Baseline Parameters

The three plots in Fig. 4 show all data and examine robustness across increasing complexity measured by the numbers of several different circuit elements.

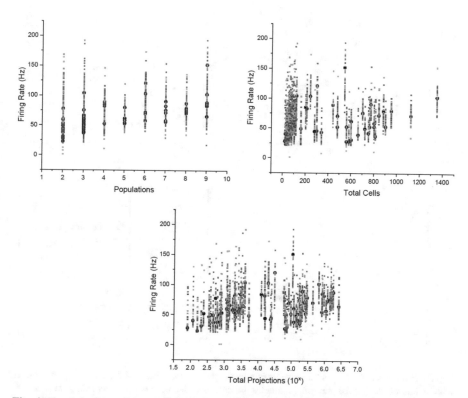

Fig. 4 Top left: firing rate vs number of neural cell *populations*. Top right: firing rate vs *total number of neurons in the system*. Bottom: firing rate vs *total system projections*. Each point is a 1-second run with a unique set of 10 neural cell parameter values (colored dots) vs baseline values (black dots), totaling 5100 sample points. In general, as complexity increases, variance decreases, i.e., the system becomes more robust

3.1.2 Robustness vs Number of Elements as Measured by Coefficient of Variance (CV)

The next plots (Fig. 5) show CV (Eq. 1) of Monte Carlo sample firing rates vs complexity parameters compared to the baseline configuration (red line). Note that we take the baseline firing rate as the "expected value" and therefore $CV < 1$ means the standard deviation of the sampled firing rates is smaller than the firing rate of the baseline (Eq. 1). Each point shows the CV over 50 population configurations, each with a randomly selected set of neuron parameters, and each graph shows 100 such points. Since the total synapses and the combined complexity are linearly related (in our study, synapses are two times the number of projections since each projection synapsed on dendritic compartments 4 & 5), their distributions of CV are similar, and thus data for a number of synapses are omitted.

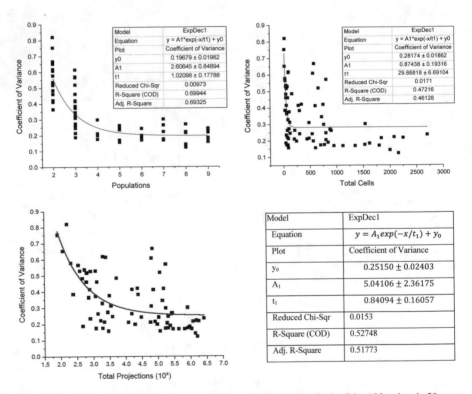

Fig. 5 Top left: coefficient of variance vs *number of populations*. Each of the 100 points is 50 runs with a unique set of 10 neuron parameter values. *CV* decreased with increasing number of populations and converged to a constant 0.197 at 7+ populations. Top right: coefficient of variance vs. *number of total cells*. Against different numbers of cells in the system, *CV* is larger than when varying the number of populations. Here *CV* converges to 0.281. Bottom: coefficient of variance vs *number of total projections*. *CV* converges to 0.252 at ~10^5 projections, similar to when varying the number of cells

3.1.3 Robustness vs Number of Elements as Measured by Lyapunov Exponent (LE)

Here we used a different measure of robustness, the Lyapunov exponent (LE, Eq. 2), based on how the firing rates of sampled population-group-neuron parameter sample points diverged over time from those of the baseline neuron parameter case. Yet the results for different levels of complexity were quite similar to the previous results using CV as the metric. Figure 6 shows LE against numbers of populations, cells, and projections.

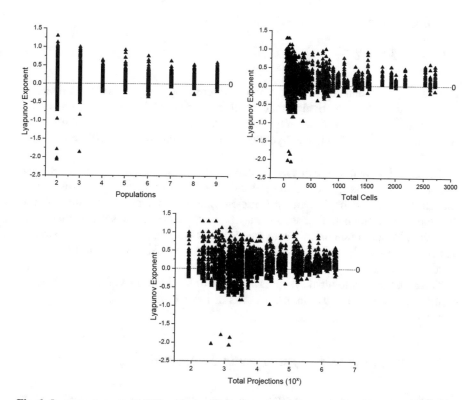

Fig. 6 Lyapunov exponent (LE) analysis of firing rates over time vs system complexity. Each point is a run with a unique set of 10 neuron parameter values in the 50 combinations of population and group numbers, totaling 5000 sample points. The LE measures how the firing rates of sample configurations deviated from the baseline case over time. $LE > 0$: sample firing rates > baseline. $LE < 0$ sample firing rates < baseline. $LE = 0$: no deviation from baseline. Top left: LE of firing rate vs *populations*. Top right: LE of firing rate vs *number of total cells*. Bottom: LE of firing rate vs *number of total projections*. Divergence decreases with increasing numbers of populations. Generally firing rates decrease over time as steady states are reached in the circuits and those of sample points are less than those of baseline cases. Firing rates of sampled configurations were generally higher than those of baseline cases and divergence decreased against total projections in the system. Divergence decreases with increasing system complexity

3.1.4   Robustness vs Number of Elements as Measured by Cumulative Firing Rate (CFR)

The *cumulative firing rate* (*CFR*) can be compared to momentary firing rate as a global reflection of the circuitry behavior over time instead of at a given point in time.

Figure 7 shows *CFR* for total populations, total cells, total projections, total projections' divergence from baseline (*LE*), and total populations' divergence from baseline (*CV*). Results were similar to simulations shown above.

4   Discussion

4.1   Key Results

The data suggests that neural circuits become increasingly robust as their complexity is increased, with a relatively small amount of complexity as compared to the number of neurons in various organisms. Synapse totals in the study ranged from 160 to 5.18×10^6 – recall that a 3-year-old human has an estimated 10^{15} synapses in their entire brain.

Thus, the inherent nature of a moderately complex and sufficiently interconnected neural circuit results in local parameter changes being unlikely to have significant changes in circuit behavior. Given sufficient connectivity, robustness increases as the number of elements increases and may be sustained once it reaches a window of transition. In our simulations, where the elements were neurons, and other parameters being equal, this transition window to maximal robustness was in the range of ~100–1000 cells.

The number of individual synapses appeared to have the least effect on achieving robustness compared to the number of cells or axonal projections.

4.2   Robustness and Degeneracy in Biological Systems

Our pragmatic interest is in applying the results of the study and related work to faithfully modeling neurological disorders, such as Parkinson's disease, epilepsy, neuropathic pain, and others, and how to treat them with neuromodulation.

Biological systems have evolved adaptive homeostasis to increase their robustness; conversely, loss of adaptive homeostasis can result in disease and disorder and is a hallmark of aging [1, 2, 40–45]. More broadly, the term *degeneracy* in complex

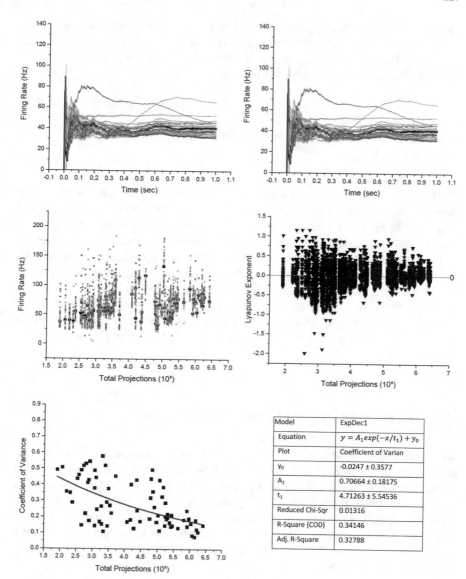

Fig. 7 *Cumulative firing rate (CFR)* is the dynamic cumulative firing rate from 100 ms at each time step divided by the elapsed time in the simulation. Plots show *CFR* for total populations (top left), total cells (top right), total projections (middle left), *LE* of total projections' *CFR* divergence from that of baseline parameters' model (middle right), and *CV* of *CFR* of total populations' divergence from that of baseline (bottom left with legend bottom right). In all cases, the *CFR* of sampled configurations decreases with increasing number of system elements

systems theory has been applied to biological systems to describe a decline in their ability to maintain functional integrity [46], and the application of degeneracy analysis to biology and neural circuits in particular is likely to play a major role in neurological disorders and aging and their treatment [47].

4.3 Robustness and Degeneracy in Functional Connectivity Brain Networks

In the past two decades, functional connectivity (FC) – the correlated activity between brain regions as identified by imaging and monitoring the brain's electrical activity – has become a major paradigm for understanding healthy and diseased states. The "resting state" and time course of FC networks can constitute signatures of healthy vs diseased conditions. Degeneracy of the healthy FC is identified with the diseased state. In many cases, modelers of neurological disorders and disease may find that system complexity may underlie diseased states. For instance, the reason why, in Parkinson's disease, up to 80% of the dopaminergic neurons of the *substantia nigra* are lost before movement disorders manifest is unknown. Degeneracy via loss of complexity and connectedness in the basal ganglia may be part of the mechanism.

4.4 Inadvertent Modeling Error Due to Scaling

Robustness helps illuminate possible errors in modeling connectomes when scaling the number of elements. When building a complete model of the human spinal cord connectome [34], we found that Rexed lamina II had ~1300 times the number of neurons that lamina I had. Thus, in scaling down the total number of neurons to make a more computationally efficient model, we inadvertently reduced the robustness of the model, which showed up as high firing rates in lamina I, and had to scale it back up to correct for that. Such an inadvertent effect would equally invalidate many neural circuit models, notably of seizure.

5   Conclusion

The complexity of biological systems, notably neural circuitry, as defined by the number of element (neurons) and connectivity (synaptic topology) is a critical element in their robustness – the degree to which, when perturbed from healthy states toward diseased or disordered states, they inherently attempt to return to the healthy state. We simulated neural circuitry complexity and connectivity in a large parameter space via Monte Carlo sampling. Our results bear out the theory that more complex, connected systems can be inherently more stable than simpler systems. These results have implications for modeling neurological disorders, such as Parkinson's disease, chronic pain, seizure, and age-related cognitive decline, and how to treat these conditions with neuromodulation.

Acknowledgment   Funded in part by the Sydney Family Foundation.

References

1. Lehar, J., et al. (2008). High-order combination effects and biological robustness. *Molecular Systems Biology, 4*, 215.
2. Davies, K. J. (2016). Adaptive homeostasis. *Molecular Aspects of Medicine, 49*, 1–7.
3. Bernard, C. (1974). *Lectures on the Phenomena of Life Common to Animals and Plants*. Springfield: Charles C Thoma.
4. Cannon, W. B. (1926). Physiological regulation of normal states: Some tentative postulates concerning biological homeostatics. In A. Pettit (Ed.), *A Charles Riches amis, ses collègues, ses élèves (in French)* (p. 91). Paris: Les Éditions Médicales.
5. Felix, M. A., & Barkoulas, M. (2015). Pervasive robustness in biological systems. *Nature Reviews. Genetics, 16*, 483–496.
6. Demongeot J., & Demetrius, L. (2015). Complexity and stability in biological systems. *International Journal of Bifurcation and Chaos, 25*.
7. Demongeot, J., et al. (2010). Attraction basins as gauges of robustness against boundary conditions in biological complex systems. *PLoS One, 5*, e11793.
8. Li, Y., & Lin, Z. (2013). Multistability and its robustness of a class of biological systems. *IEEE Transactions on Nanobioscience, 12*, 321–331.
9. Kaluza, P., et al. (2008). Self-correcting networks: Function, robustness, and motif distributions in biological signal processing. *Chaos, 18*, 026113.
10. Alcalde Cuesta, F., et al. (2016). Exploring the topological sources of robustness against invasion in biological and technological networks. *Scientific Reports, 6*, 20666.
11. Apri, M., et al. (2010). Efficient estimation of the robustness region of biological models with oscillatory behavior. *PLoS One, 5*, e9865.
12. Barkai, N., & Shilo, B. Z. (2007). Variability and robustness in biomolecular systems. *Molecular Cell, 28*, 755–760.
13. Chambers, A. R., & Rumpel, S. (2017). A stable brain from unstable components: Emerging concepts and implications for neural computation. *Neuroscience, 357*, 172–184.
14. Anafi, R. C., & Bates, J. H. (2010). Balancing robustness against the dangers of multiple attractors in a Hopfield-type model of biological attractors. *PLoS One, 5*, e14413.
15. Ay, N., & Krakauer, D. C. (2007). Geometric robustness theory and biological networks. *Theory in Biosciences, 125*, 93–121.
16. Radde, N. (2009). The impact of time delays on the robustness of biological oscillators and the effect of bifurcations on the inverse problem. *EURASIP Journal on Bioinformatics and Systems Biology*. https://doi.org/10.1155/2009/327503.
17. Khammash, M. (2016). An engineering viewpoint on biological robustness. *BMC Biology, 14*, 22.
18. Kwon, Y. K., & Cho, K. H. (2008). Quantitative analysis of robustness and fragility in biological networks based on feedback dynamics. *Bioinformatics, 24*, 987–994.

19. O'Leary, T. (2018). Can engineering principles help us understand nervous system Robustness? In C. T. Wolfe, P. Huneman, T. A. C. Reydon, M. Bertolaso, S. Caianiello, & E. Serrelli (Eds.), *Biological Robustness: Emerging perspectives from within the life sciences*. Cham: Springer.

20. Mensi, M., & Oliva, G. (2018). Robustness vs. control in distributed systems. In C. T. Wolfe, P. Huneman, T. A. C. Reydon, M. Bertolaso, S. Caianiello, & E. Serrelli (Eds.), *Biological Robustness: Emerging Perspectives from within the life sciences*. Cham: Springer.

21. Wu, Y., et al. (2009). Identification of a topological characteristic responsible for the biological robustness of regulatory networks. *PLoS Computational Biology, 5*, e1000442.

22. Von Dassow, G., et al. (2000). The segment polarity network is a robust developmental module. *Nature, 406*, 188–192.

23. Whitacre, J. M. (2012). Biological robustness: Paradigms, mechanisms, and systems principles. *Frontiers in Genetics, 3*, 67.

24. Carlson, K. W., et al. (2020). Functional requirements of small- and large-scale neural circuitry connectome models. In S. Makarov, M. Horner, & G. Noetscher (Eds.), *Brain and human body modeling: Computational human modeling at EMBC 2019*. Cham: Springer. In press.

25. Melin, P., & Castillo, O. (2019). *Modelling, simulation and control of non-linear dynamical systems : An intelligent approach using soft computing and fractal theory*. CRC Press.

26. S.J. Guastello, et al. (2008). *Chaos and complexity in psychology : The theory of nonlinear dynamical systems,*

27. Pusnik, Z., et al. (2019). Computational analysis of viable parameter regions in models of synthetic biological systems. *Journal of Biological Engineering, 13*, 75.

28. Calin-Jageman, R. J., & Cumming, G. (2019). *eNeuro*. https://doi.org/10.1523/ENEURO.0205-19.2019.

29. Gutenkunst, R. N., et al. (2007). Extracting falsifiable predictions from sloppy models. *Annals of the New York Academy of Sciences, 1115*, 203–211.

30. Gutenkunst, R. N., et al. (2007). Universally sloppy parameter sensitivities in systems biology models. *PLoS Computational Biology, 3*, 1871–1878.

31. Soltis, A. R., & Saucerman, J. J. (2011). Robustness portraits of diverse biological networks conserved despite order-of-magnitude parameter uncertainty. *Bioinformatics, 27*, 2888–2894.

32. Arle, J. E. (1992). *Neural modeling of the cochlear nucleus* (PhD Thesis), University of Connecticut.

33. Arle, J. E., et al. (2008). Modeling parkinsonian circuitry and the DBS electrode. I. Biophysical background and software. *Stereotactic and Functional Neurosurgery, 86*, 1–15.

34. Arle, J. E., et al. (2018). Dynamic computational model of the human spinal cord connectome. *Neural computation*. https://doi.org/10.1162/neco_a_01159 1–29.

35. Barreira, L. (2017). *Lyapunov exponents*. New York/Berlin/Heidelberg: Springer.

36. Politi, A. (2016). *Lyapunov exponents: A tool to explore complex dynamics*. Cambridge: Cambridge University Press.

37. Bijsterbosch, J., et al. (2017). *Introduction to resting state fMRI functional connectivity*

38. Arle, J. E., & Carlson, K. W. (2016). The use of dynamic computational models of neural circuitry to streamline new drug development. *Drug Discovery Today: Disease Models, 19*, 69–75.

39. Arle, J. E., et al. (2014). Mechanism of dorsal column stimulation to treat neuropathic but not nociceptive pain: Analysis with a computational model. *Neuromodulation, 17*, 642–655.

40. Pomatto, L. C. D., et al. (2019). To adapt or not to adapt: Consequences of declining adaptive homeostasis and proteostasis with age. *Mechanisms of Ageing and Development, 177*, 80–87.

41. Pomatto, L. C. D., et al. (2019). Limitations to adaptive homeostasis in an hyperoxia-induced model of accelerated ageing. *Redox Biology, 24*, 101194.

42. Pomatto, L. C. D., & Davies, K. J. A. (2018). Adaptive homeostasis and the free radical theory of ageing. *Free Radical Biology & Medicine, 124*, 420–430.

43. Pomatto, L. C. D., & Davies, K. J. A. (2017). The role of declining adaptive homeostasis in ageing. *The Journal of Physiology, 595*, 7275–7309.

44. Lomeli, N., et al. (2017). Diminished stress resistance and defective adaptive homeostasis in age-related diseases. *Clinical Science (London, England), 131*, 2573–2599.
45. Whitacre, J. M. (2010). Degeneracy: A link between evolvability, robustness and complexity in biological systems. *Theoretical Biology & Medical Modelling, 7*, 6.
46. Mason, P. H., et al. (2015). Hidden in plain view: Degeneracy in complex systems. *Biosystems, 128*, 1–8.
47. Rathour, R. K., & Narayanan, R. (2019). Degeneracy in hippocampal physiology and plasticity. *Hippocampus, 29*, 980–1022.

Insights from Computational Modelling: Selective Stimulation of Retinal Ganglion Cells

Tianruo Guo, David Tsai, Siwei Bai, Mohit Shivdasani,
Madhuvanthi Muralidharan, Liming Li, Socrates Dokos,
and Nigel H. Lovell

1 Introduction

Retinal prostheses aim to restore patterned vision to those with retinitis pigmentosa by electrically stimulating surviving neurons in the degenerate retina. Despite a significant global interest in the "race" to develop a high-resolution implant, commercialization of three devices in this space, and numerous human trials having demonstrated the ability of devices to restore some functional vision, the experience for most implanted patients has been largely underwhelming. The phosphenes evoked by all implants tested to date have remained complex, with human subjects reporting evoked percepts that resembled halos, blobs, wedges, streaks, or other shapes [1–4]. As a result, current devices are prescribed only to patients with profound blindness and until the vision quality has significantly improved. Many patients with residual vision, who would have benefited from the uptake of such technology, remain "waiting" for an alternative appropriate treatment.

T. Guo (✉) · D. Tsai · M. Muralidharan · S. Dokos · N. H. Lovell
Graduate School of Biomedical Engineering, UNSW, Sydney, Australia
e-mail: t.guo@unsw.edu.au

S. Bai
Graduate School of Biomedical Engineering, UNSW, Sydney, Australia

Department of Electrical and Computer Engineering, Technical University of Munich, Munich, Germany

Munich School of Bioengineering, Technical University of Munich, Garching, Germany

M. Shivdasani
Graduate School of Biomedical Engineering, UNSW, Sydney, Australia

Bionic Institute, Melbourne, Australia

L. Li
School of Biomedical Engineering, Shanghai Jiao Tong University, Shanghai, China

© The Author(s) 2021
S. N. Makarov et al. (eds.), *Brain and Human Body Modeling 2020*,
https://doi.org/10.1007/978-3-030-45623-8_13

One possible primary reason for the inability of existing retinal neuroprostheses to provide better visual perception may be the indiscriminate activation of different neuronal types across large regions of the retina, providing conflicting information to higher visual centers. To address this problem, we require an improved understanding of how different functional retinal ganglion cell (RGC) types respond to artificial electrical stimulation. In particular, if different RGC types can be selectively or differentially activated in a desired temporospatial sequence, the elicited signals may be interpreted more accurately by the brain, giving rise to visual percepts of greater meaning and utility.

Previous reports have indicated that 1–6 kHz high-frequency electrical stimulation (HFS) may elicit differential excitation of different RGCs in a manner similar to RGC responses to light stimuli in a healthy retina [5–8]. These studies suggest a great promise in eliciting RGC responses that parallel RGC encoding: one RGC type exhibited an increase in spiking activity during electrical stimulation, while another exhibited decreased spiking activity, given the same stimulation parameters. To test whether a larger range of HFS parameters can improve or even maximize the differential excitation of ON and OFF RGC pathways, we began with in silico investigations using biophysically and morphologically detailed computational models of ON and OFF RGCs using the NEURON computational environment, to evaluate the performance of a range of electrical stimulation amplitudes (10–70 µA) and frequencies (1–10 kHz) on RGC responses.

In addition, in order to investigate the effect of ON and OFF RGC dendritic morphologies on HFS-induced responses, we developed a neural morphology generator, capable of generating RGCs with tunable morphological properties, including the dendritic field radius, total dendritic length, and stratification level. Neuronal morphology has been reported to play a vital role in shaping response properties as well as the integration of neuronal inputs in many cell types throughout the central nervous system (CNS) [9, 10]. It is therefore likely that similar morphological dependence is also present in RGCs.

Finally, we used a population-based computational model of ON and OFF layers to explore the performance of electrical stimulation with clinically relevant electrode sizes and locations, as well as stimulation duration.

2   Materials and Methods

2.1   Computational Model of ON and OFF RGC Clusters

ON and OFF RGC clusters were implemented using the computational software NEURON [11]. Techniques used to model individual RGCs have been described in detail previously [12–14].

Firstly, the morphological structures of different RGCs were simulated by a customized neural morphology generator [13, 15] (see Fig. 1A). The RGC soma was initially defined as a point at the origin. With the soma as the center, a number of

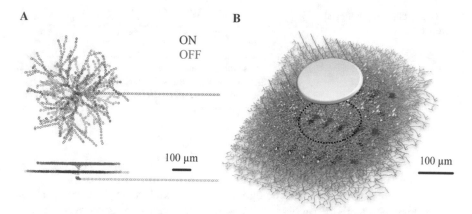

Fig. 1 (**a**). Reconstructed RGC morphologies of ON (red) and OFF (blue) cells. (**b**). A zoomed area of the population-based model with ON (red) and OFF (blue) RGC layers. RGCs were uniformly distributed on a square grid with 40-μm lateral and 40-μm longitudinal distances between neighboring cells. In total, 21 × 21 cells were simulated in each layer. A subretinal stimulation electrode ($\Phi = 200$ μm) was placed 200 μm above the RGC soma array

random carrier points, which serve as the basis of dendritic growth, were distributed within a circular planar region having a user-defined radius. An algorithm, based on the minimum spanning tree algorithm [16], generated dendritic branches by connecting unconnected carrier points to nodal points of the tree. At each step, a sweep through all nodes starting from the soma was undertaken to find the carrier point closest to the tree. A cost function was used to calculate the weighted distance $\tilde{d}$ between a carrier point and a node in the tree, as follows:

$$\tilde{d} = d_e \cdot (1 - bf) + d_p \cdot bf \qquad (1)$$

where d_e is the Euclidean distance between a carrier point and a node in the dendritic tree; d_p is the length of the path along the corresponding branch from the soma to the carrier point, which is the sum of d_e and the length of the branch from the soma to the corresponding node; and bf is a balance factor, which weighs d_e and d_p against each other in the cost function. The carrier point with the shortest $\tilde{d}$ was chosen as the candidate point to be connected to the corresponding node. After creating the dendritic tree, the soma was then extended into a 15-μm segment. A 50-μm-long axonal hillock and a 1000-μm-long axon were added subsequently. The vertical distance between the axon and the soma was set to 10 μm, and the first 50-μm segment of the axon was defined as the axon initial segment (AIS). RGC dendritic morphological parameters [17], including dendritic field radius (μm), total dendritic length (μm), and stratification level (μm), were adjusted based on published data for ON and OFF RGC morphologies in guinea pig retina (see Table 1).

Secondly, the ionic model used in this study can be represented by the equivalent cable equation:

Table 1 RGC morphological parameters

RGC	R (μm)	L (μm)	S (μm)
ON	287	7300	10
OFF	218	6700	50

R radius of dendritic field area, L total length of dendrites, S vertical distance between the soma and the dendritic tree layer. All parameters were estimated based on published data of ON and OFF RGC morphologies [17]

$$\sigma \frac{\partial^2 V_m}{\partial x^2} = A\left(C_m \frac{\partial V_m}{\partial t} + J_{ion} \right) \tag{2}$$

where V_m represents membrane potential, x is the axial cable distance, σ is the intracellular conductivity (mS·cm^{-1}), A is the local cell surface to volume ratio (cm^{-1}), and membrane capacitance (C_m) per unit membrane area was set to 1 μF·cm^{-2}. The intracellular axial resistivity ($1/\sigma$) was set to 110 Ω·cm. The simulation temperature was 37 °C. J_{ion} (mA·cm^{-2}) represents the total cell membrane ionic current, consisting of seven time-dependent currents and one leakage current:

$$J_{ion} = I_{Na} + I_K + I_{KA} + I_{Ca} + I_{KCa} + I_h + I_{CaT} + I_L \tag{3}$$

I_{Na}, I_K, I_{KA}, I_{Ca}, and I_{KCa} were defined in a previous RGC model [18], while the hyperpolarization-activated non-selective cationic current (I_h) and the low-threshold voltage-activated calcium current (I_{CaT}) are newer additions, with both latter currents known to contribute to RGC excitation [19–21]. All gating variables, except those for I_{CaT}, satisfied the following first-order ordinary differential equation (ODE):

$$dx/dt = \alpha_x(1 - x) - \beta_x x \tag{4}$$

where x is a gating variable, with α_x, β_x its opening and closing rates, respectively, which are typically functions of membrane potential V_m.

For I_{CaT}, second-order dynamics were used:

$$mT : \frac{dm_T}{dt} = m_T(1 - \alpha_{mT}) - \beta_{mT}\alpha_{mT}$$

$$hT : dh_T/dt = \alpha_{hT}(1 - h_T - d_T) - \beta_{hT}h_T$$

$$dT : d(d_T)/dt = \beta_{dT}(1 - h_T - d_T) - \alpha_{dT}d_T \tag{5}$$

where the inactivation process for I_{CaT} was modelled using two transition steps: h_T and d_T [22].

In this chapter, ON and OFF RGC models shared the same kinetic-defining parameters for all ionic currents. Ionic channel distributions across different cellular regions were set to be cell-specific to reproduce the stimulus dependency of recently published in vitro whole-cell patch-clamping data [14, 23]. The estimated kinetic-defining parameters of each rate and maximum membrane conductance values (mS/cm^2) per region in each cell are listed in Tables 2 and 3.

2.2 ON and OFF Layer Simulation

To explore the generalizability of HFS-induced differential activation with more clinically relevant electrode size, location, and stimulation duration, we conducted in silico investigations using population-based neuronal models. ON and OFF RGCs were uniformly distributed on a square grid with 40-μm lateral and 40-μm longitudinal distances between neighboring cells. In total, 21×21 cells were simulated in each layer (see Fig. 1B).

2.3 Extracellular Electrical Stimulation and Electrode Settings

To simulate extracellular stimulation, we used a mono-polar circular electrode disk with ground located at infinity and approximate the extracellular domain to be homogeneous. The extracellular potential at each spatial point was adapted based on an analytic formula [24–26]:

$$V_e = \frac{2V_o}{\pi} \sin^{-1}\left(\frac{2R}{\sqrt{(r-R)^2 + z^2} + \sqrt{(r+R)^2 + z^2}}\right) \qquad (6)$$

where r and z are the radial and axial distance, respectively, from the center of the disk for $z \neq 0$ and R is the radius of the disk. The disk potential V_o can be determined by the electric stimulation current I and extracellular resistivity ρ_e (500 Ω·cm) [25]:

$$V_o = \frac{I\rho_e}{4R} \qquad (7)$$

For simulation of single ON and OFF RGC stimulation, we defined a 3D Cartesian (x, y, z) coordinate system, with the soma as the origin, so that the upper surface of the RGC dendritic field was aligned in the x-y plane and the RGC axon was aligned with the y-axis. A hexapolar electrode array (each disk electrode of 15-μm radius, with a center-to-center distance of 60 μm) was positioned at the location (0, −40, −50) μm, where (0, 0, 0) μm was the local 3D coordinates of the soma [23]. Cathodic-first, charge-balanced, symmetric, constant-current biphasic stimuli, each with a pulse width of 50 μs per phase, were used without an interphase interval. The extracellular stimulus amplitude ranged from 10 to 70 μA in 5-μA

Table 2 Ionic current formulations and kinetic parameters

Membrane current	Formulation	
I_{Na}	$I_{Na} = \bar{g}_{Na} m^3 h (V_m - 35)$	
	$\alpha_m = -0.6(V_m + 30)/\left(e^{-0.1(V_m+30)} - 1\right)$	$\beta_m = 20e^{-(V_m+55)/18}$
	$\alpha_h = 0.4e^{-(V_m+50)/20}$	$\beta_h = 6/\left(1+e^{-0.1(V_m+20)}\right)$
I_{Ca}	$I_{Ca} = \bar{g}_{Ca} c^3 (V_m - V_{Ca})$	
	$\dfrac{d[Ca^{2+}]_i}{dt} = -\left(\dfrac{3}{2Fr}I_{Ca}\right) - \dfrac{[Ca^{2+}]_i - 0.0001}{55}$	$V_{Ca} = \dfrac{RT}{2F}\ln\left(\dfrac{1.8}{[Ca^{2+}]_i}\right)$
	$\alpha_c = -0.15(V_m + 13)/\left(e^{-0.1(V_m+13)} - 1\right)$	$\beta_c = 10e^{-(V_m+38)/18}$
I_K	$I_K = \bar{g}_K n^4 (V_m + 68)$	
	$\alpha_n = -0.02(V_m + 40)/\left(e^{-0.1(V_m+40)} - 1\right)$	$\beta_n = 0.4e^{-(V_m+50)/80}$
I_{KA}	$I_{KA} = \bar{g}_{KA} A^3 h_A (V_m + 68)$	
	$\alpha_A = -0.003(V_m + 90)/\left(e^{-0.1(V_m+90)} - 1\right)$	$\beta_A = 0.1e^{-(V_m+30)/10}$
	$\beta_{hA} = 0.04e^{-(V_m+70)/20}$	$\beta_{hA} = 0.6/\left(1+e^{-0.1(V_m+40)}\right)$
I_{KCa}	$I_{KCa} = \bar{g}_{KCa}(V_m + 68)$	
	$g_{KCa} = \bar{g}_{KCa}\left[\left(\dfrac{[Ca^{2+}]_i}{0.001}\right)^2 \Big/ \left(1 + \left(\dfrac{[Ca^{2+}]_i}{0.001}\right)^2\right)\right]$	
I_h	$I_h = \bar{g}_{Na}\, y\, (V_m + 26.8)$	
	$y_\infty = 1/(1+e^{(V+75)/5.5})$	$\tau_h = 588.2\, e^{0.01(V_m+10)}/\left(1 + e^{0.2(V_m+10)}\right)$
I_{CaT}	$I_{CaT} = \bar{g}_{Na}\, m_T^3 h_T\, (V_m - V_{Ca})$	
	$\alpha_{mT} = 1/\left(1 + e^{-(V_m+28.8)/13.5}\right)$	$\beta_{mT} = \left(1 + e^{\frac{V_m+63}{7.8}}\right)/\left(1.7 + e^{\frac{V_m+28.8}{13.5}}\right)$
	$\alpha_{hT} = e^{-(V_m+160.3)/17.8}$	$\beta_{hT} = \alpha_{hT}\sqrt{0.25 + e^{\frac{V_m+83.5}{6.3}}} - 0.5$
	$\alpha_d = \left(1 + e^{\frac{V_m+37.4}{30}}\right)\Big/\left(240\left(0.5 + \sqrt{0.25 + e^{\frac{V_m+83.5}{6.3}}}\right)\right)$	$\beta_d = \alpha_d\sqrt{0.25 + e^{\frac{V_m+83.5}{6.3}}}$
I_L	$I_L = \bar{g}_L(V_m + 70.5)$	

Table 3 Ionic channel distributions

Channel	Regional maximum membrane conductances (mS/cm^2)				
	Soma	Axon	AIS	Hillock	Dendrites
ON					
I_{Na}	68.4	68.4	254.1	68.4	7.2
I_K	45.9	45.9	68.85	45.9	42.83
I_{KA}	18.9	–	18.9	18.9	13.86
I_{Ca}	1.6	–	1.6	1.6	2.133
I_{KCa}	0.0474	0.0474	0.0474	0.0474	–
I_h	0.0286	0.0286	0.0286	0.0286	0.0572
I_{CaT}	–	–	–	–	–
I_L	0.2590	0.2590	0.2590	0.2590	0.2590
OFF					
I_{Na}	45.6	45.6	165.9	45.6	4.818
I_K	45.9	45.9	68.85	45.9	42.83
I_{KA}	18.9	–	18.9	18.9	13.86
I_{Ca}	1.6	–	1.6	1.6	2.133
I_{KCa}	0.0474	0.0474	0.0474	0.0474	–
I_h	2.1	2.1	2.1	2.1	4.2
I_{CaT}	0.1983	0.1983	0.1983	0.1983	0.9915
I_L	0.0519	0.0519	0.0519	0.0519	0.0519

I_{Na}. sodium current, I_K. delayed rectifier potassium current, I_{KA}. A type potassium current, I_{Ca}. calcium current, I_{KCa}. Ca-activated potassium, I_h. Hyperpolarization-activated current, I_{CaT}. Low-threshold voltage-activated calcium current, I_L. leakage current

steps, and stimulus frequencies ranged from 1.0 to 10 kHz in 0.5-kHz steps. All pulse trains were 200 ms in duration.

For population-based simulations, ON and OFF layers were stimulated using a subretinally placed large diameter electrode ($\Phi = 200$ μm), positioned 200 μm above the RGC soma layer. The stimulus frequency was set to 10 kHz. The extracellular stimulus amplitude ranged from 10 to 350 μA in 10-μA steps. All pulse trains were reduced to 40 ms in duration. All elicited spikes were observed and counted at the soma. Differential activation was determined from the difference in averaged total spike numbers between one cell cluster and the other.

3 Results

3.1 Differential Activation of Individual ON and OFF RGCs Using a Large HFS Parameter Space

Figure 2A1-2 illustrates the stimulus-dependent (from 20 to 140 μA) ON and OFF RGC action potential spike counts (spikes/200 ms) in response to a large range of stimulation frequencies (from 1 to 10 kHz). The elicited spikes were observed and counted at the soma. Our model predicted that the elicited RGC spike counts were highly dependent on stimulating frequency and amplitude. The colors in Fig. 2A1–2 denote the number of evoked spikes from ON or OFF cells for a given stimulation frequency and pulse amplitude.

The total spike count of the ON cell (panel A1) reached a plateau as the stimulus current surpassed a certain threshold at frequencies up to 6.5 kHz. However, with frequencies higher than 7 kHz, the total spike number increased initially with stimulus amplitude, followed by a sudden decline with further amplitude increases, creating a non-monotonic surface in the frequency-amplitude topological space. In contrast, the OFF cell (panel A2) exhibited a non-monotonic profile at stimulation frequencies higher than 2 kHz. Nevertheless, both activation maps indicated a decreasing stimulation threshold trend with respect to HFS pulse trains with increasing stimulation frequency, where the threshold was defined as the stimulation amplitude capable of eliciting 10% of the maximal spike number of each non-monotonic spike-stimulus profile.

As shown in Fig. 2B, with increasing stimulation frequency, both ON and OFF RGCs exhibited an increased slope of the rising phase in spikes/µA (the epoch in which spike counts increased with increasing stimulation current) and, concomitantly, an earlier onset of the falling phase (in which the total spike numbers saturated or declined). Interestingly, the stimulus-dependent response of the ON cell became relatively stable only at stimulation frequencies higher than 9 kHz, while the OFF cell response tended to be unchanged already at frequencies higher than 5 kHz, thus indicating the ON/OFF-cell-specific frequency dependency.

The differential activation map shown in Fig. 2C provides an alternative visualization of differential activation of RGC types at each stimulation frequency and amplitude. Each grid point was defined as the difference of total spike number (spikes/200 ms) of ON and OFF cells. Our model suggested that differential activation of the ON RGCs was maximized at high stimulation amplitudes (>45 µA) and frequencies (between 3 and 10 kHz). In contrast, HFS pulse trains across all tested frequencies induced robust differential activation of OFF RGCs with different stimulation amplitudes ranging from 10 to 50 µA. Moreover, in Fig. 1C, the threshold at which differential activation began for both cell types gradually reduced as the stimulus frequency was increased from 1 to 6 kHz. The stimulation current range for preferentially activating ON RGCs increased when the stimulus frequency increased from 2 to 5 kHz, then gradually decreased when the stimulation frequency increased from 7 to 10 kHz. In contrast, the stimulation current range for preferentially activating OFF RGCs was mostly stable across all frequencies.

Based on the information provided by Fig. 2, potential optimal stimulation parameter combinations can be chosen to selectively excite ON and OFF cells. For example, with 7-kHz stimuli of 60 µA, the ON RGC was strongly activated, while simultaneously blocking the OFF RGC spontaneous spikes. In contrast, with 1-kHz stimuli of 30 µA, the OFF RGC was strongly activated, while the ON RGC remained silent.

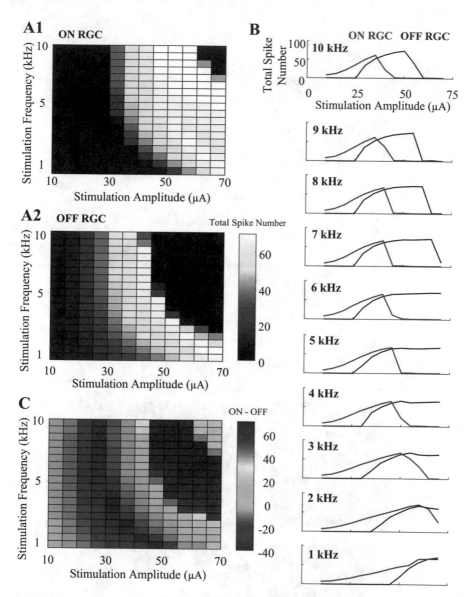

Fig. 2 Computational models of ON and OFF RGCs using an epiretinal hexapolar electrode array. (A1-A2). Activation maps showing the total spike number (spikes/200 ms) elicited in ON and OFF RGCs in response to a range of stimulation amplitudes (10–70 μA) and frequencies (1–10 kHz) delivered over a 200-ms interval. (**b**). Juxtaposition of the ON and OFF spike count against stimulating amplitude, at frequencies ranging from 1 to 10 kHz. (**c**). Differential RGC excitability map, defined as the difference of total spike count between ON and OFF RGCs, indicating stimulation parameters which can preferentially activate one cell type, while minimally activating the other type

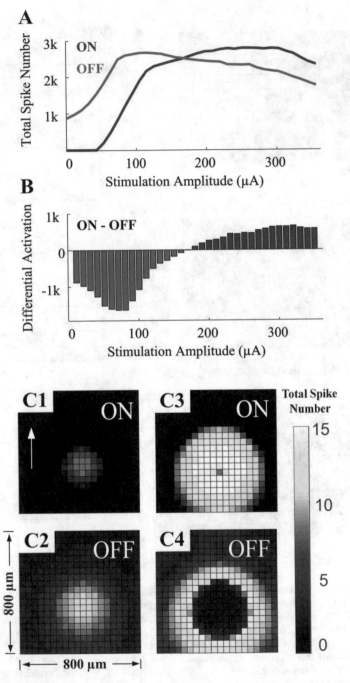

Fig. 3 In silico population-based *RGC* responses using 21 × 21 pairs of ON and OFF cells.
Each grid point represents the elicited total spike number (spikes/40 ms) of the cell at the given X-Y

3.2 Simulating Population-Based RGC Activity Under Clinically Relevant Conditions

Figure 3 shows the ability of the model to predict differential activation in a population of 21×21 pairs of ON and OFF cells using a subretinally placed large diameter electrode (200 μm), positioned at 200 μm above the RGC layers. Figure 3A demonstrates the total spike number (spikes/40 ms) elicited among the whole simulated ON and OFF populations. Our model suggested that 10-kHz HFS pulse trains were still able to induce differential activation of ON and OFF RGC populations with different stimulating amplitude parameters. Differential excitability shown in Fig. 3B was determined by the difference in the total spike numbers between the ON population and OFF population. Our model suggested that the activation of the ON RGC population was maximized at higher stimulation amplitudes (>200 μA). In contrast, excitation of OFF population is maximized at amplitudes ranging from 20 to 80 μA.

Excitation maps shown in Fig. 3C provide the differential stimulus dependency of ON and OFF populations during HFS. For example, when stimulated at a small stimulation amplitude (60 μA), a large OFF population (C2) was strongly activated, while only local ON RGCs (C1) located close to the electrode were weakly excited. However, when stimulation amplitude was gradually increased above a certain level (>120 μA), local OFF RGCs below the electrode started to be inhibited while a large population of ON RGCs were still being increasingly excited.

4 Discussion and Conclusion

The simulation results illustrated in this chapter suggest the possibility of translating recent laboratory advances in differential neural activation to large-scale, clinically relevant conditions by (1) relaxing the constraint requiring stimulation electrodes to be near the RGC cell bodies, which are impossible to locate under clinical conditions, and (2) translating the HFS-based differential activation from epiretinal to subretinal stimulation. Computational RGC models provide the ability to investigate neural modulation by changing key stimulation parameters. One advantage of the

Fig. 3 (continued) location. A stimulus electrode was located at the center. (**a**). Juxtaposition of the total elicited spike number recorded from all ON and OFF cells against stimulation amplitude. (**b**). Differential excitability is determined as the difference of the total spike numbers of ON and OFF populations. (**c**). Examples of activation maps of ON and OFF populations stimulated at different stimulus amplitudes. (**C1-C2**): When stimulated using a small amplitude (60 μA), a large OFF population was strongly activated, while only local ON RGCs were weakly excited. The white arrow indicates RGC axonal directions. (**C3-C4**): When stimulation amplitude was gradually increased to certain level (>120 μA), local OFF RGCs below the electrode started being inhibited while ON RGCs were still increasingly excited

computational approach is that the model-generated response space map can be made arbitrarily large and fine-grained for thorough exploration of stimulus parameters. This is difficult, if not impossible, to achieve through biological experiments due to the invasiveness of intracellular recordings. Moreover, simulation can be used to guide in vitro experimental design. For example, it is worthwhile to investigate population-based RGC responses to HFS predicted by our model using high-density multielectrode arrays [27] or using a calcium imaging technique [28].

Discrimination between ON and OFF RGCs with electrical stimulation is an initial step toward improving artificial vision. Until recently, retinal stimulation has not been able to provide differential activation of ON and OFF RGCs. Such co-activation is highly unnatural, providing conflicting information to higher visual centers, and potentially degrading the efficacy of retinal implants. This chapter, built on previous in vitro [5, 8, 23, 29] and in silico [6, 7, 26, 30] studies, demonstrated that preferential or differential activation of individual and population-based RGC types could be achieved. Here, we further showed that the effect was possible over a wide range of HFS parameters. In particular, the ON RGC could be targeted at relatively higher stimulation amplitudes and frequencies, while the OFF RGC could be targeted with lower stimulation amplitudes across all tested frequencies. The precise mechanism underlying differential RGC activation remains largely unknown. Further modeling and in vitro studies are still required to better understand the factors that shape the response of a retinal neuron to biphasic HFS. In particular, efforts should be devoted to assessing the contribution of intrinsic RGC properties including cell-specific ionic channel distributions in shaping RGC spiking profiles. To the best of our knowledge, no studies have identified the distribution of different ionic channel subtypes between the RGC types. Experimental studies in spinal sensory neurons reported that different sodium channel subtypes may respond differentially to high stimulus frequencies [31]. It is therefore likely that a similar frequency dependence is also present in RGCs. Further experiments based on variable ionic channel identifiers [32–34] will help us to better understand the reason for the unique stimulus dependency of each RGC type.

The HFS-based stimulation strategy described here may be useful for closely mimicking the natural encoding of RGC visual patterns. Specifically, the ON ganglion cells showed an increase in spike counts (spikes/200 ms) as the stimulus current was increased, while the OFF RGC responses were inhibited by the increased stimulus. In addition, our results suggest that differential activation of the ON RGC may be maximized within stimulation frequencies of 5–10 kHz, as shown in Fig. 2. However, it should be noted that higher frequencies can degrade stimulation efficacy [29, 35]. Therefore, a balance between current amplitude and HFS frequency may be necessary for a practical stimulation strategy.

In summary, the modelling approach can predict where the optimal stimulation parameter space is likely to be without detailed experimental investigations, providing insights into stimulation strategies that may contribute further to the development of retinal prostheses.

Acknowledgments This study was funded by the National Health and Medical Research Council (APP1087224), Retina Australia (RG181282), and SJTU-UNSW Collaborative Research Fund (RG173376). S. Bai is sponsored by the German Research Foundation (DFG) under the D-A-CH program (HE6713/2-1).

References

1. Rizzo, J. F., Wyatt, J., Loewenstein, J., Kelly, S., & Shire, D. (2003). Perceptual efficacy of electrical stimulation of human retina with a microelectrode array during short-term surgical trials. *Investigative Ophthalmology & Visual Science, 44*, 5362–5369.
2. Zrenner, E., Bartz-Schmidt, K. U., Benav, H., Besch, D., Bruckmann, A., Gabel, V. P., et al. (2011). Subretinal electronic chips allow blind patients to read letters and combine them to words. *Proceedings of the Royal Society B: Biological Sciences, 278*, 1489–1497.
3. Humayun, M. S., Weiland, J. D., Fujii, G. Y., Greenberg, R., Williamson, R., Little, J., et al. (2003). Visual perception in a blind subject with a chronic microelectronic retinal prosthesis. *Vision Research, 43*, 2573–2581.
4. Sinclair, N. C., Shivdasani, M. N., Perera, T., Gillespie, L. N., McDermott, H. J., Ayton, L. N., et al. (2016). The appearance of phosphenes elicited using a suprachoroidal retinal prosthesis. *Investigative Ophthalmology & Visual Science, 57*, 4948–4961.
5. Twyford, P., Cai, C., & Fried, S. (2014). Differential responses to high-frequency electrical stimulation in ON and OFF retinal ganglion cells. *Journal of Neural Engineering, 11*, 025001.
6. Kameneva, T., Maturana, M. I., Hadjinicolaou, A. E., Cloherty, S. L., Ibbotson, M. R., Grayden, D. B., et al. (2016). Retinal ganglion cells: Mechanisms underlying depolarization block and differential responses to high frequency electrical stimulation of ON and OFF cells. *Journal of Neural Engineering, 13*, 016017.
7. Guo, T., Lovell, N. H., Tsai, D., Twyford, P., Fried, S., Morley, J. W., et al. (2014). Selective activation of ON and OFF retinal ganglion cells to high-frequency electrical stimulation: A modeling study. In: Presented at the 36th annual international conference of the IEEE engineering in medicine and biology society, Chicago, US.
8. Cai, C., Twyford, P., & Fried, S. (2013). The response of retinal neurons to high-frequency stimulation. *Journal of Neural Engineering, 10*, 036009.
9. Vetter, P., Roth, A., & Hausser, M. (2001). Propagation of action potentials in dendrites depends on dendritic morphology. *Journal of Neurophysiology, 85*, 926–937.
10. Spruston, N. (2008). Pyramidal neurons: Dendritic structure and synaptic integration. *Nature Reviews. Neuroscience, 9*, 206–221.
11. Hines, M. L., & Carnevale, N. T. (1997). The NEURON simulation environment. *Neural Computation, 9*, 1179–1209.
12. Guo, T., Tsai, D., Morley, J. W., Suaning, G. J., Kameneva, T., Lovell, N. H., et al. (2016). Electrical activity of ON and OFF retinal ganglion cells: A modelling study. *Journal of Neural Engineering, 13*, 025005.
13. Bai, S. W., Guo, T. R., Tsai, D., Morley, J. W., Suaning, G. J., Lovell, N. H., et al. (2015). Influence of retinal ganglion cell morphology on neuronal response properties – A simulation study. In: Presented at the 2015 7th international IEEE/EMBS conference on neural engineering, Paris, France.
14. Guo, T., Tsai, D., Yang, C. Y., Al Abed, A., Twyford, P., Fried, S. I., et al. (2019). Mediating retinal ganglion cell spike rates using high-frequency electrical stimulation. *Frontiers in Neuroscience, 13*, 413.

15. Cuntz, H., Forstner, F., Borst, A., & Hausser, M. (2010). One rule to grow them all: A general theory of neuronal branching and its practical application. *PLOS Computational Biology, 6,* e1000877.

16. Prim, R. C. (1957). Shortest connection networks and some generalizations. *At&T Technical Journal, 36,* 1389–1401.

17. Ratliff, C. P., Borghuis, B. G., Kao, Y. H., Sterling, P., & Balasubramanian, V. (2010). Retina is structured to process an excess of darkness in natural scenes. *Proceedings of the National Academy of Sciences of the United States of America, 107,* 17368–17373.

18. Fohlmeister, J. F., & Miller, R. F. (1997). Impulse encoding mechanisms of ganglion cells in the tiger salamander retina. *Journal of Neurophysiology, 78,* 1935–1947.

19. Henderson, D., & Miller, R. F. (2003). Evidence for low-voltage-activated (LVA) calcium currents in the dendrites of tiger salamander retinal ganglion cells. *Visual Neuroscience, 20,* 141–152.

20. Margolis, D. J., Gartland, A. J., Euler, T., & Detwiler, P. B. (2010). Dendritic calcium signaling in ON and OFF mouse retinal ganglion cells. *The Journal of Neuroscience, 30,* 7127–7138.

21. Tabata, T., & Ishida, A. T. (1996). Transient and sustained depolarization of retinal ganglion cells by Ih. *Journal of Neurophysiology, 75,* 1932–1943.

22. Wang, X. J., Rinzel, J., & Rogawski, M. A. (1991). A model of the T-type calcium current and the low-threshold spike in thalamic neurons. *Journal of Neurophysiology, 66,* 839–850.

23. Guo, T., Yang, C. Y., Tsai, D., Muralidharan, M., Suaning, G. J., Morley, J. W., et al. (2018). Closed-loop efficient searching of optimal electrical stimulation parameters for preferential excitation of retinal ganglion cells. *Frontiers in Neuroscience, 12,* 168.

24. Tsai, D., Chen, S., Protti, D. A., Morley, J. W., Suaning, G. J., & Lovell, N. H. (2012). Responses of retinal ganglion cells to extracellular electrical stimulation, from single cell to population: Model-based analysis. *PLoS One, 7,* e53357.

25. Mueller, J. K., & Grill, W. M. (2013). Model-based analysis of multiple electrode array stimulation for epiretinal visual prostheses. *Journal of Neural Engineering, 10,* 036002.

26. Guo, T., Lovell, N. H., Tsai, D., Twyford, P., Fried, S., Morley, J. W., et al. (2015). Optimizing retinal ganglion cell responses to high-frequency electrical stimulation strategies for preferential neuronal excitation. In: Presented at the 2015 7th international IEEE/EMBS conference on neural engineering, Paris, France.

27. Tsai, D., Sawyer, D., Bradd, A., Yuste, R., & Shepard, K. L. (2017). A very large-scale microelectrode array for cellular-resolution electrophysiology. *Nature Communications, 8,* 1802.

28. Weitz, A. C., Nanduri, D., Behrend, M. R., Gonzalez-Calle, A., Greenberg, R. J., Humayun, M. S., et al. (2015). Improving the spatial resolution of epiretinal implants by increasing stimulus pulse duration. *Science Translational Medicine, 7,* 318–203.

29. Cai, C. S., Ren, Q. S., Desai, N. J., Rizzo, J. F., & Fried, S. I. (2011). Response variability to high rates of electric stimulation in retinal ganglion cells. *Journal of Neurophysiology, 106,* 153–162.

30. Guo, T., Barriga-Rivera, A., Suaning, G. J., Tsai, D., Dokos, S., Morley, J. W., et al. (2017). Mimicking natural neural encoding through retinal electrostimulation. In: Presented at the 2017 8th international international IEEE/EMBS conference on neural engineering, Shanghai, China.

31. Rush, A. M., Dib-Hajj, S. D., & Waxman, S. G. (2005). Electrophysiological properties of two axonal sodium channels, Na(v)1.2 and Nav1.6, expressed in mouse spinal sensory neurones. *Journal of Physiology-London, 564,* 803–815.

32. Van Wart, A., Trimmer, J. S., & Matthews, G. (2007). Polarized distribution of ion channels within microdomains of the axon initial segment. *The Journal of Comparative Neurology, 500,* 339–352.

33. Fjell, J., Dib-Hajj, S., Fried, K., Black, J., & Waxman, S. (1997). Differential expression of sodium channel genes in retinal ganglion cells. *Brain Research. Molecular Brain Research, 50,* 197–204.

34. Boiko, T., Van Wart, A., Caldwell, J. H., Levinson, S. R., Trimmer, J. S., & Matthews, G. (2003). Functional specialization of the axon initial segment by isoform-specific sodium channel targeting. *Journal of Neuroscience, 23*, 2306–2313.
35. Hadjinicolaou, A. E., Savage, C. O., Apollo, N. V., Garrett, D. J., Cloherty, S. L., Ibbotson, M. R., et al. (2015). Optimizing the electrical stimulation of retinal ganglion cells. *IEEE Transactions on Neural Systems and Rehabilitation Engineering, 23*, 169–178.

Functional Requirements of Small- and Large-Scale Neural Circuitry Connectome Models

Kristen W. Carlson, Jay L. Shils, Longzhi Mei, and Jeffrey E. Arle

1   Introduction

We have truly entered the Age of the Connectome due to a confluence of advanced imaging tools, methods such as the flavors of functional connectivity analysis and interspecies connectivity comparisons, and computational power to simulate neural circuitry [1–8]. The interest in connectomes is reflected in the exponentially rising number of articles on the subject (Fig. 1). What are our goals? What are the "functional requirements" of connectome modelers? We give a perspective on these questions from our group whose focus is modeling neurological disorders, such as neuropathic back pain, epilepsy, Parkinson's disease, and age-related cognitive decline, and treating them with neuromodulation.

K. W. Carlson (✉) · L. Mei
Department of Neurosurgery, Beth Israel Deaconess Medical Center, Boston, MA, USA
e-mail: kwcarlso@bidmc.harvard.edu

J. L. Shils
Department of Anesthesiology, Rush Medical Center, Chicago, IL, USA

J. E. Arle
Department of Neurosurgery, Beth Israel Deaconess Medical Center, Boston, MA, USA

Department of Neurosurgery, Harvard Medical School, Boston, MA, USA

Department of Neurosurgery, Mount Auburn Hospital, Cambridge, MA, USA

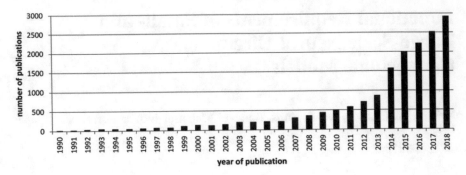

Fig. 1 Number of publications on network neuroscience per year between 1990 and 2018. Reprinted from Douw et al. [1]

2 Goals and Means

2.1 Electroceuticals and Neuromodulation

The ultimate goal of electroceuticals and neuromodulation is to use electromagnetic fields to modify any component of the central and peripheral nervous system in a predictable way to restore or enhance its normal functionality. Neuromodulation focuses more on restoring functionality from a diseased state, while electroceuticals' emphasis is using electromagnetic fields to replicate pharmaceutical effects and thereby provide a far less expensive and time-consuming path than the drug development and regulatory route via an alternative medical device route cutting cost and time as much as 90%.

In the past decade, as numerical modeling and simulation have become more sophisticated, their role in reducing research and development time and expense has become increasingly clear and valuable [9, 10]. The Food and Drug Administration and the American Society of Mechanical Engineers have led the way to formalize the simulation paradigm so that its results and their level of validity are satisfactorily transparent [11, 12].

2.2 Benefits of Numerical Modeling

The benefits of numerical modeling begin with the ability to predict electromagnetic field distributions and the resulting forces and energy imparted by the field to targeted and non-targeted structures. Modeling provides significant advantages over empirical studies particularly under the following conditions, which apply pervasively to the nervous system:

1. Where it is difficult or impossible to measure values empirically
2. Where preclinical studies are expensive or impossible
3. In inhomogeneous, anisotropic materials
4. Where material parameters are imprecisely established

Further, modelers can perform extensive "what if" explorations in large parameter spaces either interactively or in an automated batch mode. Using standards such as those mentioned above, modelers can benefit by exchanging their models to compare results or to use another's model as a starting point for new explorations.

2.3 The Role of Simple Versus Complex Models

Freddie Hansen of Abbott Laboratories built over 300 simulation models in 6 years – how did he do it? For most of his models, Hansen uses the COMSOL finite element modeling (FEM) software as a "FEM pocket calculator" with which he builds quick-and-dirty FEMs in a few hours to a few days [13]. These models are designed to answer a single, simple question. Roughly defined parameters and large error bars can be addressed with parameter sweeps across the relevant parameters to outline possible range of responses and determine sensitivities.

In the field, such simple models are often called "sub-models," i.e., they are a study of a component from a larger model. The nervous system – even in simpler model organisms than humans – is so complex that sub-modeling will be an important technique for calibration, validation, and dissecting functionality of connectomes.

More complex and sophisticated models, such as Hansen's principal model of a heart pump, take weeks and months to build, calibrate, and validate. Generally, much greater emphasis is placed on validation in complex vs. simple models, and far greater time is required to beat the desired calibration and validation behavior out of the model. Hence, calibration of sub-models within connectomes can be an important route to efficiency and model control.

An example of a connectome sub-model is the H-reflex, which is the relatively simple spinal cord circuit triggered when the doctor hits your knee with a rubber mallet. The connectivity of the monosynaptic H-reflex is well-known (Fig. 2), which is not generally the case with more complex circuits, hence the need to start with what is known and get that calibrated before venturing into unknown territory. Calibration of the H-reflex involves balancing the connection strength between the Ia excitatory sensory fiber, Renshaw cell (RC) inhibitory feedback loop, and alpha motor neuron circuit such that this mini-circuit replicates its reported ~50 ms refractory time (Fig. 3).

While one or two such mini-calibrations may suffice for a simple model, many may be necessary in a complex model in order to impose sufficient constraints to validate the model. The growing sophistication of new top-down modeling methods is greatly improving the calibration/validation process (see Sect. 2.5).

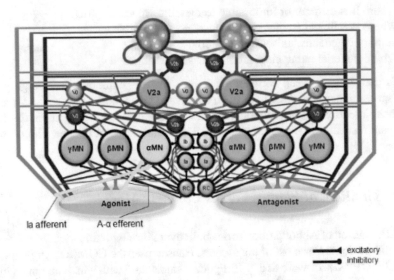

Fig. 2 Schematic of the Hoffmann reflex (H-reflex), a monosynaptic relay from afferent sensory fiber Ia to an alpha motor neuron (αMN) and back to the muscle via a large A-α efferent fiber. The refractory time of the circuit is principally mediated by a Renshaw cell (RC) clock

2.4 Ockham's Razor Drives All Modeling

"What can be accounted for by fewer assumptions is explained in vain by more" was a principle frequently invoked by the influential medieval Scholastic thinker, William of Ockham (1285–1349) [14]. Similarly, in modeling, a guiding principle is to make the model no more complex than is required to capture the desired phenomena. In modeling the human brain and spinal cord, one has little choice but to invoke this principle frequently, because the systems are so complex and intertwined. For example, one cannot model the entire brain to capture a given disorder, such as movement disorder due to loss of dopaminergic neurons in the *substantia nigra pars compacta* in Parkinson's disease, where the central basal ganglia loop sends and receives signals from external centers. In such a model, vast regions of, e.g., cortex and thalamus may be represented as single groups, or single excitatory and inhibitory pairs [15].

2.5 Capturing the Required Level of Detail

Similar to the principle of Ockham's razor, modelers must make a decision about the level of detail they wish to capture and the developmental and computational cost required to capture the target level [16, 17].

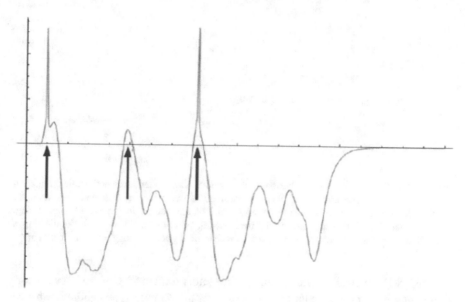

Fig. 3 Recorded transmembrane potential in a neural circuitry model calibration of the H-reflex spinal motor circuit to replicate its ~50 ms circuit refractory time. *x-axis:* time in ms. *y-axis:* transmembrane potential in mV. Action potentials (AP) are initiated at 12, 47, and 82 ms (red arrows) in Ia sensory fibers, triggering the reflex. The timing of alpha motor neuron cell response is modulated by a double-Renshaw cell (RC) inhibitory "clock" (Fig. 1). With proper calibration of axonal delays and connection strengths, the alpha motor neuron fires in response to the Ia fiber at 12 ms and 82 ms APs but, inhibited by the RC clock, does not fire at 47 ms, evidencing the refractory period

Biology is fundamentally a multisystems level discipline, and accordingly, modelers must decide at which systems level to focus [18–20]. Concomitantly, though, the systems level approach gives modelers some flexibility with regard to "axiomatizing" or "black-boxing" elements at the underlying systems level to the one at hand, thereby simplifying the model and rendering its behavior more understandable. This approach goes back as far as von Neumann's earliest thoughts on how to model the nervous system [21].

2.6 Which Neural Circuitry Software?

Once a decision is at least tentatively made on the systems level and extent of detail, a modeling tool is selected that is designed to capture the target level of detail [22].

By way of example, our connectome models have used the following tools, each designed to model a different neural systems level and to be computationally efficient for the required tasks:

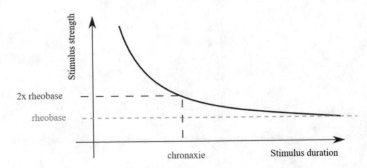

Fig. 4 Rheobase, the minimum threshold of a nerve fiber under sustained stimulation, and chronaxie, the time constant of the fiber standardized at twice the rheobase stimulus strength, together determine the fiber threshold under stimuli of any length in passive fiber modeling. Rheobase may be relative to a given empirical model, or absolute, measured by replicating a model in finite element software and measuring electric field gradient along a virtual fiber [26]

1. UNCuS (Universal Neural Circuitry Simulator) written in C++ and Java incorporating electrotonic dendritic compartments and neuron type calibration based on 12 parameters [6, 16, 23, 24]
2. Active nerve fiber cable models in C++ and Java incorporating ion channel gating at 0.25 ms timesteps [25]
3. Simpler passive fiber models based on relative threshold or an absolute version of the Weiss equation using the second finite difference of electric field potential along the fiber as predicted in a finite element model (Fig. 4) [26, 27]
4. Finite element models in COMSOL Multiphysics™ of electromagnetic neuromodulation devices and their generation of electric potential, current density, etc., in heterogeneous biological tissue [26, 27]
5. A special-purpose numerical model written in *Mathematica* (WRI, Champaign, IL, USA) of fiber threshold changes due to cathodic and anodic stimulation phases to elucidate a hypothesis on traditional low-frequency, "burst," and high-frequency spinal cord stimulation [28, 29]

In recent years, neural circuitry software is often categorized into three different model paradigms, all using coupled differential equations. The kind of considerations used to select which paradigm is appropriate for modeling the basal ganglia in Parkinson's disease, for instance, is described by Rubin [30]:

1. Activity-based or firing-rate models
2. Integrate-and-fire models
3. Conductance-based models

An older tool, NEURON, has often been used for small neural circuits as well as individual neuron and axon behavior and offers the advantage of transparency – for those who can march up the learning curve required to learn its interface [31].

However, as the need to model larger connectomes and longer time series is growing rapidly, newer methods are being developed [32–36]. Other driving forces of high efficiency model paradigms are "biomimetic" or "bio-inspired" systems and the prospect of interfacing neural prostheses with biological neural circuits [37–40]. Such paradigms seek to marry software and hardware requirements ("algorithm-hardware co-design") to produce ultimate computational efficiency [41, 42]. These newer approaches are likely to obsolete the current methods within a decade.

2.7 Initial Conditions

There are no firm guidelines for setting up the initial conditions for a neural circuit, which remains more of an art than a science. One can start with randomized connectivity and connection strength, for instance, the absence of any knowledge of specific connectivity and connection strength, i.e., a high entropy configuration, and impose calibrations that constrain the circuit, reducing its entropy, until it sends its "message," i.e., behaves according to a desired specification.

An example of a "middle ground" initial conditions approach would be specifying coarse-grained connectivity that is known from the literature and using heuristics to ensure some reasonable performance or more likely avoid unreasonable performance. One such heuristic is setting connection strength from inhibitory to excitatory groups and excitatory to inhibitory groups incrementally stronger than from excitatory to excitatory groups and from inhibitory to inhibitory groups, such that the circuit does not initially slip into hyperactivity. A similar heuristic is lowering the ratio of excitatory to inhibitory connections such that stability is initially generated. A third technique is imposing a hypothesized exogenous "black box" inhibitory input to stabilize the circuit. Any highly recurrent circuit topology requires stabilizing, negative-feedback heuristics of this genre [6]. We used the first and third techniques in our human spinal cord connectome, in which the exogenous black box represented known but unquantified descending inhibitory control from higher brain centers.

Our human spinal cord connectome model entailed an ultimate attempt to perform a bottom-up calibration by culling from the literature all topology including source peripheral fibers or neural groups, their targets (specific Rexed laminae, i.e., the gray matter of the spinal cord), connection strength, and neurotransmitters [6]. The end result was that the overall model was under-constrained due to the dearth of complete data, in particular on connection strengths. Our conclusion was twofold. First, the model was valuable as a starting template for calibrating the innumerable local circuit models that embody the spinal cord connectome. Second, while advances in imaging techniques foreshadowed the possibility of complete bottom-up calibration at some point in the future, for practical purposes, top-down calibration, e.g., by replicating circuit behavior (input-output specifications) would be required in connectome modeling.

2.8 Calibration and Validation

The general calibration/validation procedure is to cull a set of calibrations for a given circuit from the literature and add calibrations one by one until the model predicts one or more of the remaining calibrations with reasonable accuracy, i.e., is validated against the remaining empirical criteria. Calibration can be done iteratively by hand, which can be painstakingly slow and tedious, but quite educational as to how neural circuitry behaves in general, how the specific software implementing the circuitry behaves, and how the specific circuit and its components affect each other.

On the other hand, calibration of a neural circuit can be performed automatically by an optimization program using one or more objective functions and error tolerance in the objective criteria as terminating conditions of a "while" loop. Such automated batch processing may be the only expeditious way to program evenly a moderately sized circuit.

As stated, rather than using bottom-up calibration, the field is moving toward top-down calibration utilizing a variety of empirical techniques (e.g., imaging techniques, electro- and magnetoencephalography), which, in conjunction, constrain models more than what has been possible to date, and notably in diseased state modeling [43, 44]. The lower-level connectivity, connection strength, variety of individual neural processing, and axonal delays are reflected in the emergent high-level behavior [45, 46].

For example, to uncover connectivity and connection strength, fiber-tracing studies and identifying where on the dendritic tree an axon synapsed have been replaced with diffusion-weighted magnetic resonance image (MRI) and blood-oxygen-level-dependent (BOLD) data from which the "functional" connectivity of the network is inferred [47, 48]. In a rapidly evolving paradigm, variations on this technique are utilized with older methods to offer the modeler a menu of possible calibration/validation datasets (Tables 1, 2, 3, and 4). Disease "signatures" (biomarkers) in the new paradigms can be used to calibrate models of diseased vs. healthy states and measure the effects of neuromodulation, electroceutical, or pharmaceutical techniques to restore the healthy state [24, 49–52].

Table 1 Evolving methods of top-down calibration [43]

Resting-state functional connectivity network behavior
Dynamic functional connectivity network behavior
Resting-state fMRI oscillations
Brain rhythm relationships (e.g., inverse α-rhythms)
Excitation-inhibition balance
Spike-firing patterns and fMRI on short- and long-time scales
fMRI power-law scaling
fMRI functional magnetic resonance imaging

Table 2 Type of modeling software used for the relevant neural systems level and time scale

Model software	Time scale	Typical application
Molecular dynamics	$10^{-12}-10^{-9}$	Fine-grained ion channel dynamics
Active fiber	$10^{-6}-10^{0}$	Nerve fiber activation and blocking thresholds detailed at µs/ms ion-channel level
Passive fiber	$10^{-3}-10^{3}$	Fast, black-box nerve fiber activation and blocking thresholds
Neural circuitry[a]	$10^{1}-10^{3+}$	Resting state and dynamic functional connectivity of normal, disease, and neuromodulated circuits

[a]For more details on neural circuitry software types, see Sect. 2.6, "Which Neural Circuitry Software?

Table 3 Modeling software functional requirements for efficient circuit assembly

Batch processing
Parameter sweep
Automatic calibration via optimization criteria:
While$[f_{objective}[u]$ = false, loop $(parameter_1, \ldots parameter_n)]$
Import user-specific waveform
Import user-specific voltage-current (current-voltage) curve for neuron types
Import calibration-validation connectivity data

Table 4 Idealized requirements for connectome models designed to inform medical device and drug development

Peripheral fiber groups involved, their receptor type, diameter, conduction speed, numbers, and activation and blocking thresholds
Source and target neuron groups, their neurotransmitters and receptors, known connectivity, and axonal delays
Resting state, dynamic state, and their variational methods of measuring functional connectivity in healthy and diseased circuits
Tissue parameters: geometry, conductivity, permittivity, and permeability
Medical device parameters: geometrical electrode array and waveform characteristics including anodic, cathodic, and rest phase pulse width and shape, frequency, and duty cycle
Drug effects on neural system targets such as peripheral fiber types and central neural groups

3 The Functional Requirements

At a high level, the following are required to efficiently make valuable connectome models:

1. User-friendly modeling software for the correct systems level
2. Standardized parts list (axon and neuron types)

3. Assembly instructions (topography and weights)
4. Methods of calibration, validation, and their associated datasets to use in those processes

4 Conclusion

The "connectome" has come of age and is now in a similar stage as the genome was 15 years ago. The great advances and benefits loom ahead. Yet the functional requirements to build connectomes are now known. Connectome simulation is fast and cheap compared to hardware prototyping, or in vitro and in vivo investigation of how the neural system works. Connectome modeling of neuromodulation and electroceutical action on the nervous system will lead to an explosion of targets and means for greater control over disease and disorder treatment and enhanced neural behavior.

Acknowledgments We are grateful for what we have learned from the following experts in simulation and modeling: Freddie Hansen (Abbott Laboratories), Socrates Dokos (Dept. of Biomedical Engineering, UNSW), Nirmal Paudel, Thomas Dreeben (Osram, Inc.), Jack Tuszynski (Dept. of Physics, University of Edmonton), Ze'ev Bomzon (Novocure Ltd.). Funded in part by the Sydney Family Foundation.

References

1. Douw, L., et al. (2019). The road ahead in clinical network neuroscience. *Network Neuroscience, 3*, 969–993.
2. Ardesch, D. J., et al. (2019). The human connectome from an evolutionary perspective. *Progress in Brain Research, 250*, 129–151.
3. Ito, T., et al. (2019). Discovering the computational relevance of brain network organization. *Trends in Cognitive Sciences.* https://doi.org/10.1016/j.tics.2019.10.005.
4. Ito, T., et al. (2017). Cognitive task information is transferred between brain regions via resting-state network topology. *Nature Communications, 8*, 1027.
5. Rilling, J. K., & van den Heuvel, M. P. (2018). Comparative primate connectomics. *Brain, Behavior and Evolution, 91*, 170–179.
6. Arle, J. E., et al. (2018). Dynamic computational model of the human spinal cord connectome. *Neural Computation,* 1–29. https://doi.org/10.1162/neco_a_01159.
7. Van Essen, D. C., et al. (2013). The WU-Minn human connectome project: An overview. *NeuroImage, 80*, 62–79.
8. Kuan, L., et al. (2015). Neuroinformatics of the Allen mouse brain connectivity atlas. *Methods, 73*, 4–17.
9. Glotzer, S. C., et al. (2009). *International assessment of research and development in simulation-based engineering and Science.* Baltimore: World Technology Evaluation Center, Inc.
10. Forster, M. (2019). Multiphysics simulation illuminates the future. In *IEEE spectrum: Multiphysics modeling.* Piscataway: IEEE.
11. U.S. FDA. (2016). *Reporting on computational modeling studies in medical device submissions.* Rockville: U. S. Food and Drug Administration.

12. ASME. (2018). *Assessing credibility of computational modeling through verification and validation: Application to medical devices.* New York: American Society of Mechanical Engineers.
13. Hansen, F. (2018). *COMSOL Inc. Designing improved heart pumps with simulation.* https://www.comsol.com/blogs/keynote-video-designing-improved-heart-pumps-with-simulation/. Accessed 20 Dec 2018.
14. Moody, E. A. (1974). Ockham. In C. C. Gillispie (Ed.), *Dictionary of scientific biography* (Vol. X, pp. 171–175). New York: Charles Scribner's Sons.
15. Humphries, M. D., et al. (2018). Insights into Parkinson's disease from computational models of the basal ganglia. *Journal of Neurology, Neurosurgery, and Psychiatry, 89,* 1181–1188.
16. Arle, J. E., et al. (2008). Modeling parkinsonian circuitry and the DBS electrode. I. Biophysical background and software. *Stereotactic and Functional Neurosurgery, 86,* 1–15.
17. Igarashi, J., et al. (2011). Real-time simulation of a spiking neural network model of the basal ganglia circuitry using general purpose computing on graphics processing units. *Neural Networks, 24,* 950–960.
18. Olivier, B. G., et al. (2016). Modeling and simulation tools: From systems biology to systems medicine. *Methods in Molecular Biology, 1386,* 441–463.
19. Tretter, F. (2010). *Systems biology in psychiatric research: From high-throughput data to mathematical modeling.* Weinheim: Wiley-VCH.
20. Wellstead, P. (2012). *Systems biology of Parkinson's disease.* New York: Springer.
21. von Neumann, J. (1956). The general and logical theory of automata. In J. R. Newman (Ed.), *The world of mathematics* (Vol. 4, pp. 2070–2098). New York: John Wiley & Sons.
22. Herz, A. V., et al. (2006). Modeling single-neuron dynamics and computations: A balance of detail and abstraction. *Science, 314,* 80–85.
23. Arle, J. E. (1992). *Neural modeling of the cochlear nucleus (PhD Thesis),* University of Connecticut.
24. Arle, J. E., & Carlson, K. W. (2016). The use of dynamic computational models of neural circuitry to streamline new drug development. *Drug Discovery Today: Disease Models, 19,* 69–75.
25. Arle, J. E., et al. (2014). Mechanism of dorsal column stimulation to treat neuropathic but not nociceptive pain: Analysis with a computational model. *Neuromodulation, 17,* 642–655.
26. Arle, J. E., et al. (2014). Modeling effects of scar on patterns of dorsal column stimulation. *Neuromodulation, 17,* 320–333.
27. Arle, J. E., et al. (2016). Investigation of mechanisms of vagus nerve stimulation for seizure using finite element modeling. *Epilepsy Research, 126,* 109–118.
28. Arle, J. E., et al. (2019 (in press)). Neuromodulation. https://doi.org/10.1111/ner.13076
29. Breakspear, M., et al. (2003). Modulation of excitatory synaptic coupling facilitates synchronization and complex dynamics in a biophysical model of neuronal dynamics. *Network, 14,* 703–732.
30. Rubin, J. E. (2017). Computational models of basal ganglia dysfunction: The dynamics is in the details. *Current Opinion in Neurobiology, 46,* 127–135.
31. Hines, M. L., & Carnevale, N. T. (2001). NEURON: A tool for neuroscientists. *The Neuroscientist, 7,* 123–135.
32. Olin-Ammentorp, W., & Cady, N. (2019). Biologically-inspired neuromorphic computing. *Science Progress, 102,* 261–276.
33. Thakur, C. S., et al. (2018). Large-scale neuromorphic spiking array processors: A quest to mimic the brain. *Frontiers in Neuroscience, 12,* 891.
34. Rozenberg, M. J., et al. (2019). An ultra-compact leaky-integrate-and-fire model for building spiking neural networks. *Scientific Reports, 9,* 11123.
35. Breslin, C., & O'Lenskie, A. (2001). Neuromorphic hardware databases for exploring structure-function relationships in the brain. *Philosophical Transactions of the Royal Society of London. Series B, Biological Sciences, 356,* 1249–1258.

36. Sanz-Leon, P., et al. (2015). Mathematical framework for large-scale brain network modeling in the virtual brain. *NeuroImage, 111*, 385–430.
37. Lee, Y., & Lee, T. W. (2019). Organic synapses for neuromorphic electronics: From brain-inspired computing to sensorimotor Nervetronics. *Accounts of Chemical Research, 52*, 964–974.
38. Park, H. L., et al. (2019). Flexible neuromorphic electronics for computing, soft robotics, and neuroprosthetics. *Advanced Materials*, e1903558. https://doi.org/10.1002/adma.201903558.
39. Tang, J., et al. (2019). Bridging biological and artificial neural networks with emerging neuromorphic devices: Fundamentals, progress, and challenges. *Advanced Materials, 31*, e1902761.
40. Chiolerio, A., et al. (2017). Coupling resistive switching devices with neurons: State of the art and perspectives. *Frontiers in Neuroscience, 11*, 70.
41. Roy, K., et al. (2019). Towards spike-based machine intelligence with neuromorphic computing. *Nature, 575*, 607–617.
42. Neftci, E. O. (2018). Data and power efficient intelligence with neuromorphic learning machines. *iScience, 5*, 52–68.
43. Schirner, M., et al. (2018). Inferring multi-scale neural mechanisms with brain network modelling. *eLife, 7*, e28927.
44. Garbarino, S., et al. (2019). Differences in topological progression profile among neurodegenerative diseases from imaging data. *eLife, 8*, e49298.
45. Beim Graben, P., et al. (2019). Metastable resting state brain dynamics. *Frontiers in Computational Neuroscience, 13*, 62.
46. Noble, S., et al. (2019). A decade of test-retest reliability of functional connectivity: A systematic review and meta-analysis. *NeuroImage, 203*, 116157.
47. Smitha, K. A., et al. (2017). Resting state fMRI: A review on methods in resting state connectivity analysis and resting state networks. *The Neuroradiology Journal, 30*, 305–317.
48. Chen, L. M., et al. (2017). Biophysical and neural basis of resting state functional connectivity: Evidence from non-human primates. *Magnetic Resonance Imaging, 39*, 71–81.
49. Hughes, L. E., et al. (2019). Biomagnetic biomarkers for dementia: A pilot multicentre study with a recommended methodological framework for magnetoencephalography. *Alzheimer's & Dementia (Amst), 11*, 450–462.
50. Lopez-Sanz, D., et al. (2019). Electrophysiological brain signatures for the classification of subjective cognitive decline: Towards an individual detection in the preclinical stages of dementia. *Alzheimer's Research & Therapy, 11*, 49.
51. Zhang, C., et al. (2019). Dynamic alterations of spontaneous neural activity in Parkinson's disease: A resting-state fMRI study. *Frontiers in Neurology, 10*, 1052.
52. Siekmeier, P. J., & vanMaanen, D. P. (2013). Development of antipsychotic medications with novel mechanisms of action based on computational modeling of hippocampal neuropathology. *PLoS One, 8*, e58607.

Part VI
High-Frequency and Radiofrequency Modeling

Simplifying the Numerical Human Model with *k*-means Clustering Method

Kyoko Fujimoto, Leonardo M. Angelone, Sunder S. Rajan,
and Maria Ida Iacono

1 Introduction

Computational modeling is widely used to assure patient safety with respect to radio-frequency (RF) related concerns during magnetic resonance imaging (MRI). It allows for evaluation of RF power absorption and specific absorption rate (SAR) in anatomically detailed numerical human models. Such evaluations are especially important for safety of patients with implantable devices.

RF-induced heating depends on the physical characteristics of the patient [1], and as such, it is expected that numerical human models are sufficiently detailed to be able to estimate the differences across the patient popupation; however, the time-consuming model generation process prevents achieving a realistic safety evaluation.

A limited number of anatomically realistic numerical human models are available for research and development use. Currently available whole-body numerical models have 26–77 anatomical structures [2–5]. A previous study showed that three different dielectric properties (muscle, fat, and lung) were sufficient to estimate SAR with a 5-mm resolution [6]; yet there are implantable devices with dimensions of less than 5 mm that can be implanted in very thin anatomical structures such as arteries and veins. Thus, a more detailed model may be needed for robust RF-induced heating evaluation. Moreover, a study using a higher resolution model showed that the blood vessel SAR can be up to ten times higher than the maximum standard gel phantom SAR value [7]. Knowing the limits of simplification in a numerical model can help not only to reduce the time needed for segmentation for

K. Fujimoto (✉) · L. M. Angelone · S. S. Rajan · M. I. Iacono
Center for Devices and Radiological Health, US Food and Drug Administration, Silver Spring, MD, USA
e-mail: kyoko.fujimoto@fda.hhs.gov

© The Author(s) 2021
S. N. Makarov et al. (eds.), *Brain and Human Body Modeling 2020*,
https://doi.org/10.1007/978-3-030-45623-8_15

model generation but also to fabricate a standard phantom which can accurately reflect the in vivo result.

In this study, we used the k-means clustering method, one of the commonly used vector quantitization methods for cluster analysis, to reduce the number of anatomical structure types with different dielectric properties in a detailed human model. Then we investigated the resulting differences in SAR with respect to the number of clusters. The simplified models were then used to simulate a test senario for RF-exposure by calculating background tangential electric field (E_{tan}) along five stent trajectories in selected arteries.

2 Methods

2.1 *k-means Clustering*

The k-means clustering was applied on the electrical conductivity and permittivity of anatomical structures of the AustinMan model [5] with 61 different anatomical structures in MATLAB (The MathWorks Inc., Natick, MA, United States). Ten model variations with different dielectric property configurations were used in the simulations: one full model with the original 61 anatomical structures with 51 dielectric properties and nine models with varying number of dielectric property clusters ($k = 33, 30, 27, 24, 21, 18, 15, 12,$ and 9). The total sum of distances was used as a distance measure, and the model with 33 clusters was chosen as a starting point because the distance was less than 0.5 for numbers of clusters above 34. The example of clustering for $k = 9$ is shown in Fig. 1(a).

2.2 *Computational Modeling Setups*

The computational modeling setups were implemented using the commercially available finite-difference time-domain platform Sim4Life (Zurich Med Tech, Switzerland). A 32-port 16-rung birdcage coil, 700 mm in length and 650 mm in diameter, with idealized excitation was modeled. All the current carrying coil structures were modeled as perfect electric conductors (PEC). Electromagnetic simulations were performed by feeding the coil with a continuous sinusoidal wave at 128 MHz.

The Huygens' approach [8] was applied to facilitate a fair comparison between full and simplified models. The incident field was calculated with an unloaded coil first, then used to compute electromagnetic fields within body models, with 1-mm isotropic grid. The modeling software could estimate electomagnetic fields with a coarser grid, especially with the clustered models, which would reduce computational burdens. However, to facilitate fair comparisons between the full model and

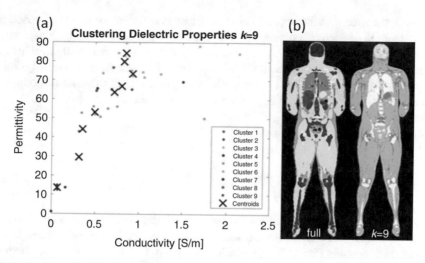

Fig. 1 (**a**) Example of the *k*-means clustering based on the sum of absolute differences. Different colors represent different clusters. Calculated centroid values were used in the simulation for anatomical structures within clusters. (**b**) The coronal view of the segmented model with full properties and nine properties is shown

the clustered models, the same resolution of the AustinMan model was used to discretize the models in this study. All the models were simulated at the hip bone imaging landmark.

2.3 SAR Calculation

The single-voxel SAR (SAR_{raw}), 1 g-averaged SAR (SAR_{1g}), and 10 g-averaged SAR (SAR_{10g}) results were compared by calculating the mean and maximum percentage difference between the full model and the clustered models. Voxel-wise comparison of each pair was performed by linear regression of the SAR values. All SAR values were computed with original mass density values of the model and normalized to a whole-body averaged SAR equal to 2 W/kg [9]. All the analyses were performed in MATLAB.

2.4 Electric Field Tangential to Stents in Blood Vessels

To simulate a test case of RF heating assessment, stent trajectories were chosen in the five locations in the arteries of AustinMan as described in Fujimoto et al. [10]. Five case studies were analyzed for the ascending aorta, the brachial, the femoral, the iliac, and the popliteal arteries. Stent trajectories were created based on the centerline of each blood vessel that was calculated in MATLAB by binarizing a selected vessel and determining the centroid of the consecutive axial slices of the model. The centerline was then imported into Sim4Life to create a smooth trajectory. The E_{tan} value was calculated along each trajectory using the IMSAFE module in Sim4Life. The magnitudes of E_{tan} values were calculated offline for each number of clusters.

3 Results

The values of dielectric properties for each k-clustered model were determined by the k centroids. The example of clustering plot and coronal slices of full and $k = 9$ models are shown in Fig. 1. The centroids as shown in Fig. 1(a) are distributed across the range of the original permittivity values from 1 (air) and 90 (kidney) and the original conductivity values from 0 (air) to 2.1 (cerebrospinal fluid). For example, with these centroids, skin, muscle, diaphragm, and liver became one cluster in the $k = 9$ clustered model. The maximum intensity projection of each SAR map showed that the SAR_{1g} and SAR_{10g} maps were qualitatively similar among different models regardless of numbers of dielectric properties used (Fig. 2).

The mean and maximum percentage difference (Table 1) revealed that there were up to 15.3% mean difference in SAR. The example cross-sectional (transverse slice) SAR maps are shown in Fig. 3. The 12-clustered models estimated higher SAR_{raw} values compared to the full model on the skin, whereas the values for SAR_{1g} and SAR_{10g} were similar. This trend was observed when compared between the rest of the simplified models and the full model. The SAR values from the full model plotted against the SAR values from each clustered model revealed that all the clustered models showed high correlations with the full model (Fig. 4). The E_{tan} values calculated in selected stent paths were similar among the full and clustered models (Fig. 5).

4 Discussion

The clustered analysis showed that reducing the number of dielectric properties from 51 (original) to 30 has less than 0.2% effect on the mean SAR results. Further reduction in the number of dielectric properties was not linearly correlated with mean and maximum SAR differences. The greatest mean SAR difference was 15.3% for $k = 18$. Each voxel pair between the full and the 30-clustered SAR values was highly correlated. Our results suggest not only reducing the segmentation time on

Fig. 2 Maximum intensity projection of SAR_{1g} and SAR_{10g} from the simulations with the models with full, 33-clustered, and nine-clustered properties. No subjective differences were observed among the three SAR_{1g} and SAR_{10g} maps

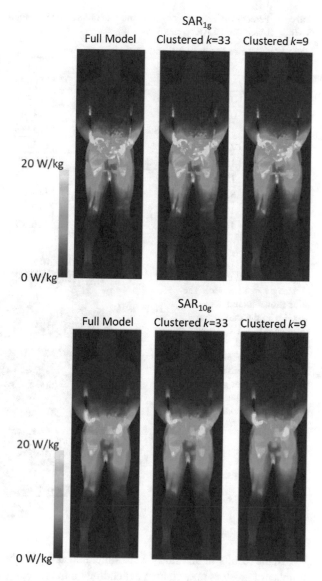

generating models but also using existing models with less anatomical structures which result in reduction in computational time.

The E_{tan} results (Fig. 5) showed that they were not linearly correlated with the number of clusters. In other words, as shown with the ascending aorta, the iliac, and the femoral artery trajectories, the models with small numbers of clusters can estimate the E_{tan} results as the full model simulation does.

Table 1 Percent differences for mean and maximum values computed between full and clustered SAR_{raw}, SAR_{1g}, and SAR_{10g} maps

Percent difference between full and clustered SAR maps						
	SAR_{raw}		SAR_{1g}		SAR_{10g}	
# Clusters	Mean	Maximum	Mean	Maximum	Mean	Maximum
33	0.0%	54.9%	0.0%	114.8%	0.0%	152.7%
30	0.2%	56.9%	0.2%	114.8%	0.1%	152.7%
27	3.0%	177.8%	2.8%	115.0%	2.3%	152.6%
24	2.7%	53.8%	2.5%	115.1%	2.1%	152.6%
21	15.1%	151.8%	11.7%	117.4%	9.4%	156.3%
18	15.3%	168.9%	11.8%	116.5%	9.3%	156.6%
15	8.1%	171.8%	6.3%	115.4%	4.8%	152.6%
12	12.0%	183.7%	10.0%	118.0%	8.0%	153.9%
9	10.4%	183.5%	8.4%	116.6%	6.7%	153.6%

Fig. 3 Cross-sectional SAR maps for the full model and clustered $k = 12$ model. The slice shown includes the voxel with the maximum difference in SAR_{raw} maps between full and 12-clustered models (see Table 1). The maximum difference resides on the skin of AustinMan where the arm and torso are in contact

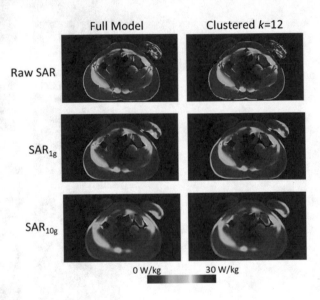

All the results in this study were simulated at the hip bone landmark. The optimal number of clusters may change depending on the imaging landmarks as electric field distribution varies depending on the exposed mass. Another limitation of this study was that only one set of dimensions of RF coil and one field strength were studied. As a previous study showed [11], both can affect the field distribution and the resulting optimal number of clusters.

Our k-means clustering approach was only based on dielectric properties. Incorporating the location of the dielectric property may help improve the clustered models. Different approaches such as the Gaussian hidden Markov random field models may be able to help the generation of clustered model based on not only the dielectric properties but also spatial constraints based on neighboring voxels.

Fig. 4 Scatter plots between full and clustered SAR_{1g} and SAR_{10g} values. The linear fit is shown with the lines and R^2 values. All of the slope values showed close to 1. Therefore, there was a one-to-one correlation between each pair. The 33-, 30-, and 15-clustered models showed especially high correlation

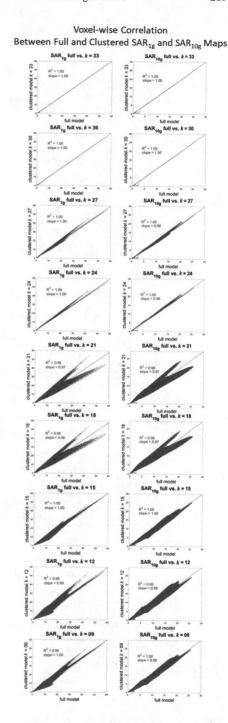

Voxel-wise Correlation
Between Full and Clustered SAR_{1g} and SAR_{10g} Maps

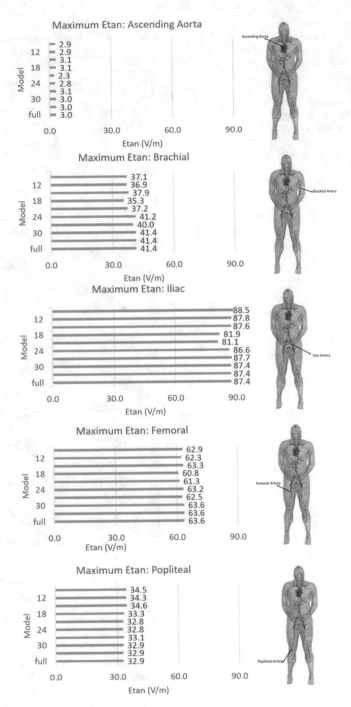

Fig. 5 E$_{tan}$ on selected blood vessel paths where short stents are commonly implanted are calculated

5 Conclusion

Simplified numerical models based on dielectric properties can show equivalent SAR result. Further investigation of the clustering method may enable efficient MRI safety assessment by simplifying the model generation and reducing computational time.

Acknowledgments This work was supported by the FDA Office of Women's Health and the Research Participation Program at the Center for Devices and Radiological Health administered by the Oak Ridge Institute for Science and Education through an interagency agreement between the US Department of Energy and the US Food and Drug Administration. The authors would like to thank Tayeb A. Zaidi, Trent V. Robertson, Drs. Brian B. Beard, David A. Soltysik, and Eriko S. Yoshimaru for their helpful comments and discussion.

Disclaimer The mention of commercial products, their sources, or their use in connection with material reported herein is not to be construed as either an actual or implied endorsement of such products by the Department of Health and Human Services.

References

1. Murbach, M., Cabot, E., Neufeld, E., et al. (2010). Local SAR enhancements in anatomically correct children and adult models as a function of position within 1.5 T MR body coil. *Progress in Biophysics and Molecular Biology, 107*(3), 428–433.
2. Christ, A., Kainz, W., Hahn, E. G., et al. (2010). The virtual family—Development of surface-based anatomical models of two adults and two children for dosimetric simulations. *Physics in Medicine and Biology, 55*, N23–N38.
3. Nagaoka, T., Watanabe, S., Sakurai, K., et al. (2004). Development of realistic high-resolution whole-body voxel models of Japanese adult males and females of average height and weight, and application of models to radio-frequency electromagnetic-field dosimetry. *Physics in Medicine and Biology, 49*, 1–15.
4. Visible Human Project.: https://www.nlm.nih.gov/research/visible/visible_human.html.
5. Massey, J. W., & Yilmaz, A. E. (2016). AustinMan and AustinWoman: High-Fidelity, anatomical voxel models developed from the VHP color images. *Conference Proceedings: Annual International Conference of the IEEE Engineering in Medicine and Biology Society*, 3346–3349.
6. Homann, H., Börnert, P., Eggers, H., et al. (2011). Toward individualized SAR models and in vivo validation. *Magnetic Resonance in Medicine, 66*(6), 1767–1776.
7. Fujimoto, K., Angelone, L. M., & Lucano, E. et al. (2017) Conf. Proc. BMES/FDA.
8. Benkler, S., Chavannes, N., & Kuster, N. (2009). Novel FDTD Huygens source enables highly complex simulation scenarios on ordinary PCs. *Conf Proc IEEE APSURSI, 2009*, 1–4.
9. IEC 60601-2-33. (2010). *Magnetic resonance equipment for medical diagnosis testing.*

10. Fujimoto, K., Angelone, M. L., Lucano, E., et al. (2018). Radio-frequency safety assessment of stents in blood vessels during magnetic resonance imaging. *Frontiers in Physiology, 9*, 1439.
11. Fujimoto, K, Angelone, L. M., & Lucano, E., et al. (2017). Effect of simulation settings on local specific absorption rate (SAR) in different anatomical structures. In Proceedings of ISMRM Workshop on Ensuring RF Safety in MRI: Current Practices & Future Directions.

Using Anatomical Human Body Model for FEM SAR Simulation of a 3T MRI System

Alexander Prokop, Tilmann Wittig, and Abhay Morey

1 Introduction

Specific absorption rate (SAR), the dissipated power per tissue mass, is used to quantify the human exposure to electromagnetic fields in frequency ranges between 100 kHz and 6 GHz. To approximate the temperature rise distribution, it is usually averaged over masses of 1 g (SAR-1g) or 10 g. The averaging procedure to be used after computational determination of the electromagnetic fields using the finite difference time domain (FDTD) method or finite integration theory (FIT) in rectilinear hexahedral grid meshes has been defined in the IEC/IEEE 62704-1 standard [1].

However, with the finite element method (FEM), electromagnetic fields are usually calculated on unstructured, i.e. tetrahedral or curved element meshes, for which the 62704-1 averaging procedure is not directly applicable. Therefore, in the IEC/IEEE 62704-4 standard [2], an iterative sampling procedure has been defined. Its output can be used with the 62704-1 averaging procedure.

In the following, we show how this sampling procedure, which was developed for the standard application to wireless communication devices, is also applicable to other field distributions from devices like magnetic resonance imaging (MRI) systems for which SAR is also an essential quantity to evaluate patient safety.

A. Prokop (✉) · T. Wittig
Dassault Systèmes SIMULIA, Darmstadt, Germany
e-mail: alexander.prokop@3ds.com

A. Morey
Dassault Systèmes SIMULIA, Pune, India

© The Author(s) 2021
S. N. Makarov et al. (eds.), *Brain and Human Body Modeling 2020*,
https://doi.org/10.1007/978-3-030-45623-8_16

273

2 Simulation Setup

A generic MRI RF coil for 3 Tesla resonating at 128 MHz is simulated in full 3D together with an anatomically correct human body model (HBM), the Female Visible Human [3, 4]; tissue properties are based on the typical Gabriel [5] parameters with background tissue modelled as fat (Fig. 1). Coil capacitors are represented by ports in the 3D model, so the coil can easily be tuned by an EM-circuit co-simulation in post-processing. During the full simulation run, a relatively coarse mesh is sufficient, as no details of the HBM need to be resolved. The resulting fields are recorded on a Huygens box enclosing the HBM.

As a next step, the obtained fields are used as an equivalent field source (EFS) for a simulation where only the human body model is contained. Agreement of the results with the full simulation have been verified as presented in [6]. The EFS has been applied for both FIT hexahedral time domain simulation with a 2-mm mesh step and a FEM frequency domain simulation with a mesh of about 1 million tetrahedra.

SAR-1g has been calculated for FIT on the 2-mm discretization mesh and with different sampling steps for FEM, scaled to an accepted power of 1 W.

As an initial sampling step, 62704-4 [2] suggests to stay below the cubic root of the quotient of the averaging mass divided by the maximum occurring mass density, in our case 2000 kg/m^3 for bone resulting in a maximum initial mesh step of 7.94 mm. We choose an initial mesh step of 5 mm.

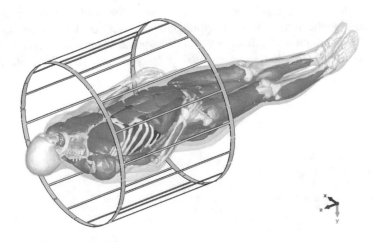

Fig. 1 MRI coil setup with HBM

3   Results

The following figures show the SAR-1g distributions in coronal cross sections. With 5-mm sampling, the maximum SAR-1g is 0.264 W/kg. Figure 2a shows the plane with the maximum; see the red spot between stomach and intestine. With 2-mm sampling, maximum SAR-1 g is 0.37 W/kg, but at a different location. Figure 2b

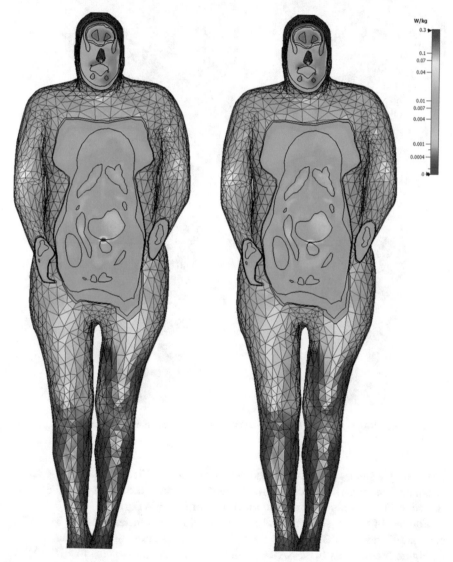

Fig. 2 (**a**) left: SAR-1g with 5-mm sampling at y = −65; (**b**) right: SAR-1g with 2-mm sampling at y = −65

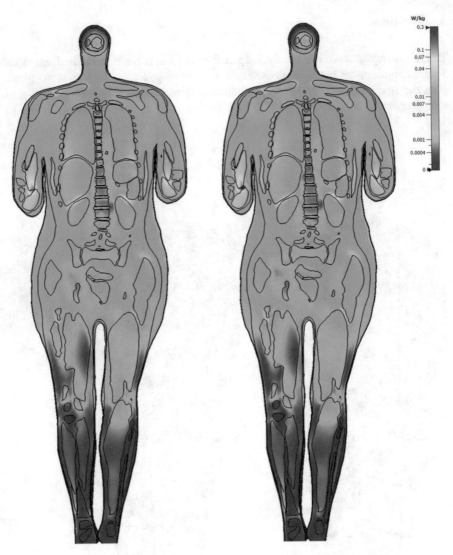

Fig. 3 (**a**) left: SAR-1g with 5-mm sampling at y = 63; (**b**) right: SAR-1g with 2-mm sampling at y = 63

shows only a small rise at the stomach/intestine spot. In Figs. 3a and 3b, we see the actual maximum SAR-1g position in the right arm, where the 5-mm sampling results shown in Fig. 3a only have a local maximum of 0.123 W/kg.

Figures 4a and 4b show a good agreement of the 1-mm sampling (0.341 W/kg) and the FIT results (0.338 W/kg) for maximum SAR-1g in value and location.

Figure 5 compares FEM SAR-1g with different sampling steps to FIT SAR-1g along the line shown dashed in Fig. 6.

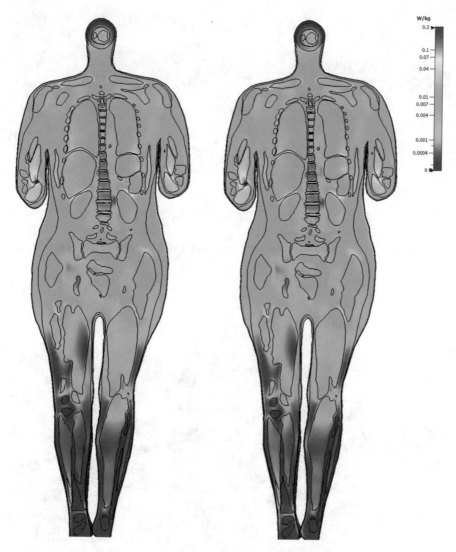

Fig. 4 (**a**) left: SAR-1g with 1-mm sampling at y = 63; (**b**) right: SAR-1g from FIT according to 62704-1 at y = 63

As in Fig. 2a, the 5-mm sampling shows a peak around z = 680 mm, which does not exist in the finer sampling and the FIT results. There are some minor deviations of the 1-mm and 2-mm sampling results from FIT with reasonable relative error or in regions where SAR is very low.

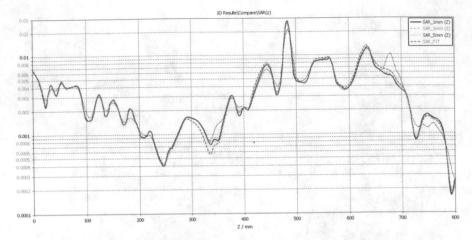

Fig. 5 FEM SAR-1g for different sampling steps with FIT SAR-1 g compared along a line through $x = 30, y = 50$

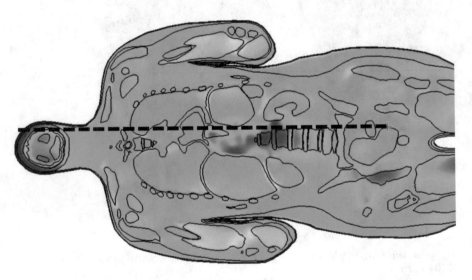

Fig. 6 Dashed line used for comparison in plane $y = 50$

4 SAR Profiles

Figures 7, 8, and 9 show the maximum SAR-1g per plane for each coordinate direction, i.e., Fig. 7 shows the maximum SAR-1g in each sagittal plane at coordinates x from −250 to 250 (right to left body part), clearly showing the maximum in

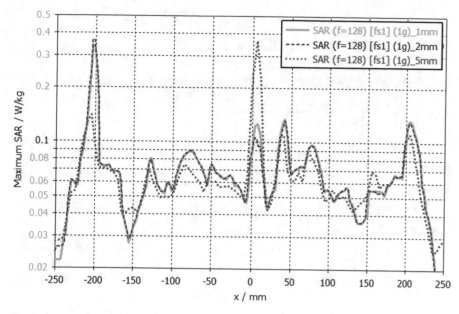

Fig. 7 SAR profile showing the plane-wise maximum along x

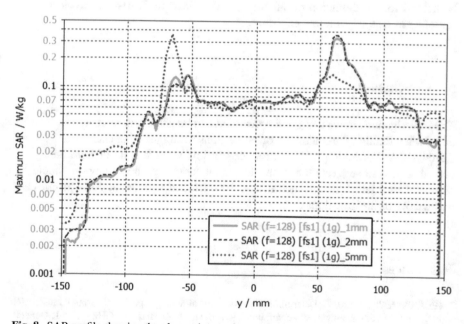

Fig. 8 SAR profile showing the plane-wise maximum along y

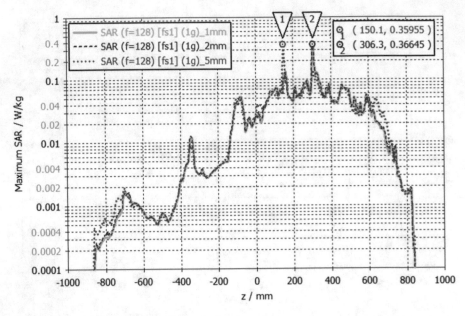

Fig. 9 SAR profile showing the plane-wise maximum along z

the right arm at $x = -200$. Such representations can be called SAR profiles and can be helpful for the definition of subvolumes described in 62704-4 to avoid refined sampling in regions with low exposure.

5   Conclusion

The investigation shows that good agreement can be achieved, but care needs to be taken when limiting the evaluation on subregions as suggested in clause 6.2.2.2 e) of 62704-4 [2]. The SAR profiles introduced in this work can be of help here. The investigated case suggests that from the initial sampling, the subregion should be chosen at least large enough such that in the excluded regions the SAR profiles are below 10% of the global maximum.

References

1. IEC/IEEE 62704-1. (2017). Determining the peak spatial-average specific absorption rate (SAR) in the human body from wireless communications devices, 30 MHz to 6 GHz – Part 1: General requirements for using the finite-difference time-domain (FDTD) method for SAR calculations. www.iec.ch

2. IEC/IEEE 62704-4. (2020). Determining the peak spatial-average specific absorption rate (SAR) in the human body from wireless communications devices, 30 MHz to 6 GHz – Part 4: General requirements for using the finite-element method for SAR calculations. www.iec.ch
3. Yanamadala, et al. (2016). Multi-purpose VHP-female version 3.0 cross-platform computational human model. In EuCAP16, Davos, Switzerland, April 2016, pp. 1–5.
4. Massey, J. W., Prokop, A., & Yılmaz, A. E. (2017). A comparison of two anatomical body models derived from the female Visible Human Project data. In: EMBC'17, Jeju Island, Korea, July 2017.
5. Andreuccetti, D., Fossi, R., & Petrucci, C. (1997). An Internet resource for the calculation of the dielectric properties of body tissues in the frequency range 10 Hz–100 GHz. In IFAC-CNR, Florence (Italy). Based on data published by C. Gabriel et al. in 1996. [Online]. Available: http://niremf.ifac.cnr.it/tissprop/
6. Prokop, A., Wittig, T., & Levine, S. (2018). Efficient computational investigation of implant RF safety with anatomical human models in MRI systems. In IEEE EMBC, Honululu, USA, July 2018.

RF-Induced Unintended Stimulation for Implantable Medical Devices in MRI

James E. Brown, Rui Qiang, Paul J. Stadnik, Larry J. Stotts, and Jeffrey A. Von Arx

1 Introduction

As the preferred imaging modality for soft tissue imaging, there has been significant interest in recent years to provide access to magnetic resonance imaging (MRI) to patients with active implantable medical devices (AIMDs). Historically, these patients have been denied access to this important diagnostic tool due to several potentially hazardous interactions with AIMDs [1]. Examples of AIMDs include pacemakers, cochlear implants, implantable glucose monitors, spinal cord stimulators, and deep brain stimulators.

AIMD manufacturers, in cooperation with MRI manufacturers, regulatory bodies, and academia, have developed a technical specification which outlines test methods for assessing the risk of several hazards [2]. In test methods, hazards are separated according to the field component (static, gradient, or radio frequency (RF)), and conservative test conditions are derived to stress the device in a laboratory setting beyond what is possible in the complex MR environment.

Among the potential hazards of MRI for patients with active implantable medical device is RF-induced rectified voltage. RF energy incident on the device may be rectified by internal active components and cause unintended stimulation of tissue near the device electrodes. In order to assess the risk to the patient, device manufacturers use computational human models to quantify the incident RF on the device and perform benchtop testing to determine the likelihood of unintended stimulation.

For a cardiac implantable electronic device (CIED) such as a pacemaker, if this rectified voltage exceeds the patient's physiological threshold, it could induce unintended cardiac stimulation (UCS) leading to tachycardia. A standard test for leaded CIEDs is presently being developed to quantify the risk of UCS for these

J. E. Brown (✉) · R. Qiang · P. J. Stadnik · L. J. Stotts · J. A. Von Arx
Micro Systems Engineering, Inc., Lake Oswego, OR, USA
e-mail: james.brown@biotronik.com

© The Author(s) 2021
S. N. Makarov et al. (eds.), *Brain and Human Body Modeling 2020*,
https://doi.org/10.1007/978-3-030-45623-8_17

devices [3]. The safety assessment for RF-induced UCS utilizes computational human models (CHMs).

In this work, the general process of using CHMs for the assessment of RF-induced unintended stimulation is discussed in the next section. Then, the process is applied to a pacemaker as an example in the following section for the specific hazard of UCS. Next, the impact of the model development on the results of this analysis is discussed. Finally, the overall impact of this work and areas which should be considered for future work are presented.

2 Evaluation of RF-Induced Unintended Stimulation Using Computational Human Models

There are many advantages of using CHMs for risk-based safety assessments for AIMD in MRI [4, 5]. The RF-induced energy incident on a leaded AIMD is typically calculated through the use of the well-known transfer function method [6, 7] as

$$V_{DUT} = A \int_0^L S(\tau) \cdot E_{tan}(\tau) d\tau.$$

Here, V_{DUT} is the voltage at the AIMD under test, S is the transfer function, and E_{tan} is portion of the incident electric field which is tangent to the lead pathway. These lead models are combined with RF field distributions within CHMs derived through electromagnetic simulation to conservatively estimate the induced RF level.

The variability of the predicted induced RF level for leaded devices is quite large due to the presence of resonant phenomena. These resonance effects have been extensively studied in the literature [8–11], usually in terms of RF-induced heating, but the same principle applies to the RF level induced at the implantable pulse generator (IPG). RF heating is evaluated at the distal end of the lead, while RF-induced energy is quantified at the proximal end, but otherwise the phenomena are very much related. Both may be influenced differently by the terminating impedance at the IPG [12].

2.1 3-D Field Distribution Within the CHM

A library of CHMs spanning the population in terms of height and BMI in different body positions, MR coils, landmark positions is used to study the distribution of expected electromagnetic fields along the lead pathway. The overall procedure is shown in the flowchart of Fig. 1.

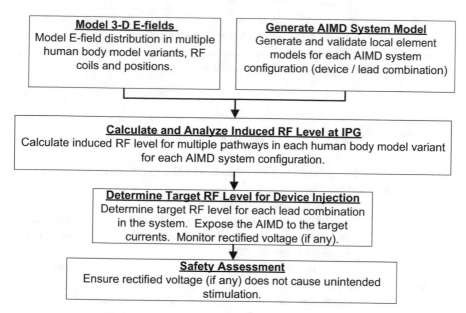

Fig. 1 Flow chart illustrating workflow for the assessment of protection from harm to the patient caused by RF-induced unintended stimulation

2.2 AIMD System Model

Lead models are developed numerically or experimentally [13–17], in one or more tissue-simulating media (TSM). The homogenous TSM should be chosen to accurately compute the induced RF level once the transfer function is applied in the human body (via CHM). The variability of computed RF level with the TSM used during lead model development is discussed in Sect. 4.

The geometric accuracy of the CHM along the lead trajectory, including the continuity of organs through which the leads are placed, is paramount to the accuracy of the model. Variations in critical parameters such as surrounding anatomy and tissue parameters must be included in the set of simulations used to generate the worst-case predicted RF-induced energy.

2.3 Assessment of AIMD Response

A risk assessment is then performed to determine the set of target exposure values for a benchtop RF injection test. Due to the nonlinear nature of rectified voltage, the test must be performed using the large-signal RF level (i.e., results cannot be scaled). Observed rectified voltage can be compared with a patient's expected physiological response, which is dependent on the specific tissues that are near the electrodes.

As an example of the evaluation of this hazard for a specific implantable device, this procedure is detailed for a pacemaker system in the next section. Analogous methods may be applied to extend the analysis to other types of AIMDs, including spinal cord stimulators and deep brain stimulators.

3 Approach Applied to a Pacemaker System

In order to apply the procedure from Sect. 2 to CIEDs, the incident fields in the CHMs are extracted for clinically relevant pathways. An example orientation for a dual-lead pacemaker system is shown in Fig. 2.

Additionally, lead models must be developed for all lead combinations which are to be evaluated. The induced RF level for multiple lead systems has been observed to be higher than for single lead systems [3].

A probability distribution of induced RF levels is created using each lead model and set of extracted fields from the library of CHMs. Additionally, if the device includes an RF antenna, the induced voltage on the antenna must be quantified [18, 19]. Then, benchtop testing is performed to expose the pacemaker to a suitable range of target RF levels so as to appropriately characterize the risk to the patient. Any observed rectified voltage can then be compared with the statistical likelihood that the waveform will cause cardiac stimulation, for example, through the strength duration curve [19]. In [3], the criterion is that the probability of unintended cardiac stimulation, $P(UCS)$, shall be less than 1 in 10,000. This probability is assessed by computing the integral

Fig. 2 Example orientation of a dual-chamber pacemaker in the human body

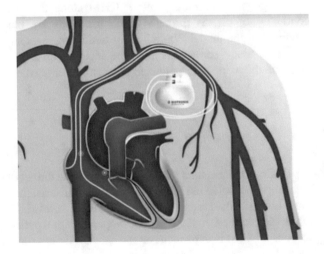

Fig. 3 Computing the probability of unintended stimulation

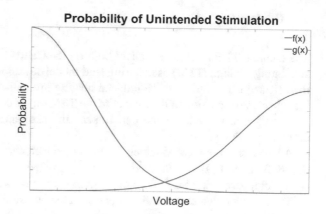

Fig. 4 The strength-duration curve

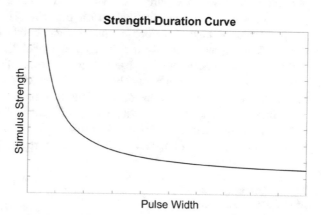

$$P(UCS) = \int\limits_0^\infty f(x) \int\limits_0^x g(y) dy \, dx.$$

Here, $f(x)$ is the probability that a given rectified voltage is induced during the MRI, and $g(x)$ is the probability that a given voltage is the minimum required to stimulate the patient's cardiac tissue. This is illustrated in Fig. 3, where the y-axis is probability and the x-axis is voltage. The voltage is the voltage monitored across a tissue-simulating resistor and thus is directly proportional to current. Current is a surrogate for charge, which is the true figure of merit for evaluating the probability of stimulation.

The pulse width of the rectified voltage is important to assess the probability of unintended stimulation. The well-known strength duration curve [20], shown in Fig. 4, can be used to determine the dependence of the analysis on rectified pulse width.

4   Variability of Induced RF Level

The computed RF level is dependent both on the CHMs being used and the tissue-simulating medium (TSM) used during lead model development. This topic has been investigated extensively for RF-induced heating in the literature [21–25]. Here, we include some discussion of the impact of the TSM on the computation of the induced RF level, which would in the end impact the assessment of the probability of unintended stimulation.

A basic structure approximating an implanted medical device lead is described in [2] as SAIMD-1. Here, the lead length is changed to 45 cm in order to better represent a pacemaker lead. The device geometry is shown in Fig. 5.

The incorporation of the lead model uses a method similar to [7], where a voltage is introduced between the wire and device ground, and the transfer function is then proportional to the current distribution on the wire. The surrounding medium is swept through a range of dielectric constants and conductivities. For example, the transfer function magnitude for 0.2 S/m, 1.2 S/m, and 2.2 S/m are given in Fig. 6.

From these plots, it can be observed that a lower dielectric constant of TSM pushes the lead below resonance. The resonant behavior is more pronounced for the higher conductivity of TSM due to wavelength compression effects.

Naturally, these lead models will lead to different predictions for the induced RF level in the patient and thus varied risk of unintended stimulation. The lead model is inserted into the VHP-Female v3.0 [26] along a representative pathway, and the

Fig. 5 Cross section of the insulated wire. The wire is 0.8 mm diameter with insulation of 0.5 mm wall thickness

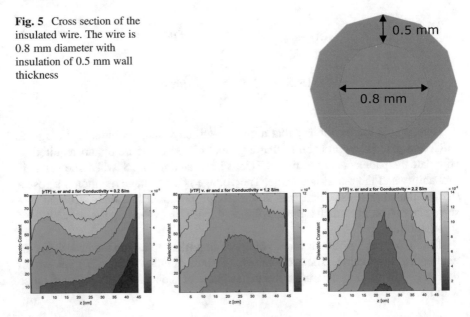

Fig. 6 Transfer function magnitude (v. length) versus dielectric constant of the TSM for three selected TSM conductivities

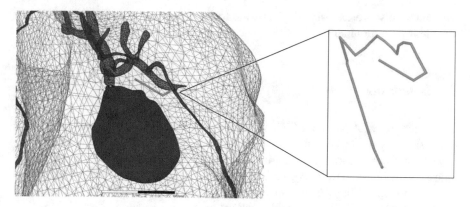

Fig. 7 Representative cardiac pathway within VHP-female v3.0 (only selected anatomical features are visible)

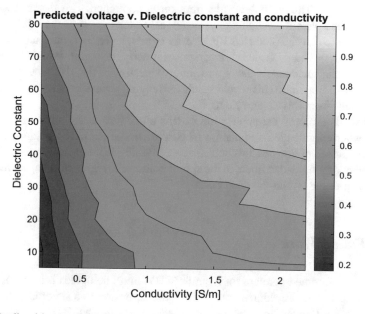

Fig. 8 Predicted in vivo MRI RF-induced voltage as a function of TSM dielectric constant and conductivity

incident field values are extracted. The pathway and its insertion into the model are shown in Fig. 7.

For this pathway, the RF-induced energy at the device is calculated for each of the transfer functions with different TSM. The normalized voltages are shown in Fig. 8. Notably, the higher the dielectric constant and the conductivity, the higher the predicted induced RF level which must be used for device testing. Care should be taken that an appropriate TSM is chosen such that the device test is conservative without wildly overexposing the hardware to RF levels beyond what is expected in the clinical environment.

5 Discussion

Manufacturers of AIMDs use CHMs in order to protect the patient from harm due to the risk of RF-induced unintended stimulation. The use of CHMs enables the investigation of millions of scenarios of scan parameters, patient sizes and anatomies, and MR system technologies. Therefore, CHMs allow AIMD manufacturers to quantify low probability events such as (for cardiac implantable electronic devices) UCS that represent a high risk to the patient.

The predicted RF-induced energy incident on the device is used for benchtop testing in order to convert a probability of an incident RF level to a probability of rectifed voltage. This probability distribution is combined with the probability of stimulating the patient through a physiological model.

It is notable that the predicted RF level is model dependent; thus, an appropriate TSM should be used. A TSM that shows overly resonant lead behavior will overpredict the incident RF level and thus overestimate the risk to the patient of unintended stimulation. This would result in denying access to MRI to patients that could benefit from this diagnostic tool.

This work explored the application of this procedure to the example of a pacemaker, where unintended stimulation of cardiac tissue could result in tachycardia. For other classes of devices, this could mean anything from muscular discomfort, to recruitment of unintended fibers in the spinal column, to significant impact related to stimulating tissue within the deep brain.

5.1 Future Work

The results presented here are for a single CHM and one device orientation within the MRI field. The procedures are valid for the extension to a set of CHMs and for many devices, enabling the safety assessment of AIMDs by identifying low-probability worst-case-based assessments.

Future work in this area could allow for co-simulation of the incident RF field and circuit model of internal device hardware, especially active components that could rectify the incident RF level. One option for this is to combine the circuit model with a Tier 4 model [2] of the realistic medical device lead geometry in the patient, with the MR coil. Additionally, computational methods have been shown to be effective in evaluating a range of shimming conditions [27, 28].

Further, the incorporation of a physiological model into the electromagnetic CHM could eliminate the need for clinical data to assess the relationship between rectified waveforms and the probability of stimulating tissue. An additional benefit of this technique would be the evaluation of waveforms not often used in the clinical setting, including cross-channel rectification which may not mimic a therapy waveform.

References

1. Kalin, R., & Stanton, M. S. (Apr 2005). Current clinical issues for MRI scanning of pacemaker and defibrillator patients. *Pacing and Clinical Electrophysiology, 28*(4), 326–328.
2. ISO/TS 10974:2018 (E). (2018). Assessment of the safety of magnetic resonance imaging for patients with an active implantable medical device.
3. AAMI PC76 (Draft), "Requirements and Test Protocols for Safety of Patients with Pacemakers and ICDs Exposed to MRI", to be published.
4. Wilkoff, B. L., et al. (2013). Safe magnetic resonance imaging scanning of patients with cardiac rhythm devices: A role for computer modeling. *Heart Rhythm, 10*(12), 1815–1821.
5. Brown, J. E., et al. (2016). MR conditional safety assessment of implanted medical devices: Advantages of computational human phantoms. In: Proceedings of 38th Annual International Conference IEEE EMBC, Orlando, FL, pp. 6465–6468.
6. Park, S.-M., et al. (2007). Calculation of MRI-induced heating of an implanted medical Lead wire with an electric field transfer function. *Journal of Magnetic Resonance Imaging, 26*, 1278–1285.
7. Feng, S., et al. (2015). A technique to evaluate MRI-induced electric fields at the ends of practical implanted lead. *IEEE Transactions on Microwave Theory and Techniques, 63*(1), 305–313.
8. Brown, J. E. (2012). Radiofrequency heating near medical devices in magnetic resonance imaging. Ph.D. dissertation, Bobby B. Lyle School of Engineering, Southern Methodist University, Dallas, TX.
9. Brown, J. E., & Lee, C. S. (2013). Radiofrequency resonance heating near medical devices in magnetic resonance imaging. *Microwave and Optical Technology Letters, 55*, 2–299.
10. McCabe, S. O., & Scott, J. B. (2014). Cause and amelioration of MRI-induced heating through medical implant lead wires. In 21st electrical New Zealand conference, Hamilton, New Zealand.
11. McCabe, S. O., & Scott, J. B. (2015). Technique to assess the compatibility of medical implants to the RF field in MRI. In Asia-Pacific Microwave Conference 2015, pp. 6–9.
12. Liu, J., et al. (2019). On the relationship between impedances of active implantable medical devices and device safety under MRI RF emission. In IEEE Transactions, EMC, Accepted (to be published).
13. Yao, A., et al. (2019). Efficient and reliable assessment of the maximum local tissue temperature increase at the electrodes of medical implants under MRI exposure. *Bioelectromagnetics, 40*(6), 422–433.
14. Kozlov, M., & Kainz, W. (2017). Comparison of lead electromagnetic model and 3D EM results for helix and straight leads. In Proceedings of 19th International Confernce on Electromagnetic Advances and Applications, pp. 649–652.
15. Kozlov, M., & Kainz, W. (2018). Lead electromagnetic model to evaluate RF-induced heating of a coax lead: A numerical case study at 128 MHz. *IEEE Journal of Electromagnetics. RF and Microwaves in Medicine and Biology, 2*(4), 286–293.
16. Zastrow, E., Capstick, M., & Kuster, N. (2016).Experimental system for RF-heating characterization of medical implants during MRI. In Proceedings of 24th Annual Meeting ISMRM, Singapore.
17. Zastrow, E., Yao, A., & Kuster, N. (2017). Practical considerations in experimental evaluations of RF-induced heating of leaded implants. In 32nd URSI GASS, Montreal, Canada.

18. Brown, J. E., et al. (2018). Calculation of MRI RF-induced voltages for implanted medical devices using computational human models. In S. Makarov et al. (Eds.), *Brain and human body modeling: Computational human modeling at EMBC 2018* (pp. 283–294). Cham: Springer Nature.

19. Brown, J. E., et al. (2018). Calculating RF-induced voltages for implanted medical devices in MRI using computational human models. In Proceedings of 40th annual international conference, IEEE EMBC, Honolulu, HI, pp. 3866–3869.

20. Coates, S., & Thwaites, B. (2000). The strength-duration curve and its importance in pacing efficiency: A study of 235 pacing leads in 229 patients. *PACE, 23*, 1273–1277.

21. Kurpad, K., et al. (2018) MRI RF safety of active implantable medical devices (AIMDs): Effect of conductivity of tissue simulating media on device model accuracy. In Proceedings of 26th annual meeting international society of magnetic resonance in medicine, June 16–21, 2018, Paris, France, pp. 4075.

22. Min, X., & Sison, S. (2017). Impact of mixed media on transfer functions with a pacemaker system for estimation of RF heating during MRI scans. *Computers in Cardiology, 44*, 1–4.

23. Min, X., & Sison, S. (2018). Transfer functions of a spinal cord stimulation systems in mixed media and homogeneous media for estimation of RF heating during MRI Scans. In Proceedings of 40th annual international conference, IEEE EMBC, Honolulu, HI, pp. 2048–2051.

24. Kurpad, K. N., et al. (Accepted). MRI rf safety of active implantable medical devices (AIMDs): Tissue simulating medium properties for accurate modeling of MRI-conditional AIMDs. In 28th annual meeting international society of magnetic resonance in medicine.

25. Brown, J. E., et al. (Accepted). MRI safety of active implantable medical devices: Numerical study of the effect of lead insulation thickness on the RF-induced tissue heating at the lead electrode. In 42nd annual international conference, IEEE EMBC.

26. Noetscher, G. M., et al. (2016). Computational human model VHP-female derived from datasets of the National Library of Medicine. In Proceedings of 38th annual international conference, IEEE EMBC, Orlando, FL, pp. 3350–3353.

27. Ibrahim, T. S., et al. (2000). Application of finite difference time domain method for the design of birdcage RF head coils using multi-port excitations. *Magnetic Resonance Imaging, 10*, 733–742.

28. McElcheran, C. E., Golestanirad, L., Iacono, M. I., et al. (2019). Numerical simulations of realistic lead trajectories and an experimental verification support the efficacy of parallel radiofrequency transmission to reduce heating of deep brain stimulation implants during MRI. *Scientific Reports, 9*, 2124.

Estimating Electric Field and SAR in Tissue in the Proximity of RF Coils

Rosti Lemdiasov, Arun Venkatasubramanian, and Ranga Jegadeesan

1 Introduction

Inductive systems for delivering power and performing communications have become ubiquitous in medical implants [1–3]. In modern literature [4–9], there are numerous design approaches for building robust induction-based wireless links, and some of them include constraints based on exposure regulations as part of their design methodology. We find that there is a huge reliance on full-wave simulation tools such as Ansys HFSS, Remcom XFdtd, and Zurich Med Tech SIM4LIFE for exposure assessments. However, in some cases, this may be the only means to get some insights on the electromagnetic field distribution in tissues, a result of nonhomogeneous dispersive behavior of human tissues and its associated irregular stratified geometry. However, we lose the ability to get an intuitive understanding of the electromagnetic phenomena in the human body from the point of view of the basic electromagnetic principles, such as those outlined in [10, 11]. Additionally, working with full-wave simulation tools is very involved, and it requires familiarity with the tool and has a sizeable setup time and computational effort. In this paper, we present our earnest attempt to simplify exposure assessment so that the effort needed can be significantly reduced while not compromising on the degree of accuracy of results. We approach this problem by trying to estimate electric field in human body due to a nearby inductive system that has an RF coil adjacent to the human body. The

R. Lemdiasov (✉)
Cambridge Consultants, Boston, MA, USA
e-mail: rosti.lemdiasov@cambridgeconsultants.com

A. Venkatasubramanian · R. Jegadeesan
Cambridge Consultants, Singapore, Singapore
e-mail: arun.venkat@cambridgeconsultants.com; ranga.jegadeesan@cambridgeconsultants.com

© The Author(s) 2021
S. N. Makarov et al. (eds.), *Brain and Human Body Modeling 2020*,
https://doi.org/10.1007/978-3-030-45623-8_18

293

specific absorption rate (SAR) and induced currents can then be computed from first principles using well-established tissue properties [12] and compared with regulatory limits.

2   SAR in the Tissue

The magnetic field generated by the inductive systems used in medical implants induces currents in the tissues that are exposed to it. These currents generate heat and can cause tissue damage if left unchecked. To address this issue and regulate the use of induction technology so that safe operating conditions can be ensured in medical devices, exposure restrictions are enforced by the Food and Drug Administration (FDA) regulations in the United States and CE regulations in the EU. Both regulatory bodies use SAR metric as the key yardstick to evaluate field exposure. To cite the SAR regulatory limits for general exposure, the FDA has coordinated with the Federal Communications Commission (FCC) to set a limit of 1.6 W/kg averaged over a volume containing a mass of 1 g of tissue for the torso and head [13]. The corresponding limit in the EU is 2 W/kg averaged over a volume containing 10 g of tissue [14].

In a typical inductive charging system, the receive coil is part of the implant which is located at a certain depth inside the tissue (Fig. 1). The implant charging rate depends on the distance between the transmit coil and receive coil. At low frequencies (when skin depth in tissue is large compared to tissue size), the effect of tissue on the generated magnetic field is negligible. Hence, for purposes of estimating charging rate, tissue needs not be considered, and the system can be considered as though present in air. This also implies that the losses in tissue due to the magnetic

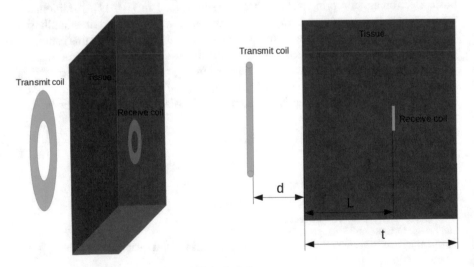

Fig. 1 Transmit coil, receive coil, and biological tissue

field are only a tiny fraction of the power that is delivered to the implant. As a result, for various distances from the transmit coil, it is simply possible to calculate magnetic field of the transmit coil as though in air. However, the tissue losses, though small, are a crucial factor in exposure assessment.

To assess these losses, we look at the electric field generated by the time-varying magnetic field. The electric field leads to circulating eddies and causes heating. It is this heating that needs to be kept within limits, and the metric used in this regard is the specific absorption rate which is defined as the amount of heat energy absorbed in the tissue per kg and is expressed in W/kg.

To estimate SAR due to a certain transmit coil located near tissue, it is normally required to do full-wave simulations of the coil loaded with biological tissue using the software package such as Ansys HFSS. Furthermore, such simulations would need to be performed for a range of distances between the transmitter and the exposed tissue so that the safe standoff distance can be estimated. For each distance, one can calculate the maximum allowable current and input RF power that would correspond to maximum allowable SAR in the tissue. To do this work, it requires significant computational resources and time. Is there an easier way to estimate electric field and SAR in the tissue?

3 Tissues

There are four abundant tissues of the human body that are of particular interest for exposure assessment: skin, fat, muscle, and bone. Each one of these tissues exhibits different frequency-dependent material characteristics such as electric permittivity, conductivity, and mass density. Depending on the tissue properties at a given frequency of interest, we can identify the most lossy tissue type that will exhibit the highest SAR per unit field strength.

As an example, let us choose an ISM frequency, say 13.56 MHz. For the chosen frequency, the muscle tissue dissipates most power. For simulation purposes, we use a block of muscle tissue next to the transmit coil and study effect of the heating. The relative electric permittivity for the muscle tissue for the frequency that we chose is 138 [14] which indicates how much electric field is attenuated inside the body compared to the electric field in air. Additionally, conductivity of muscle tissue is 0.628 S/m. Conductivity also contributes to attenuation of the electric field inside the tissue. The following chapter discusses the electric field and its components.

4 Two Components of Electric Field Inside the Tissue

It is known from electromagnetics that electric field can be represented as a sum of two components:

$$\mathbf{E} = -\nabla\varphi - j\omega\mathbf{A}$$

where φ is the scalar electric potential (electric charge as a source) and $\mathbf{A}$ is the magnetic vector potential (electric current as a source). For our analysis, let us call $-\nabla\varphi$ as "charge" electric field and $-j\omega\mathbf{A}$ as "current" electric field. Then, the following holds good:

$$\mathbf{E} = \mathbf{E}_{charge} + \mathbf{E}_{current}$$

Transmit coil in a wireless charging system generates both types of electric field when powered. As the transmit coil approaches biological tissue, the "charge" and "current" components of electric field behave quite differently inside the tissue.

4.1 "Charge" Electric Field

Lines of "charge" electric field have beginning (positive charges) and end (negative charges) (Fig. 2).

As the electric charges accumulate on the transmit coil, the field lines originate and end on the coil (outside of the biological tissue). That is, every line of "charge" electric field enters the block of tissue and then exits it. This "charge" electric field is significantly attenuated (by a complex factor of $\varepsilon_r - j\frac{\sigma}{\omega\varepsilon_0} = 138 - j833$) by charges that accumulate on surfaces of the block of tissue (Fig. 3).

Fig. 2 "Charge" component of electric field

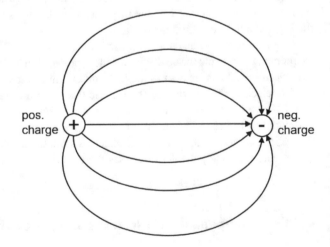

pos. charge

neg. charge

Fig. 3 Attenuation of "charge" electric field in the tissue

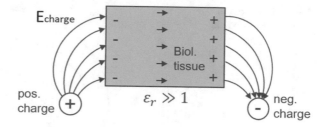

Fig. 4 Loop of current

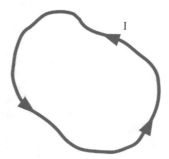

4.2 "Current" Electric Field

Generally speaking, the "current" electric field is not divergence-free, because the current in the coil is not the same at different points of the loop. However, if the self-resonance frequency of a transmit coil is high (compared to the chosen frequency), then the current can be considered to be the same throughout the coil. Let us consider a loop where the current I is the same throughout as shown in the following figure (Fig. 4).

The "current" electric field of the loop can be written as follows:

$$\mathbf{E}(\mathbf{r}) = -\frac{j\omega\mu_0}{4\pi} I \oint \frac{\exp(-jk|\mathbf{r}' - \mathbf{r}|)}{|\mathbf{r}' - \mathbf{r}|} \vec{dl}'$$

It can be shown that in free space, the divergence of "current" electric field of the loop of current is zero. This means that lines of this electric field have no beginning and no end, that is, they terminate into themselves. Just as magnetic field **B**, the "current" electric field of the loop of current consists of self-terminating field lines (Fig. 5).

Some of these lines are located entirely inside the block of tissue. These lines are *not* attenuated by surface charges as these lines never cross the surface. If a round transmit coil is positioned parallel to the block of tissue, then the "current" component of the electric field is *not attenuated* inside the block of tissue. Hence, it is the "current" electric field that contributes to tissue heating (SAR).

Fig. 5 Field lines of "current" electric field

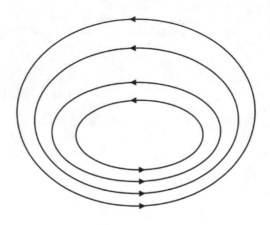

Table 1 Comparing loaded and unloaded simulations

Distance, cm	HFSS, unloaded simulation			HFSS, loaded simulation	
	Electric field of round coil			Electric field of round coil	
	Free space	Value V/m		Value V/m	Loaded with human tissue
10		5.62		1.63	

4.3 Comparing Effects of Tissue Loading Using HFSS Simulations of Transmit Coil at 10 cm from Tissue

When there is no tissue (unloaded), the electric field in the plane located 10 cm away from the coil is plotted. When tissue is present (loaded), the electric field on the surface of the Ø40 cm-diameter and 6 cm-tall cylinder of muscle tissue is plotted with the transmitting coil still 10 cm away.

From Table 1, we observe that the electric field generated by a transmit coil at a plane 10 cm from it is quite different if tissue is introduced. The presence of the tissue significantly attenuates overall electric field. Based on our previous definitions of types of electric fields, it is the "charge" electric field that is attenuated inside the tissue.

5 "Uniform Current" Approximation

When an RF coil is excited, current develops in the coil and there is charge accumulation on different parts of the coil. Currents give rise to "current" electric field and charges produce "charge" electric field. As it is clear from the previous sections, it is current flowing in the RF coil that is responsible for "current" electric field that in turn contributes to SAR. If the coil is operating at the frequency that is several times lower than self-resonance frequency, it makes good sense to assume uniform current flowing throughout the RF coil. The coil can be modelled as a combination of short current elements (Fig. 6).

"Current" electric field from the short current element is as follows:

$$\mathbf{E} = -\frac{j\omega\mu_0}{4\pi} \cdot I \cdot \frac{\vec{l}}{r} \cdot \exp(-jkr)$$

Each element contributes to the "current" electric field. Adding up the contributions of the elements, "current" electric field can be calculated and plotted, thereby providing an estimate for SAR. The following figure demonstrates how an RF coil can be formed from an arrangement of small current elements (Fig. 7).

There are several common coil geometries that we use in practice when performing the power transfer calculations. It is possible to write a code (MATLAB, C++, or others) to generate several commonly used coil geometries from a combination of short wire elements (Fig. 8).

Given the input parameters (radius, pitch, etc.), most common coil geometries can be generated. Furthermore, for such coils, the values of magnetic field and "current" electric field can be calculated at any point in space.

We are interested in the effects of such coils on nearby human tissues (skin, muscle, fat, bone). As has been discussed, the "current" electric field may not

Fig. 6 Current interval and observation point

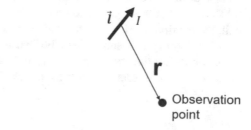

Fig. 7 Crude drawing of RF coil

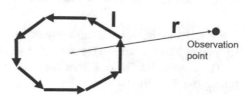

Fig. 8 Five common coils

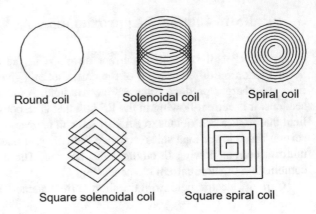

Round coil   Solenoidal coil   Spiral coil

Square solenoidal coil   Square spiral coil

attenuate when it crosses the air-to-tissue boundary. The highest electric field (and SAR) develops on the surface of the body facing the transmit coil. We calculate the electric field on the surface of the body and find the maximum value of electric field (and SAR).

6 Getting Rid of the Block of Tissue

"Current" electric field is the main contributor to tissue heating (SAR) and it does not attenuate inside the tissue. This means that we can do an RF simulation of the transmit coil in the absence of the tissue using "uniform current" simulation and then use the result of this simulation ("current" electric field) to estimate maximum SAR (Fig. 9).

We calculate maximum value of SAR per unit input RF power at several planes located at various distances from the coil. We then scale the input RF power to bring SAR to the maximum permitted value. This way, for a range of distances from the coil, the maximum RF power for a given transmit coil can be obtained.

It is important to mention that HFSS simulations are *expensive* in terms of setup time, simulation time, data processing, and license cost. "Uniform current" simulations are *cheap* for each one of these terms.

7 Maximum Allowed Current

When moving the transmit coil away from the block of tissue, we can provide more RF power into the coil (so that the SAR limit is not exceeded) (Fig. 10).

The farther the tissue is from the coil, the larger the current it can take before hitting the SAR limits.

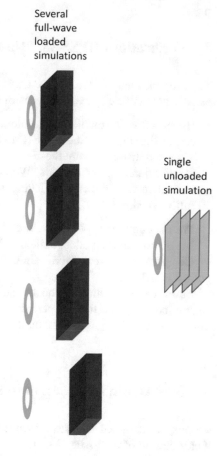

Fig. 9 Replacing several full-wave loaded simulations by a single (not full-wave) unloaded "uniform current" simulation

Several full-wave loaded simulations

Single unloaded simulation

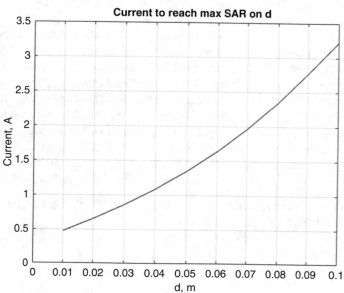

Fig. 10 Maximum allowed current on the distance to tissue, $I(d)$

8 Verification: HFSS vs. "Uniform Current"

To verify the idea that the electric field of a transmit coil inside the tissue is close to electric field in free space, two types of simulations are performed:

- HFSS full-wave simulations to plot electric field generated by a round transmit coil on the surface of the tissue (Ø40 cm diameter and 6 cm tall) by positioning it at several distances from the coil.
- "Uniform current" simulation. We define a round coil carrying uniform current throughout and plot electric field on several planes parallel to the coil. This is not full-wave simulation.

    We use one loop Ø17 cm round coil.
    Maximum values of electric field differ by about 3%.
    Comparison between HFSS simulations and "uniform current" simulations is shown in Table 2.

Full-wave simulations done by HFSS are computationally intense and they require considerable time to perform. We see that the relatively simple "uniform current" simulation produces about the same result without using HFSS.

9 Skin Depth Attenuation in the Tissue

Above, we assumed that the presence of the tissue does not significantly change fields outside of the tissue. When can we make such an assumption? Our answer is that the skin depth should be significantly larger than the thickness of the tissue: $\delta \gg t$, which happens at low frequency. In the opposite limiting case (high frequency), skin depth is small, and there is a significant attenuation of external field right near the boundary. This would correspond to the case of metal (highly conductive material), where electric field is approaching zero. In case if the skin depth is comparable to the tissue thickness, we need a formula that would describe the attenuation of the external field near the boundary. The simplest way to describe the attenuation of the external field near the tissue can be done by introducing the exponent $exp(-t/\delta)$. However, we get better agreement with full-wave simulations if we state that the attenuation is $exp(-t'/\delta)$, where t' is a weighted combination (inversely weighted) of thickness t, distance d, and radius (half size) of tissue block R: $t' = 1/\left(\frac{1}{t} + \frac{1}{d} + \frac{1}{R}\right)$. Here is our rationale for doing such weighting:

- In case if $t \ll d$ and $t \ll R$, we have a straightforward exponential attenuation of field with depth $exp(-t/\delta)$, that is, $t' \approx t$.

Table 2 Comparing loaded HFSS simulations and unloaded "uniform current" simulation

Distance, cm	Electric field of round coil from HFSS		Electric field of round coil, "uniform current" approximation	
	Plot	Value, V/m	Value, V/m	Plot
2		8.01	7.85	
4		4.80	4.71	
6		3.18	3.12	
8		2.25	2.19	
10		1.63	1.60	

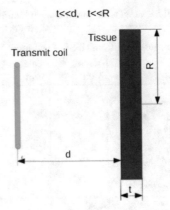

- In case if $d \ll t$ and $d \ll R$, the coil wiring is close to the tissue. Almost all induced current in the body is flowing right next to coil wiring. Effect of the tissue thickness d and tissue size $2R$ would be very small. So $t' \approx d$.

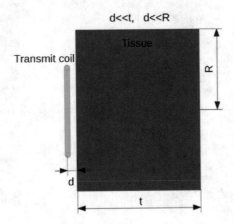

- In case if $R \ll r$ and $R \ll d$, the E-field infiltrates into the tissue from the sides. So R plays a role of thickness: $t' \approx R$.

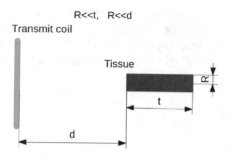

In the following figure, we plot dependence of the maximum electric field on conductivity (Fig. 11). We compare the "uniform current" approximation (theory) with HFSS simulations. In the "uniform current" approximation, we first calculate the maximum electric field and then attenuate it by a factor $exp(-t'/\delta)$. In HFSS simulation, we used a tissue disk having diameter of $2R = 20$ cm.

First of all, we see from the plot that increasing the value of ε_r leads to the flatter HFSS curve on the left side of the graph (low conductivity). This is exactly in line with our claim that high ε_r leads to significant attenuation of the "charge" electric field.

Secondly, for both HFSS and theoretical estimate ("uniform current" simulation), we observe a fall off at high conductivity ($\sigma > 1$ S/m). As a reference, the muscle conductivity at 13.56 MHz is 0.628 S/m. To our opinion, there is a satisfactory agreement between HFSS simulations and the "uniform current" estimate.

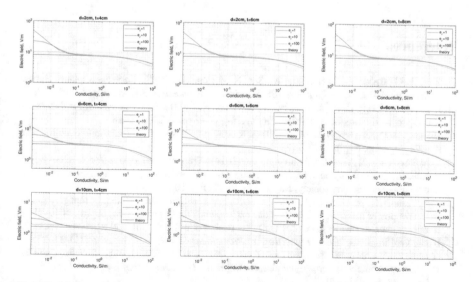

Fig. 11 Maximum electric field for a range of tissue conductivities (0.002–100 S/m), three dielectric permittivities (1, 10, 100), three coil-tissue distances (2 cm, 6 cm, 10 cm), and three tissue thicknesses (4 cm, 6 cm, 8 cm)

"Uniform current" estimate works well to describe the maximum electric field (and therefore SAR) developing on the surface of the tissue near the RF coil.

Transition between the flat response on the left (low σ) and exponential decay on the right (high σ) occurs roughly when skin depth in tissue δ is comparable to $1/\left(\frac{1}{t}+\frac{1}{d}+\frac{1}{R}\right)$.

10   Conclusions

We demonstrated that electric field inside the body can be satisfactorily estimated without full-wave simulation software using simple, time- and energy-efficient means.

Speaking about interaction of the RF coil with the human body, the following conclusions can be made:

- "Charge" component of electric field is significantly attenuated inside the biological tissue (if $\varepsilon_r \gg 1$) and generally causes negligible loss. "Current" component may not experience significant attenuation.
- It is the "current" component of electric field that is mostly responsible for heating the tissue.
- To estimate electric field and SAR inside the block of tissue located at certain distance, it may be sufficient to perform an unloaded simulation ("uniform current" simulation) that produces only "current" electric field.

References

1. Zeng, F. G., Rebscher, S., Harrison, W., Sun, X., & Feng, H. (2008). Cochlear implants: System design, integration, and evaluation. *IEEE Reviews in Biomedical Engineering, 1*, 115–142.
2. Loeb, G. E., Peck, R. A., Moore, W. H., & Hood, K. (2001). BionTM system for distributed neural prosthetic interfaces. *Medical Engineering & Physics, 23*(1), 9–18.
3. Weiland, J. D., & Humayun, M. S. (2014). Retinal prosthesis. *IEEE Transactions on Biomedical Engineering, 61*(5), 1412–1424.
4. Lemdiasov, R., & Venkatasubramanian A. (2017). Transmit coil design for wireless power transfer for medical implants. In: 39th annual international conference of the IEEE engineering in medicine and biology society, July 11–15, 2017, Jeju Island, Korea.
5. Jegadeesan, R., & Guo, Y.-X. (2012). Topology selection and efficiency improvement of inductive power links. *IEEE Transactions on Antennas and Propagation, 60*(10), 4846–4854.
6. RamRakhyani, A., & Lazzi, G. (2013). On the design of efficient multicoil telemetry system for biomedical implants. *IEEE Transactions of Biomedical Circuits and Systems, 7*(1), 11–23.
7. Jegadeesan, R., Nag, S., Agarwal, K., Thakor, N., & Guo, Y.-X. (2015). Enabling wireless powering and telemetry for peripheral nerve implants. *IEEE Journal of Biomedical and Health Informatics, 19*, 958–970.
8. Christ, A., Douglas, M. G., Roman, J. M., Cooper, E. B., Sample, A. P., Waters, B. H., Smith, J. R., & Kuster, N. (2013). Evaluation of wireless resonant power transfer systems with human

electromagnetic exposure limits. *IEEE Transactions on Electromagnetic Compatibility, 55*(2), 265–274.

9. Makarov, S., Horner, M., & Noetscher, G. (2018). *Brain and human body modeling*. Cham: Springer.

10. Jackson, J. D. (1962). *Classical electrodynamics*. New York: John Wiley & Sons.

11. Stratton, J. A. (1941). *Electromagnetic theory*. New York: John Wiley & Sons.

12. "http://niremf.ifac.cnr.it/tissprop". Accessed 6 Jan 6 2020.

13. 47 CFR 1.131, "https://www.fcc.gov/general/fcc-policy-human-exposure". Accessed 6 Jan 2020.

14. ICNIRP guidelines for limiting exposure to time-varying electric, magnetic and electromagnetic fields (up to 300 GHz), "https://www.icnirp.org/cms/upload/publications/ICNIRPemfgdl.pdf". Accessed 6 Jan 2020.

The CAD-Compatible VHP-Male Computational Phantom

Gregory M. Noetscher

1 Introduction

Computational human phantoms are an integral part of the design process in many areas of modern science and technology; this is especially true for computational electromagnetics (CEM). A review of available literature suggests that, since the very inception of CEM, its practitioners have used various model surrogates (primitive shapes, various combinations thereof, further refined models, etc.) to demonstrate the use of numerical modelling for the estimation of a body's response to external electromagnetic stimulation. The convergence of a number of disparate disciplines, including highly refined medical image collection techniques, advanced image processing, development of efficient simulation algorithms and supercomputing hardware, has resulted in computational human phantoms at a level of detail previously thought impossible.

Inspiration for the creation of the Visible Human Project (VHP)-Male model, presented herein, was the design and use of the VHP-Female model [15], which was constructed in the mid-2010s, adopted by the IEEE International Committee on Electromagnetic Safety for use as simulation of specific absorption rate (SAR), and used in a host of commercial and academic applications [2–4, 8, 12–14, 16, 17]. Throughout the development of the VHP-Female model, it became extremely apparent that compatibility with common computer-aided design (CAD) tools and interfaces was highly advantageous and enabled maximum use of the model in a variety of simulation methodologies, including the finite element method (FEM), boundary element method (BEM), finite-difference time-domain (FDTD) method and experimental methods, including the coupled boundary element-fast multipole

G. M. Noetscher (✉)
Electrical and Computer Engineering Department, Worcester Polytechnic Institute, Worcester, MA, USA
e-mail: gregn@wpi.edu

© The Author(s) 2021
S. N. Makarov et al. (eds.), *Brain and Human Body Modeling 2020*,
https://doi.org/10.1007/978-3-030-45623-8_19

method (BEM-FMM) [10]. Due to the successful implementation of the female model and the demand for a male version, the steps described below were undertaken for its construction. It is our sincere hope that the model will be adapted in a manner similar to its predecessor.

This work is organized in the following manner. Section 2 (Materials and Methods) documents the model construction process, including a description of the source data and mesh processing techniques. Section 3 (Results and Discussion) depicts the outcomes of mesh construction, global model assembly and baseline simulation. Section 4 (Conclusions) provides a summary of the work, together with plans for future work, suggestions for augmentations to the model and potential applications for which this model may be suitable.

2   Materials and Methods

2.1   Source Data

As its name would suggest, the VHP-Male model is based exclusively on medical data collected as part of the US Library of Medicine's Visible Human Project [1, 7]. Conducted during the mid-1990s to late 1990s, this effort is a collection of extremely detailed and anatomically accurate data obtained from one male and one female cadaver. The data includes magnetic resonance imagery (MRI) and computed tomography (CT) imagery together with high-definition photographs of cross-sectional cryosections. This data is provided to the greater public free of charge, enabling a staggering number of applications, from medical research to artistic endeavours.

More specifically, the male data, released in 1994, includes axial MRI data collected at 4 mm intervals throughout the majority of the body, axial CT data collected at 1 mm intervals and anatomical cryosection images collected at 1 mm intervals to coincide with the CT data. These cryosection images are 2048 by 1216 pixels, with each pixel measuring 0.33 mm in size.

2.2   Mesh Construction

The VHP cryosection images were segmented using a custom MATLAB-based segmentation tool. Each cryosection image at a given height of the cadaver and oriented along the global Z axis was imported, and a user of this tool was able to surround a given structure with points denoting the X and Y axis limits of the structure. Once all images were processed in this manner, all X, Y and Z points were assembled such that they consisted of a point cloud describing the outer surface of the structure of interest. This point cloud was then meshed using triangular surface elements such that mesh became two-manifold.

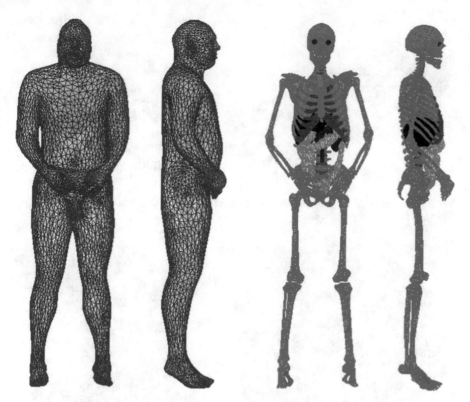

Fig. 1 The VHP-Male model. At left and mid-left: the full model including outer skin shell. At mid-right and right: the full model with the outer skin shell removed for better viewing of internal structures

Following segmentation and construction of each individual component, all components were assembled in a global reference frame and tested for intersections. All intersections were resolved using the mesh sculpting capabilities of Meshmixer. This tool is able to gradually move a triangle and its nearest neighbours along the triangle surface normal.

In certain instances, a smaller number of triangles were desired due to the need to balance simulation efficiency with model accuracy. In these cases, the quadric edge collapse decimation scheme [5] implemented in Meshlab was employed to reduce the number of triangles.

The results of these mesh manipulations are shown in Figs. 1, 2, 3 and 4. Each mesh component, the total number of triangles, mesh quality and the minimum mesh edge are given in Table 1.

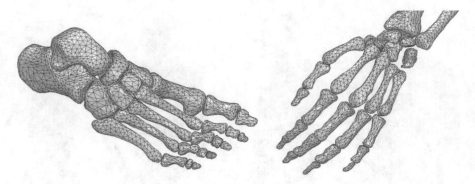

Fig. 2 Left – detailed views of the VHP-Male model foot and ankle; right – detailed views of the VHP-Male model hand and wrist

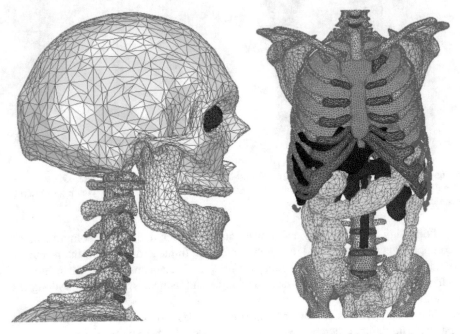

Fig. 3 Left – detailed view of the VHP-Male model skull and mandible; right – detailed views of the VHP-Male model internal organs and rib cage

2.3 Simulation Setup

Each component of the model was imported into the commercial FEM-based ANSYS Electromagnetics Suite 2019 R1 as an STL file and assigned dielectric and density material properties consistent with those published in the IT'IS Foundation [6]; this database has been widely excepted as the standard by the academic community. The excitation for this baseline simulation was a 300 MHz incident

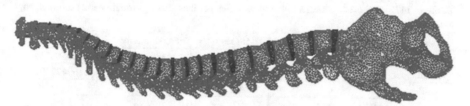

Fig. 4 Left – detailed views of the VHP-Male model spinal cord and vertebrae

plane wave with an intensity of 1 V/m originating approximately 35 mm in front of the model nose. The wave direction of propagation was toward the body, and the positive direction of the electric field was aligned with the positive vertical axis of the model.

Construction of the initial mesh was accomplished using the 'Classic' method, and mesh entity error checks were performed with the 'Strict' setting to ensure that no intersections or other mesh faults were present.

A solution setup with a target frequency of 300 MHz was applied and included adaptive mesh refinement. Following completion of the solution calculation, the volumetric mesh was refined by 30%. The resulting mesh statistics are presented in Table 2 along with simulation times and memory requirements. First-order basis functions were applied throughout. All calculations were performed by activating the HPC option and using 40 cores. The system hosting the software was running 64-bit Windows Server 2016 Standard with 64 AMD Opteron processors running at 2.66 GHz and a total of 256 GB of memory.

3 Results and Discussion

Following the initialization process described above, a triangular surface mesh composed of 313,750 elements was produced. This resulted in an initial volumetric mesh of 1,110,200 tetrahedra. Two adaptive mesh refinement steps were executed, documented in Table 2, generating a final mesh of 1,443,265 tetrahedra.

Plots of the magnitude of the electric field in the centres of the model sagittal and coronal planes are given in the top and bottom of Fig. 5, respectively. It is interesting to see the propagation of the surface wave along the skin shell in the sagittal plane. Also of note is the relatively uniform propagation of the wave within the body, shown in stark detail at the rear base of the neck; this is likely a consequence of the fact that the internal body volume was modelled with a single value of electrical permittivity and conductivity. When additional muscle structures are added, this is expected to greatly disrupt the path of the field, creating a much more inhomogeneous profile. Additional muscles will be added in the next model revision.

Table 1 Individual mesh names, number of triangles per mesh, triangle quality and minimum edge length

Mesh name	Triangles	Triangle quality	Min. edge length
'AortaLower'	1362	0.134330781	0.920544358
'AortaUpper'	652	0.222094966	1.260477698
'Bladder'	1640	0.068304298	0.812577699
'BrainWhiteMatter'	22,402	0.046567698	0.872950257
'CalcaneusLeft'	1344	0.114427155	1.452797748
'CalcaneusRight'	1412	0.097310098	1.564575094
'CapitateRight'	196	0.072770684	1.559374091
'Capitateleft'	178	0.230362472	1.800466459
'Cerebellum'	2622	0.157891988	0.494216188
'ClavicleLeft'	1258	0.045264776	1.371140432
'ClavicleRight'	1238	0.078624045	1.213447698
'Coccyx'	346	0.06891789	0.908385343
'Colon'	7260	0.001621858	0.544420676
'CuboidLeft'	506	0.202039927	1.281321767
'CuboidRight'	500	0.175812595	1.436882414
'CuniformIntermediateLeft'	252	0.162846331	1.087986024
'CuniformIntermediateRight'	210	0.196727483	1.432081059
'CuniformLateralLeft'	314	0.066617761	1.440015526
'CuniformLateralRight'	282	0.244729955	1.658142044
'CuniformMedialLeft'	390	0.138902069	1.28087339
'CuniformMedialRight'	406	0.214556367	1.330687259
'DiscC03C04'	182	0.198568394	0.896118632
'DiscC04C05'	224	0.341427842	1.366489723
'DiscC05C06'	344	0.139070512	0.551801066
'DiscC06C07'	320	0.149239492	0.490609829
'DiscC07T01'	410	0.225282064	0.949094485
'DiscL01L02'	538	0.250568351	1.366702091
'DiscL02L03'	614	0.194281326	1.534183401
'DiscL03L04'	774	0.182906967	1.272352431
'DiscL04L05'	724	0.17660326	1.232707083
'DiscL05L06'	910	0.106356352	1.239995701
'DiscL06S00'	718	0.13029574	1.206016825
'DiscT01T02'	588	0.139201049	0.465434502
'DiscT02T03'	590	0.312950053	0.630624001
'DiscT03T04'	592	0.123513989	0.874712503
'DiscT04T05'	732	0.293476947	0.580976494
'DiscT05T06'	686	0.178928426	0.910670175
'DiscT06T07'	696	0.17695128	0.658382169
'DiscT07T08'	754	0.239607771	0.768537297
'DiscT08T09'	582	0.149629995	0.925819093
'DiscT09T10'	634	0.232195174	0.979317946
'DiscT10T11'	744	0.251148608	1.013260169

(continued)

Table 1 (continued)

Mesh name	Triangles	Triangle quality	Min. edge length
'DiscT11T12'	750	0.104616503	1.140370543
'DiscT12L01'	802	0.073026765	0.906750275
'Oesophagus'	2012	0.052368945	1.018606107
'FemurLeft'	5238	0.504688464	2.29075216
'FemurRight'	6128	0.549623054	2.480630149
'FibulaLeft'	2232	0.308788783	1.02925398
'FibulaRight'	2388	0.158261673	1.125755996
'GallBladder'	490	0.085184055	1.368812877
'HipLeft'	5772	0.170565242	0.466692974
'HipRight'	5596	0.066202543	0.577033756
'HumerusLeft'	3222	0.525440799	2.029104712
'HumerusRight'	2956	0.518585689	2.544932279
'KidneyLeft'	2264	0.103633477	0.475712171
'KidneyRight'	1922	0.185812015	0.568609335
'Lens_Left'	66	0.069840435	0.151185182
'Lens_Right'	150	0.120496855	0.307647016
'Liver'	10,920	0.029780084	0.517502817
'LunateLeft'	188	0.163375398	0.70472835
'LunateRight'	194	0.142373039	0.633063654
'LungLeft'	10,034	0.00168764	0.361124034
'LungRight'	10,094	0.06703563	0.396727705
'Mandible'	5000	0.097637404	0.261716653
'MetacarpalLeft1'	382	0.33160287	0.299037544
'MetacarpalLeft2'	426	0.298271901	1.3603719
'MetacarpalLeft3'	508	0.10321164	0.556610664
'MetacarpalLeft4'	308	0.085347552	0.972818976
'MetacarpalLeft5'	264	0.176798859	0.748657523
'MetacarpalRight1'	382	0.11581314	1.086429111
'MetacarpalRight2'	434	0.05808032	1.022409002
'MetacarpalRight3'	474	0.090823214	0.988240084
'MetacarpalRight4'	310	0.262561401	1.026288459
'MetacarpalRight5'	252	0.457876386	1.050913568
'MetatarsalLeft1'	652	0.120767953	0.517845995
'MetatarsalLeft2'	452	0.141255045	1.124870802
'MetatarsalLeft3'	462	0.172389284	0.574780219
'MetatarsalLeft4'	424	0.132385844	0.558378787
'MetatarsalLeft5'	444	0.249921383	0.900482375
'MetatarsalRight1'	652	0.119668442	0.706069735
'MetatarsalRight2'	446	0.214312152	0.710267704
'MetatarsalRight3'	414	0.195978651	0.874267734
'MetatarsalRight4'	414	0.24698425	1.14134027
'MetatarsalRight5'	438	0.153710476	0.807904513

(continued)

Table 1 (continued)

Mesh name	Triangles	Triangle quality	Min. edge length
'NavicularLeft'	460	0.206977803	0.99978752
'NavicularRight'	444	0.128297957	0.964590002
'Pancreas'	2102	0.203865066	0.745068684
'PatellaLeft'	530	0.267932417	1.078763998
'PatellaRight'	532	0.335488394	1.427379689
'PhalangeDistalFootLeft1'	168	0.235004346	1.571741849
'PhalangeDistalFootLeft2'	92	0.268369191	0.831078757
'PhalangeDistalFootLeft3'	102	0.483981354	1.352923578
'PhalangeDistalFootLeft4'	96	0.464114316	1.052147256
'PhalangeDistalFootLeft5'	82	0.202013722	1.015984979
'PhalangeDistalFootRight1'	180	0.096299624	1.353143922
'PhalangeDistalFootRight2'	116	0.302680933	1.125514542
'PhalangeDistalFootRight3'	70	0.343021658	1.204871773
'PhalangeDistalFootRight4'	170	0.31481353	0.721639691
'PhalangeDistalFootRight5'	72	0.469888115	1.122329106
'PhalangeDistalHandLeft1'	178	0.268247206	1.359285138
'PhalangeDistalHandLeft2'	130	0.09784967	1.292786241
'PhalangeDistalHandLeft3'	132	0.157418063	1.098169903
'PhalangeDistalHandLeft4'	168	0.250218437	0.879250144
'PhalangeDistalHandLeft5'	100	0.294543741	1.062110587
'PhalangeDistalHandRight1'	188	0.154548236	1.155781252
'PhalangeDistalHandRight2'	174	0.082593765	0.749666824
'PhalangeDistalHandRight3'	158	0.144122875	1.006352412
'PhalangeDistalHandRight4'	120	0.051107021	1.366663165
'PhalangeDistalHandRight5'	128	0.204927838	0.672745153
'PhalangeIntermediateFootLeft2'	126	0.19671167	1.202773185
'PhalangeIntermediateFootLeft3'	102	0.292676446	1.179335756
'PhalangeIntermediateFootLeft4'	94	0.28205998	1.319829723
'PhalangeIntermediateFootLeft5'	128	0.465058544	0.712763519
'PhalangeIntermediateFootRight2'	122	0.425226618	1.229020296
'PhalangeIntermediateFootRight3'	154	0.234524916	1.159242431
'PhalangeIntermediateFootRight4'	72	0.212530112	1.190552138
'PhalangeIntermediateFootRight5'	74	0.222513152	1.097581791
'PhalangeIntermediateHandLeft2'	162	0.056994087	1.403116851
'PhalangeIntermediateHandLeft3'	244	0.090014565	1.144600211
'PhalangeIntermediateHandLeft4'	206	0.234766368	1.32785976
'PhalangeIntermediateHandLeft5'	156	0.159150685	1.287847653
'PhalangeIntermediateHandRight2'	188	0.245103352	1.179043396
'PhalangeIntermediateHandRight3'	268	0.111289314	1.036930209
'PhalangeIntermediateHandRight4'	218	0.026740507	0.927233326
'PhalangeIntermediateHandRight5'	184	0.103672908	1.272412621
'PhalangeProximalFootLeft1'	316	0.166227126	1.328956537

(continued)

Table 1 (continued)

Mesh name	Triangles	Triangle quality	Min. edge length
'PhalangeProximalFootLeft2'	210	0.191881761	1.421797168
'PhalangeProximalFootLeft3'	204	0.183090128	1.153649369
'PhalangeProximalFootLeft4'	166	0.16043228	1.435088718
'PhalangeProximalFootLeft5'	184	0.280139704	1.35571498
'PhalangeProximalFootRight1'	358	0.17528083	1.13358412
'PhalangeProximalFootRight2'	172	0.109310553	1.493342579
'PhalangeProximalFootRight3'	184	0.176204429	1.46978673
'PhalangeProximalFootRight4'	198	0.190545609	1.103740172
'PhalangeProximalFootRight5'	208	0.110182825	1.080280334
'PhalangeProximalHandLeft1'	210	0.098640872	1.447996794
'PhalangeProximalHandLeft2'	272	0.199974342	1.451075053
'PhalangeProximalHandLeft3'	312	0.121724447	1.586284823
'PhalangeProximalHandLeft4'	294	0.165683911	1.321214895
'PhalangeProximalHandLeft5'	222	0.253583311	1.355428709
'PhalangeProximalHandRight1'	212	0.222933219	1.650829938
'PhalangeProximalHandRight2'	272	0.177067742	1.478012517
'PhalangeProximalHandRight3'	360	0.121158269	1.168497
'PhalangeProximalHandRight4'	354	0.223749458	1.367527605
'PhalangeProximalHandRight5'	222	0.239075957	1.495746015
'PisiformLeft'	78	0.314852312	2.176323052
'PisiformRight'	86	0.284437859	1.511736134
'RadiusLeft'	1624	0.07812625	1.580472228
'RadiusRight'	1652	0.005233192	1.677186851
'RibLeft01'	588	0.060906359	1.059913319
'RibLeft01Cartilage'	404	0.109004079	1.054736417
'RibLeft02'	1042	0.09086824	0.932675145
'RibLeft02_Cartilage'	176	0.045449305	2.127295566
'RibLeft03'	1128	0.113561227	0.99423192
'RibLeft03_Cartilage'	260	0.026268431	1.645999605
'RibLeft04'	1302	0.084563472	0.992262712
'RibLeft04_Cartilage'	308	0.023780717	1.636533296
'RibLeft05'	1274	0.119127187	0.984158072
'RibLeft05_Cartilage'	386	0.111179556	1.617471712
'RibLeft06'	5008	0.03157469	0.243544913
'RibLeft06_09Cartilage'	1792	0.100098473	0.591946184
'RibLeft07'	2510	0.036709266	0.61452184
'RibLeft08'	2496	0.110317453	0.545243404
'RibLeft09'	5026	0.033871361	0.259944831
'RibLeft10'	5018	0.001194147	0.201178192
'RibLeft10Cartilage'	220	0.239513906	1.157833434
'RibLeft11'	1148	0.108452456	1.019565359
'RibLeft12'	530	0.1788179	0.907150243

(continued)

Table 1 (continued)

Mesh name	Triangles	Triangle quality	Min. edge length
'RibRight01'	600	0.087526113	1.204959997
'RibRight02'	1148	0.10756514	0.511703818
'RibRight02_Cartilage'	212	0.225162533	1.718796537
'RibRight03'	1236	0.099341824	0.919209026
'RibRight03_Cartilage'	340	0.040870397	1.163855878
'RibRight04'	1224	0.137964912	1.049346668
'RibRight04_Cartilage'	340	0.123182198	1.852387423
'RibRight05'	1250	0.079547932	1.155101801
'RibRight05_Cartilage'	366	0.121168084	1.80995292
'RibRight06'	1330	0.176484463	1.038825487
'RibRight06_09Cartilage'	1902	0.102953268	1.104599593
'RibRight07'	2290	0.02321827	0.436906477
'RibRight08'	2470	0.035852455	0.697197243
'RibRight09'	2488	0.007762291	0.869718583
'RibRight10'	410	0.187648969	0.881692954
'RibRight12'	872	0.12421912	0.833364362
'Sacrum'	7146	0.005022698	0.121273699
'ScaphoidLeft'	276	0.161552246	1.259282022
'ScaphoidRight'	234	0.126420481	0.919981639
'ScapulaLeft'	2118	0.002681555	0.905832545
'ScapulaRight'	2064	0.003680874	0.996482213
'Skin'	13,246	0.000609802	0.198862826
'Skull'	14,766	0.001814466	0.154270982
'SpineC1'	2460	0.00481742	0.248172617
'SpineC2'	1190	0.08315216	0.781631363
'SpineC3'	1518	0.042719709	0.252149084
'SpineC4'	1794	0.002418403	0.038447994
'SpineC5'	1704	0.019635973	0.021795812
'SpineC6'	1702	0.007202928	0.020272316
'SpineC7'	1532	0.063678597	0.188076348
'SpineL1'	2790	0.105967936	0.545063665
'SpineL2'	2214	0.169298277	0.590225123
'SpineL3'	2014	0.022869325	0.758650269
'SpineL4'	2072	0.010869297	0.55886578
'SpineL5'	1906	0.143658437	1.191150819
'SpineT1'	1356	0.033942145	0.248312562
'SpineT10'	1598	0.071890258	0.599400251
'SpineT11'	1746	0.099742236	0.411570228
'SpineT12'	1770	0.204306553	0.734127338
'SpineT2'	1546	0.004042741	0.992247621
'SpineT3'	1370	0.11645742	0.356404699
'SpineT4'	1382	0.076971261	0.159364139

(continued)

Table 1 (continued)

Mesh name	Triangles	Triangle quality	Min. edge length
'SpineT5'	1482	0.136731374	0.637760074
'SpineT6'	1298	0.098502586	0.573448349
'SpineT7'	1312	0.146509781	1.096860449
'SpineT8'	2152	0.003031485	0.395329192
'SpineT9'	1734	0.016899747	0.416672514
'Spleen'	952	0.128410792	2.317553484
'Sternum'	2336	0.017746966	0.595070383
'TalusLeft'	914	0.522314047	1.397671978
'TalusRight'	848	0.16043732	1.072112228
'TibiaLeft'	4544	0.534577008	1.721629222
'TibiaRight'	4296	0.59556915	2.341040932
'TrapeziumLeft'	178	0.199925044	0.525153998
'TrapeziumRight'	150	0.341799266	1.094954994
'TrapezoidLeft'	148	0.067468983	0.507043705
'TrapezoidRight'	128	0.130458037	0.663360569
'TriquetralLeft'	182	0.069901248	0.329812317
'TriquetralRight'	134	0.246326993	0.986282248
'UlnaLeft'	1950	0.32301674	1.023958874
'UlnaRight'	1998	0.137709703	0.985571845
'VitreousHumor_Left'	312	0.416736477	0.50974078
'VitreousHumor_Right'	268	0.266409301	1.142471306

Table 2 Individual mesh names, number of triangles per mesh, triangle quality and minimum edge length

Adaptive pass	Number of tetrahedra	Solver time (HH:MM:SS)	Memory (GB)
1	1,110,200	01:00:13	109
2	1,443,265	01:56:40	177

4 Conclusions

The work describes the construction and baseline use of the Visible Human Project (VHP)-Male computational phantom, a CAD-compatible model based on publicly available data. This model has been constructed such that it may be employed by all of the most common CEM simulation techniques in use today and easily modified to optimally fit a given application. A baseline simulation using the commercial FEM-based ANSYS Electromagnetics Suite 2019 was conducted, and the results of this simulation were presented.

Future additions to the model include major muscle groups and selected large nerves. In addition, further detail in the circulatory system will likely be required to address several simulation applications.

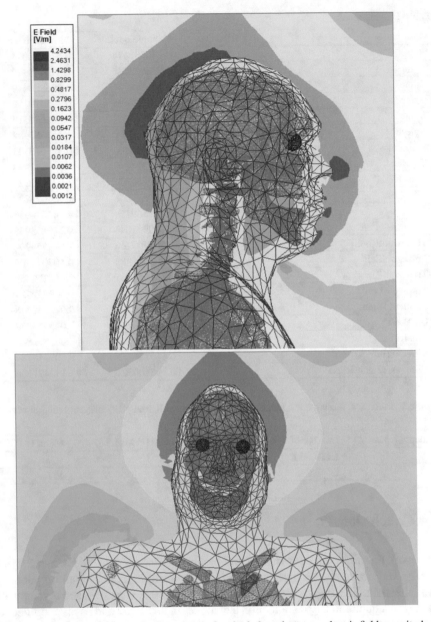

Fig. 5 Top – electric field magnitude at centre of sagittal plane; bottom – electric field magnitude at centre of coronal plane

As with the VHP-Female model, additional layers characterizing variations in skin and fat thicknesses will be included to explore the impact of body mass index on SAR. Refinements of the inner and outer ear structures are also envisioned. Inclusion of sinus cavities will also be critical to enable the highest level of accuracy possible.

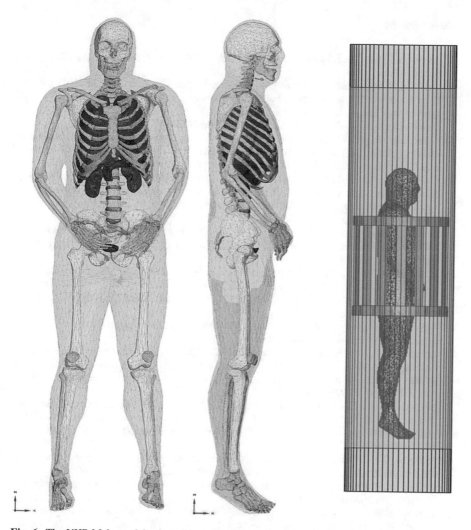

Fig. 6 The VHP-Male model oriented in a simulation of an MRI coil

Figure 6 provides one possible use of this new model: simulation of a loaded MRI coil. The VHP-Female model was used very successfully to characterize numerous MRI coil designs. There is no reason to believe this new male model would not also be highly suitable for this purpose.

Acknowledgements The author would like to thank Dr. Ali Yilmaz, Dr. Jackson Massey and the Computational Electromagnetics Group at the University of Texas at Austin for their exceptional work on the AustinMan and AustinWoman voxel models [11]. The AustinMan model in particular represented the standard against which the VHP-Male model was measured.

References

1. Ackerman, M. J. (1998, March). The visible human project. *Proceedings of the IEEE* (vol. 86. pp. 504–511).
2. Barbi, M., Garcia-Pardo, C., Cardona, N., Nevarez, N., Pons, V., & Frasson, M. (2018). Impact of receivers location on the accuracy of capsule endoscope localization. *2018 IEEE 29th annual international symposium on Personal, Indoor and Mobile Radio Communications (PIMRC)*, Bologna, (pp. 340–344). https://doi.org/10.1109/PIMRC.2018.8580862.
3. Chen, L. et al.. (2018, July). Radiofrequency propagation close to the human ear and accurate ear canal models. *40th Annual International Conference of the IEEE Engineering in Medicine and Biology* Society (EMBC 2018), Honolulu, HI (pp.17–21).
4. Garcia-Pardo, C. et al.. (2018, June). Ultrawideband technology for medical in-body sensor networks: An overview of the human body as a propagation medium, phantoms, and approaches for propagation analysis. *IEEE Antennas and Propagation Magazine* 60(3): 19–33. https://doi.org/10.1109/MAP.2018.2818458.
5. Garland, M., & Heckbert, P. S. (1997). Surface simplification using quadric error metrics. *SIGGRAPH '97 Proceedings of the 24th Annual Conference on Computer Graphics and Interactive Techniques* (pp. 209–216). New York.
6. Hasgall, P. A., Di Gennaro, F., Baumgartner, C., Neufeld, E., Lloyd, B., Gosselin, M. C., Payne, D., Klingenböck, A., & Kuster, N. (2018, May 15). *IT'IS Database for thermal and electromagnetic parameters of biological tissues*. Version 4.0. https://doi.org/10.13099/VIP21000-04-0. it is.swiss/database.
7. U.S. National Library of Medicine. The Visible Human Project® Online: http://www.nlm.nih.gov/research/visible/visible_human.html
8. Lemdiasov, R., & Venkatasubramanian, A. (2017). Transmit coil design for Wireless Power Transfer for medical implants. *2017 39th Annual International Conference of the IEEE Engineering in Medicine and Biology Society (EMBC)*, Seogwipo (pp. 2158–2161). https://doi.org/10.1109/EMBC.2017.8037282.
9. Makarov, S. N., Noetscher, G. M., Yanamadala, J., Piazza, M. W., Louie, S., Prokop, A., Nazarian, A., & Nummenmaa, A. (2017). Virtual human models for electromagnetic studies and their applications. *IEEE Reviews in Biomedical Engineering, 10*, 95–121. https://doi.org/10.1109/RBME.2017.2722420.
10. Makarov, S. N., Noetscher, G. M., Raij, T., & Nummenmaa, A. (2018). A quasi-static boundary element approach with fast multipole acceleration for high-resolution bioelectromagnetic models. *IEEE Transactions on Biomedical Engineering, 65*(12), 2675–2683. https://doi.org/10.1109/TBME.2018.2813261.
11. Massey, J. W., & Yilmaz, A. E. (2016, Aug). AustinMan and AustinWoman: High-fidelity, anatomical voxel models developed from the VHP color images. In *Proceedings of 38th Annual International Conference of the IEEE Engineering in Medicine and Biology Society (IEEE EMBC)*. Orlando.
12. Nikolayev, D. (2018). Modeling and Characterization of in-Body Antennas," *2018 IEEE 17th international conference on Mathematical Methods in Electromagnetic Theory (MMET)*, Kiev (pp. 42–46). https://doi.org/10.1109/MMET.2018.8460279.
13. Nikolayev, D., Zhadobov, M., Karban, P., & Sauleau, R. (2017, December). Conformal antennas for miniature in-body devices: The quest to improve radiation performance. *URSI Radio Science Bulletin 2017*(363):52–64. https://doi.org/10.23919/URSIRSB.2017.8409427
14. Nikolayev, D., Zhadobov, M., Le Coq, L, Karban, P., & Sauleau, R.. (2017, November). Robust Ultraminiature capsule antenna for ingestible and implantable applications. *IEEE Transactions on Antennas and Propagation 65*(11): 6107–6119. https://doi.org/10.1109/TAP.2017.2755764.
15. Noetscher, G. et al. (2016, August). Computational Human Model VHP-FEMALE Derived from Datasets of the National Library of Medicine," *38th Annual International Conference of the IEEE Engineering in Medicine and Biology Society* (EMBC 2016), Orlando. (pp. 16–20.)

16. Perez-Simbor, S., Andreu, C., Garcia-Pardo, C, Frasson, M., & Cardona, N. UWB Path Loss Models for Ingestible Devices. *IEEE Transactions on Antennas and Propagation*. https://doi.org/10.1109/TAP.2019.2891717

17. Venkatasubramanian, A., & Gifford, B. (2016). Modeling and design of antennas for implantable telemetry applications. 2016 *38th Annual International Conference of the IEEE Engineering in Medicine and Biology Society (EMBC)*, Orlando (pp. 6469–6472). https://doi.org/10.1109/EMBC.2016.7592210.

Part VII
Relevant Modeling Topics

Preprocessing General Head Models for BEM-FMM Modeling Pertinent to Brain Stimulation

William A. Wartman

1   Introduction

Transcranial magnetic stimulation (TMS) is a noninvasive neurostimulation method wherein a coil placed near the subject's head induces electric currents within the brain [7, 10, 13]. However, intermediate tissues between the coil and the cortex strongly affect the induced electric field (and thus the induced current). Numerical simulation of the interaction between the primary electric field and tissues of the head is necessary to predict the behavior of the total induced electric field and find the ultimate activation site(s). Further, wide intersubject variations cause the actual fields to deviate strongly from expected fields calculated using a generic head model. To minimize deviation between the simulated and actual fields, the simulated fields must be calculated using an accurate, high-resolution, subject-specific head model.

The TMS toolkit (complete computational code and supporting documentation) available for academic use at the Dropbox repository [2] is one such TMS simulator, which utilizes the boundary element fast multipole method (BEM-FMM) described in [4, 9]. The toolkit is written for MATLAB R2019a and has dependencies on the Image Processing Toolbox, Partial Differential Equations Toolbox/Antenna Toolbox, and Statistics and Machine Learning Toolbox. Its core FMM method is that of [3], included with permission in the redistributable software package.

This toolkit enables users to simulate TMS behavior using predefined or custom coil CAD models and subject-specific head models. These head models consist of a set of nested 3D triangular meshes, where each mesh marks the boundary between two tissues with different electrical properties (e.g., one mesh follows the skin/skull boundary, and another mesh follows the gray matter (GM)/white matter

W. A. Wartman (✉)
Electrical and Computer Engineering Department, Worcester Polytechnic Institute, Worcester, MA, USA
e-mail: wawartman@wpi.edu

© The Author(s) 2021
S. N. Makarov et al. (eds.), *Brain and Human Body Modeling 2020*,
https://doi.org/10.1007/978-3-030-45623-8_20

327

(WM) boundary) [5]. Because the BEM-FMM algorithm operates directly in terms of induced charges on these interfaces [8], it is robust against several common mesh defects that would hinder conventional volumetric finite element method (FEM) simulations, including intersecting meshes. The BEM-FMM algorithm further supports computation of the net electric field at locations arbitrarily close to tissue interfaces, where FEM routines cannot provide field resolution that exceeds the resolution of the underlying volumetric mesh.

Despite the BEM-FMM algorithm's robustness against common mesh defects, the current implementation of the software toolkit is applicable only to one specific meshing scheme: one in which each mesh represents a boundary between *exactly two* tissues. This is the standard output format of the SimNIBS v2.1 pipeline [11, 12, 14–17] in particular, and it produces meshes that are layered one inside the other. For example, the boundary between the skull and cerebrospinal fluid (CSF) completely surrounds and encloses the boundary between the CSF and gray matter (GM), which in turn completely surrounds and encloses the boundary between gray matter and white matter (WM). The goal of this exercise was to add support for a second meshing scheme, in which each mesh represents the entire outer boundary of a single tissue. One model that employs this meshing scheme is the MIDA model, produced by the IT'IS Foundation [6]

The MIDA head model comprises 115 CAD tissue models with more than 11 M triangular facets total. The model was produced from scans of a healthy 29-year-old female volunteer. Data was compiled from several medical imaging methods, including MRI, MRA, and DTI. These diverse imaging methods ensured that high-contrast images of most tissues existed in at least one of the image sets and image resolution approached 500 µm. Special care was taken to obtain high-contrast images of nerve tissue and vasculature. The entire data set was segmented independently by three experts using both manual and automated segmentation techniques, and their individual segmentations were combined to produce a highly accurate final segmentation. A triangulation algorithm was then applied to the resulting voxel model to extract triangular mesh surfaces for every tissue [6].

Several selections of model tissues are shown in Figs. 1, 2, 3, 4, 5, 6, 7, 8, 9 and 10. The mesh processing software used is open-source MeshLab v2016.12 [1]. Some characteristics of the model relevant to the task of enabling its use in the BEM-FMM toolkit are as follows:

(a) Adjacent meshes typically have coincident triangular facets at their interfaces. Observe, for example, the GM, CSF, and vasculature in Figs. 4, 5 and 6.
(b) Some meshes comprise multiple manifold surfaces. The CSF in particular includes a very large number of closed surfaces scattered throughout the cranial volume (see Figs. 5 and 10).

Fig. 1 Epidermis mesh of the MIDA head model

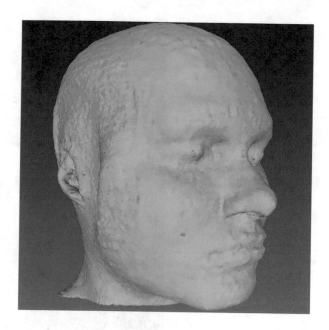

Fig. 2 Selected meshes of the MIDA model below the subcutaneous adipose tissue. Muscles are shown in pink, bones are shown in white, glands are shown in green, mucous membranes are shown in lime green, and cartilage is shown in orange

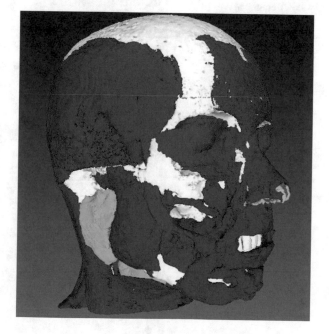

Fig. 3 The skull, vertebrae, and other bones (white); intervertebral disks (orange); veins and arteries (blue and red); and cranial nerves (yellow)

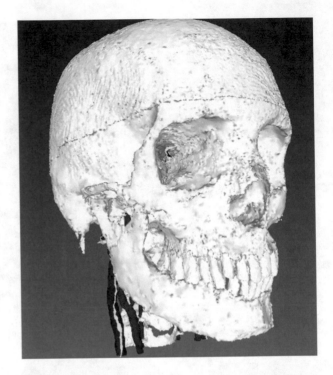

Fig. 4 Gray matter (gray), cerebrospinal fluid (light yellow), cranial nerves (dark yellow), veins (blue), and arteries (red). Note the close proximity of the CSF, GM, veins, and arteries

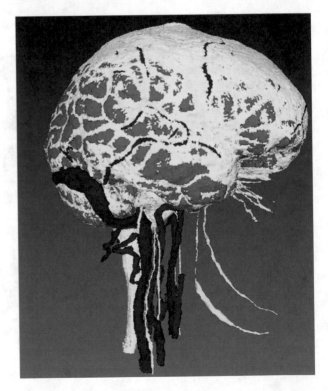

Fig. 5 The CSF mesh presented in isolation from all other tissues. Note the multitude of small, isolated compartments visible near the position of the cerebellum. Also note the tight channel for the vein near the top of the CSF mesh (compare with Fig. 4)

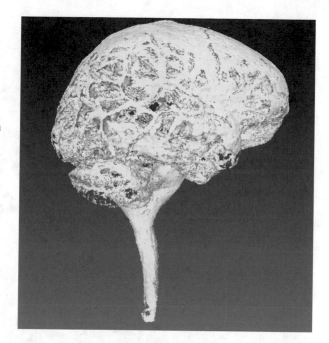

Fig. 6 Gray matter (gray), arteries (red), veins (blue), and cranial nerves and spinal cord (yellow). Other small brain components are in light gray

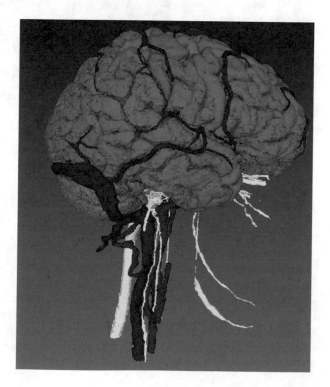

Fig. 7 White matter (white), arteries (red), veins (blue), nerves/spinal cord (yellow), and other small brain meshes (gray). Note the very fine structures of the white matter of the cerebellum

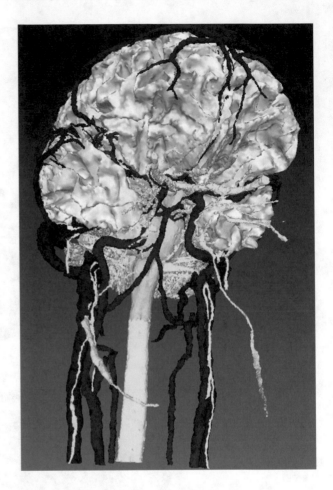

2 Methods

2.1 Mesh Preprocessing

Because the MIDA model's tissue meshes are not nested in general (i.e., a given MIDA tissue mesh explicitly segments every boundary between that mesh and any other tissue), adjacent tissue meshes each contain their own copies of the facets that form the border between them. When two adjacent tissue meshes are loaded simultaneously, their shared border comprises two sets of coincident facets, one set contributed by each tissue mesh. These coincident facets necessarily share coincident centroids, which in turn create singularities that invalidate simulation results. Figure 11 depicts this case for three hypothetical meshes, Object 1, Object 2, and Object 3. Though Object 1 and Object 2 both segment their shared boundary, Object 3 does not explicitly segment its boundary with Objects 1 and 2 for

Fig. 8 Veins (blue), arteries (red), and nerves (yellow) presented in isolation from other tissues

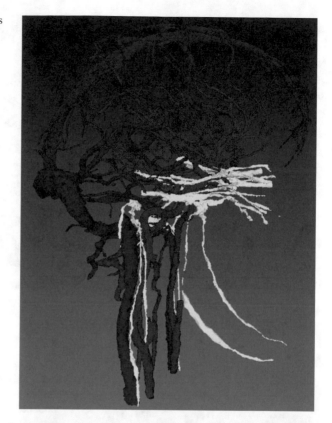

Fig. 9 Nerves of the MIDA head model, featuring the optic chiasm and optic tract. The anterior direction is toward the top of the page, and the superior direction is out of the page

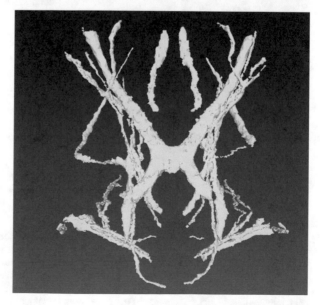

Fig. 10 View from the interior of the CSF mesh presented in isolation from all other tissues. Note the large number of isolated compartments

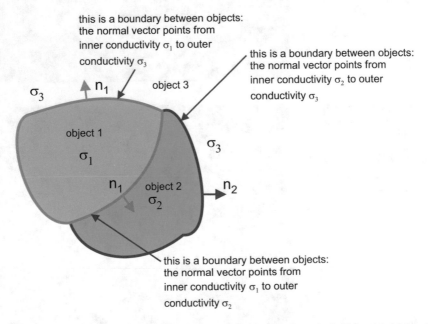

Fig. 11 Object 3 (with interior conductivity σ_3) surrounds and encloses both Object 1 (with interior conductivity σ_1) and Object 2 (with interior conductivity σ_2), so Object 1 and Object 2 initially list σ_3 as the exterior conductivity for all facets in their respective meshes. Because Object 1 and Object 2 have each explicitly segmented their mutual interface, that interface initially contains coincident facets contributed by both objects. In this example, Object 2's copies of the interface facets have been removed, and Object 1's copies of the facets remain. Object 1's facets at the interface still list σ_1 as their interior conductivity but have changed their exterior conductivity from σ_3 to σ_2

simplicity. Every mesh is assigned a default exterior conductivity derived from manual inspection of the surrounding tissues.

The function knnsearch of MATLAB's Statistics and Machine Learning Toolbox is first used to pair facets that have coincident centroids. One facet of each pair is designated as the facet to be kept, and the other is designated as the facet to be deleted. The outer conductivity of the facet to be kept is set equal to the inner conductivity of the facet to be deleted, and associated contrast information is updated for the facet to be kept. After this process has been completed for all coincident facet pairs, all information related to the facets to be deleted (e.g., centroid, area, connectivity) is removed, and any now-unreferenced vertices are cleared from the list of vertices (and face connectivity information is updated as appropriate).

3 Results

The final test setup modeled a TMS configuration intended to target the motor hand area of the precentral gyrus (the hand knob area, [18]) of the MIDA model. The coil model used was a generic figure-eight coil with circular cross-sectional wire, as shown in Fig. 12. The coil was approximated by 16,000 elementary current segments driven by time-varying current $\frac{dI}{dt} = 9.4e7 \, Amperes/sec$.

Preprocessing of the MIDA model for simulation using the BEM-FMM algorithm took approximately 525 s in total. Of those 525 s, 138 s were required to resolve coincident facets. Of the original 11 M facets, approximately 5.4 M were removed, and approximately 5.6 M remain. Table 1 lists the times associated with each preprocessing step.

The coil model was positioned above the head model according to four simple geometric rules:

1. The coil's centerline passes through a selected point on the hand knob area.
2. The coil's centerline is perpendicular to the skin surface.
3. The distance from the coil to the skin surface along the coil's centerline is 10 mm.
4. The dominant field direction (the y-axis of the coil coordinate system) is approximately perpendicular to the gyral crown and associated sulcal walls of the precentral gyrus pattern at the target point.

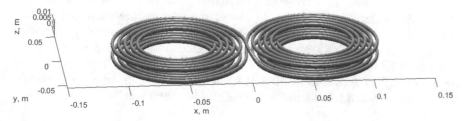

Fig. 12 The coil model employed for this test

Table 1 Preprocessing time

Step description	Step time (s)	Facet count
Load all meshes from disk	48.16	11,008,306
Calculate facet characteristics (e.g., normal vectors)	84.89	11,008,306
Assign initial conductivities	4.83	11,008,306
Find and resolve coincident facets	137.97	11,008,306
Find topological neighbors (for charge low-pass filtering)	30.08	5,632,767
Find BEM-FMM integration neighbors	17.52	5,632,767
Evaluate neighbor integrals	169.80	5,632,767
Save data to disk	30.82	5,632,767
Total preprocessing time	**524.24**	

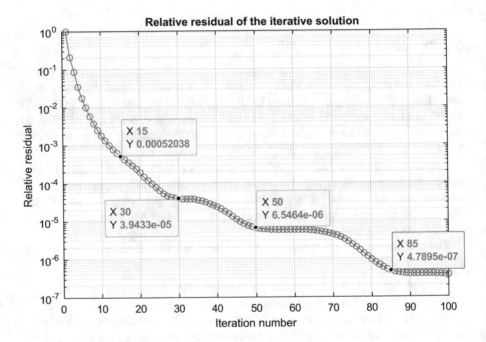

Fig. 13 Convergence curve for 100 GMRES iterations

Figure 13 shows the BEM-FMM convergence curve for this test setup after 100 GMRES iterations, and Fig. 14 shows the convergence curve for the first 15 iterations. The relative residual falls well below the threshold 10^{-3} within 15 iterations, indicating that 15 iterations produce results within an acceptable error margin. The test was run on a 32-core Intel® Xeon® E5-2683 v4 CPU operating at 2.1 GHz with 256 GB RAM. On this machine, the total computational time required with 5.6 M facets for 15 GMRES iterations was 373 s.

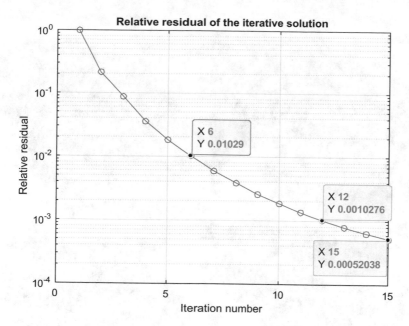

Fig. 14 Convergence curve for 15 GMRES iterations

Figure 15 shows simulation results for the electric field at the gray matter/CSF interface (a, c) and the white matter/gray matter interface (b, d). Figure 15 (a, b) shows a heat map of the magnitude of the electric field at the respective surfaces, scaled in V/m. Figure 15 (c, d) shows a focality estimate of the total electric field. In these figures, small blue balls are drawn at every facet for which the total field magnitude is within the range 80% to 100% of the maximum field magnitude observed for that particular surface. We see that the naïve geometric coil positioning rules barely stimulate the desired region at all and instead produce local maxima at distant sulci rather than the targeted motor hand area. This further reinforces the necessity of subject-specific head modeling for TMS applications.

Figures 16, 18 and 20 depict cross sections of the tissue meshes coregistered with T1 MRI data for the MIDA subject. The planes of these cross sections pass through the point on the white matter surface where the maximum E-field magnitude occurs. Pink spheres are drawn at the center of every WM facet that experiences a field with magnitude within 80–100% of the maximum E-field magnitude on this surface. Figures 17 and 19 show contour plots of the electric field magnitude in the immediate vicinity (i.e., ±10 mm) of the maximum field locations of Figs. 16 and 18, respectively.

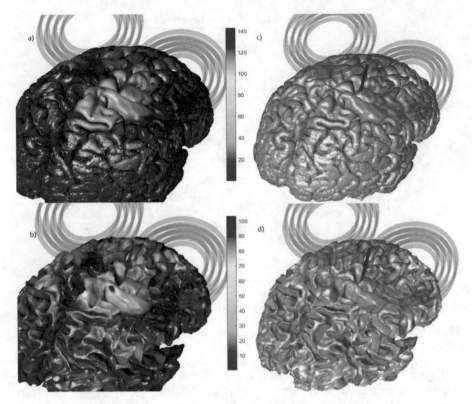

Fig. 15 Surface fields and focality. (**a**): Electric field magnitude (V/m) at the gray matter surface. (**b**): Electric field magnitude (V/m) at the white matter surface. (**c**): Locations of high field strength (80–100% of the absolute maximum field observed) at the gray matter surface. (**d**): Locations of high field strength (80–100% of the absolute maximum field observed) at the white matter surface

4   Conclusion

The BEM-FMM TMS modeling toolkit has been made compatible with a previously unsupported head mesh scheme, in which each mesh corresponds to one tissue's entire outer surface. The modifications were tested using the MIDA head model, which employs the newly supported mesh scheme. Simulation was executed successfully with the MIDA model, achieving convergence within 15 GMRES iterations. The performance penalty associated with the new mesh format occurs solely in

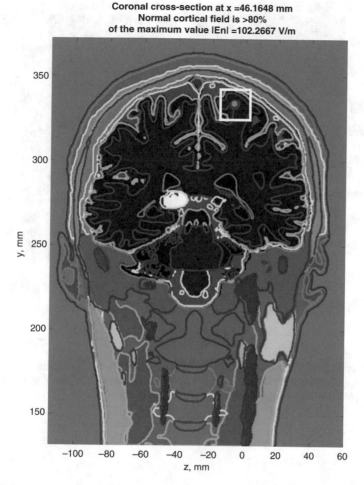

Fig. 16 Coronal cross section passing through the location of the maximum E-field in the white matter volume. Colored traces denote contours of tissue meshes passing through the cross-sectional plane. Small pink balls are drawn at the locations experiencing high field strength (80–100% of the maximum field observed in the WM volume)

the preprocessing stage, and there is little to no effect on field calculation performance. The toolkit is now applicable to a wider range of head models and is more robust against models whose meshes have coincident facets in general.

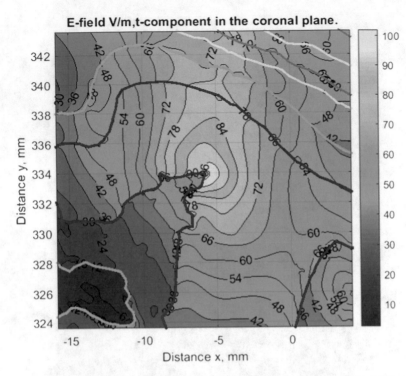

Fig. 17 Coronal-plane contour plot of electric field magnitude in the immediate vicinity of the location of the maximum electric field within the white matter volume. The boundary of this figure corresponds to the white box in Fig. 16

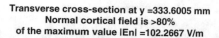

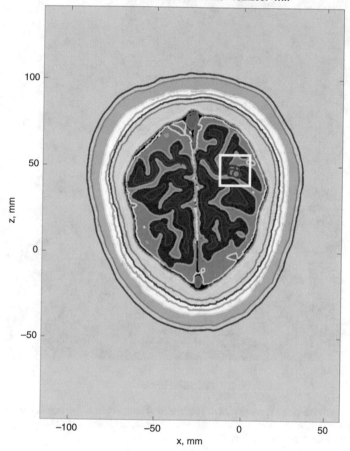

Fig. 18 Transverse cross section passing through the location of the maximum E-field in the white matter volume. Colored traces denote contours of tissue meshes passing through the cross-sectional plane. Small pink balls are drawn at the locations experiencing high field strength (80–100% of the maximum field observed in the WM volume)

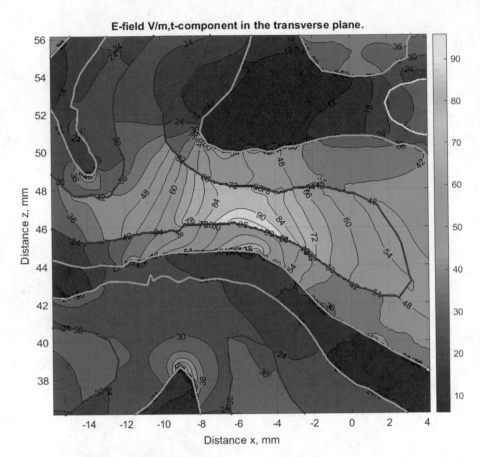

Fig. 19 Transverse-plane contour plot of electric field magnitude in the immediate vicinity of the location of the maximum electric field within the white matter volume. The boundary of this figure corresponds to the white box in Fig. 18

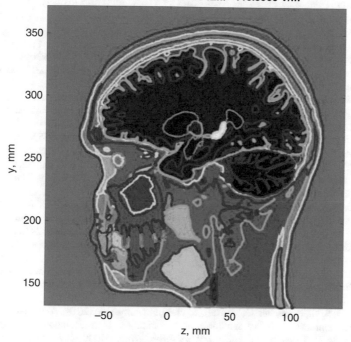

Sagittal cross-section at x =−5.8645 mm
Normal cortical field is >80%
of the maximum value IEnI =115.3966 V/m

Fig. 20 Sagittal-plane cross section passing through the location of the maximum E-field in the white matter volume. Colored traces denote contours of tissue meshes passing through the cross-sectional plane. Small pink balls are drawn at the locations experiencing high field strength (80–100% of the maximum field observed in the WM volume)

References

1. Cignoni P, Callieri M, Corsini M, Dellepiane M, Ganovelli F, Ranzuglia G. (2008). MeshLab: an open-source mesh processing tool. Sixth Eurographics Italian Chapter Conference (pp. 129–136).
2. Dropbox Repository (2019, November). TMS Modeling Package v1.1 Fall 2019. Online: https://www.dropbox.com/sh/0s0tl30a74wevr3/AAAEGu70k9Fx72hEkfdx3qfAa?dl=0.
3. Gimbutas Z, Greengard L, Magland J, Rachh M, Rokhlin V. (2019). *fmm3D documentation*. Release 0.1.0. Online: https://github.com/flatironinstitute/FMM3D.
4. Htet, A. T., Saturnino, G. B., Burnham, E. H., Noetscher, G., Nummenmaa, A., & Makarov, S. N. (2019a). Comparative performance of the finite element method and the boundary element fast multipole method for problems mimicking transcranial magnetic stimulation (TMS). *Journal of Neural Engineering, 16*, 1–13. https://doi.org/10.1088/1741-2552/aafbb9.
5. Htet, A. T., Burnham, E. H., Noetscher, G. M., Pham, D. N., Nummenmaa, A., & Makarov, S. N. (2019b). Collection of CAD human head models for electromagnetic simulations and their applications. *Biomedical Physics & Engineering Express, 6*(5), 1–13. https://doi.org/10.1088/2057-1976/ab4c76.
6. Iacono, M. I., Neufeld, E., Akinnagbe, E., Bower, K., Wolf, J., Vogiatzis Oikonomidis, I., et al. (2015). MIDA: A multimodal imaging-based detailed anatomical model of the human head and neck. *PLoS One, 10*(4), e0124126. https://doi.org/10.1371/journal.pone.0124126.
7. Kobayashi, M., & Pascual-Leone, A. (2003). Transcranial magnetic stimulation in neurology. *Lancet Neurology, 2*(3), 145–156. PMID: 12849236.
8. Makarov, S. N., Noetscher, G. M., & Nazarian, A. (2015). Low-frequency electromagnetic modeling for electrical and biological systems using MATLAB (pp. 648). New York: Wiley. ISBN-10: 1119052564.
9. Makarov, S. N., Noetscher, G. M., Raij, T., & Nummenmaa, A. (2018). A quasi-static boundary element approach with fast multipole acceleration for high-resolution bioelectromagnetic models. *IEEE Transactions on Biomedical Engineering, 65*(12), 2675–2683. https://doi.org/10.1109/TBME.2018.2813261.
10. McMullen D. (2017, November 11). *NIMH non-invasive brain stimulation E-Field modeling workshop*. Online: https://www.nimh.nih.gov/news/events/2017/brainstim/nimh-non-invasive-brain-stimulation-e-field-modeling-workshop.shtml.
11. Nielsen, J. D., Madsen, K. H., Puonti, O., Siebner, H. R., Bauer, C., Madsen, C. G., Saturnino, G. B., & Thielscher, A. (2018). Automatic skull segmentation from MR images for realistic volume conductor models of the head: Assessment of the state-of-the-art. *NeuroImage, 174*, 587–598. https://doi.org/10.1016/j.neuroimage.2018.03.001.
12. Opitz, A., Paulus, W., Will, S., Antunes, A., & Thielscher, A. (2015). Determinants of the electric field during transcranial direct current stimulation. *NeuroImage, 109*, 140–150. https://doi.org/10.1016/j.neuroimage.2015.01.033.
13. Rossi, S., Hallett, M., Rossini, P. M., Pascual-Leone, A., & Safety of TMS Consensus Group. (2009 Dec). Safety, ethical considerations, and application guidelines for the use of transcranial magnetic stimulation in clinical practice and research. *Clinical Neurophysiology, 120*(12), 2008–2039. https://doi.org/10.1016/j.clinph.2009.08.016.
14. Thielscher, A., Antunes, A., & Saturnino, G. B. (2015). Field modeling for transcranial magnetic stimulation: A useful tool to understand the physiological effects of TMS? *Conference Proceedings: Annual International Conference of the IEEE Engineering in Medicine and Biology Society*, 222–225. https://doi.org/10.1109/EMBC.2015.7318340.
15. Saturnino, G. B., Puonti, O., Nielsen, J. D., Antonenko, D., Madsen, K. H., & Thielscher, A. (2019a). SimNIBS 2.1: A comprehensive pipeline for individualized electric field modelling for transcranial brain stimulation. In S. Makarov, G. Noetscher, & M. Horner (Eds.), *Brain and human body modeling*. New York: Springer. ISBN 9783030212926.
16. Saturnino, G. B., Madsen, K. H., & Thielscher, A. (2019b). Efficient electric field simulations for transcranial brain stimulation. *bioRxiv, 541409*. https://doi.org/10.1101/541409.

17. Saturnino, G. B., Madsen, K. H., & Thielscher, A. (2019c Nov 6). Electric field simulations for transcranial brain stimulation using FEM: An efficient implementation and error analysis. *Journal of Neural Engineering, 16*(6), 066032. https://doi.org/10.1088/1741-2552/ab41ba.
18. Yousry, T. A., Schmid, U. D., Alkadhi, H., Schmidt, D., Peraud, A., Buettner, A., et al. (1997). Localization of the motor hand area to a knob on the precentral gyrus.A new landmark. *Brain, 120*, 141e57. https://doi.org/10.1093/brain/120.1.141.

Profiling General-Purpose Fast Multipole Method (FMM) Using Human Head Topology

Dung Ngoc Pham

1 Introduction

Recently, a quasistatic boundary element method (BEM) solution has been proposed [1] that combines the adjoint double-layer formulation of the boundary element method [2–4] which utilizes surface charges at the boundaries, the zeroth-order (piecewise constant) basis functions with accurate near-field integration, and the FMM accelerator [5–7]. This approach does not require explicit forming of the BEM matrix; an iterative solution with M iterations requires $O(MN)$ operations. The fast multipole method speeds up computation of a matrix-vector product of a numerical iterative solution via the boundary element method (BEM) by many orders of magnitude. In the past, it was successfully applied for modeling high-frequency electromagnetic [8, 9] and acoustic [10–12] scattering problems. It has also been applied to modeling transcranial magnetic stimulation (TMS) and demonstrated a fast computational speed and superior accuracy for high-resolution head models as compared to both the standard boundary element method and the finite element method of various orders [1, 13, 14]. The rapid increase in the use of FMM in such numerical modeling schemes calls for an accurate and thorough study of the performance of FMM in a wide range of scenarios.

The goal of this study is to benchmark the performance (both speed and memory consumption) of the fast multipole method or FMM [5, 6]. Here, we will use the established head collection and its barycentrically refined versions to perform the profiling of the FMM library provided by Z. Gimbutas and L. Greengard [7] and employed in [1, 12]. Such profiling implies running the FMM for all head geometries at different frequencies including the static case and averaging the respective results. One FMM runtime essentially corresponds to one iteration step of an iterative

D. N. Pham (✉)
ECE Department, Worcester Polytechnic Institute, Worcester, MA, USA
e-mail: dnpham@wpi.edu

BEM-FMM solution [1, 12]. Therefore, the data reported in the present study could be used to estimate the performance of a rather generic BEM-FMM algorithm if the number of iterations is approximately known or could be estimated a priori.

2 Materials and Methods

2.1 FMM Library of 2017

The core FMM algorithm is taken from the FMM library provided by Gimbutas and Greengard [7]. The latest version, last updated on November 8, 2017, is downloaded from the GitHub database to use in this study. We focus specifically in the function *fmm3d* which is used to solve Laplace and Helmholtz equations for a large number of target points. The compiled MEX versions of this function, namely, *fmm3d.mexw64* and *fmm3d.mexw64* for MATLAB compatibility in Windows and Linux, respectively, are used for all FMM calculations within the MATLAB environment. Depending on whether a solution for the Laplace or Helmholtz equation is desired, a wrapper function, either *lfmm3dpart* or *hfmm3dpart* – both available in the FMM library, is employed. A sample MATLAB command that calls *hfmm3dpart* to compute the Helmholtz equation is given by the following:

```
[U]=hfmm3dpart(iprec,k,nsource,source,ifcharge,charge,ifdipole,
dipstr,dipvec,ifpot,iffld,ntarget,target,ifpottarg,iffldtarg)
```

In the command above, the inputs parameters are as follows:

- `iprec`: precision flags for FMM
- `k`: wave number (Helmholtz parameter)
- `nsource`: number of source points
- `source`: source locations
- `ifcharge`: charge flag
- `charge`: charge values
- `ifdipole`: dipole flag
- `dipstr`: dipole magnitudes
- `dipvec`: dipole orientations
- `ifpot`: potential flag
- `iffld`: filed flag
- `ntarget`: number of targets
- `target`: target locations
- `ifpottarg`: target potential flag
- `iffldtarg`: target field flag

The output parameter is the struct U that contains the following fields:

- U.pot : the computed potential at source locations
- U.fld : the field at source location
- U.pottarg : potential at target locations
- U.fldtarg : field at target locations

In a similar manner, a sample MATLAB command that calls *lfmm3dpart* to compute the Laplace equation is given by the following:

```
[U]=lfmm3dpart(iprec,nsource,source,ifcharge,charge,ifdipole,
dipstr,dipvec,ifpot,iffld,ntarget,target,ifpottarg,iffldtarg)
```

where the input and output variables are similar to that of *hfmm3dpart*, except that for *lfmm3dpart* there is no wave number k. A more recent FMM library, developed by Flatiron Institute [15], is also investigated and compared with the library provided by Gimbutas and Greengard [7].

2.2   CAD Human Head Models

Every CAD human head model [8] has seven objects: the skin, skull, CSF, GM, cerebellum, WM, and ventricles head compartments. The models have an "onion" topology: the gray matter shell is a container for white matter, ventricles, and cerebellum objects; the CSF shell contains the gray matter shell; the skull shell contains the CSF shell; and finally, the skin or scalp shell contains the skull shell. The models have an average of 866,000 triangular facets and an average triangle quality of 0.25. The average edge length is 1.48 mm, and the average surface mesh density or resolution is 0.57 points per mm^2. A sample image of such a head model is shown in Fig. 1. Finer meshes with ~3,464,000 facets, obtained through one iteration of subdivision on the original CAD models, are also obtained for more intensive examinations on the scaling of timings and hardware resources.

2.3   Hardware Information

Windows server:

- 2 CPUs: Intel(R) Xeon(R) CPU E5–2683 v4 at 2.10GHz, 16 cores, 32 logical processors
- Physical memory (RAM): 256 GB
- OS: Microsoft Windows Server 2008 R2 Enterprise

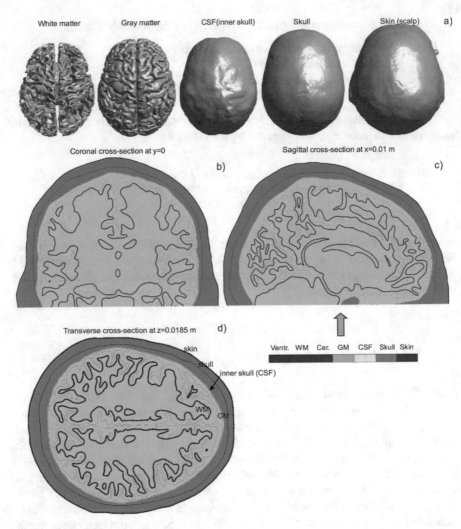

Fig. 1 Compartments of a sample brain model used in the testing of FMM software. (Image adapted from Htet et al. [8])

Linux server:

- 2 CPUs: Intel(R) Xeon(R) CPU E5–2690 0 at 2.90GHz, 64 bits
- Physical memory (RAM): 192 GB
- OS: Red Hat Enterprise Linux Server release 7.5 (Maipo)

2.4 Charge Assembly

For each of the 16 CAD models, a set of monopole charges are distributed over the surfaces of the triangular mesh so that at each triangle centers, a charge of random magnitude q is assigned. The electric potential generated at each triangle centroids, excluding self-contribution from the local charge, is given by the following:

$$\varphi(r) = \sum_i \frac{1}{4\pi\varepsilon_0} \frac{q}{|r - r_i|} e^{-jk|r-r_i|} \tag{1}$$

where ε_0 is permittivity of vacuum, q is electric charge of the source, k is the wave number, r is the target location at which the potential is sought, and r_i is the source location. The resultant electric field is given by the following:

$$E(r) = -\nabla\varphi = \sum_i + \frac{q}{4\pi\varepsilon_0} \left(\frac{r - r_i}{|r - r_i|^3} + jk\frac{r - r_i}{|r - r_i|^2} \right) e^{-jk|r-r_i|} \tag{2}$$

As a measure of FMM's performance, both the potential and the electric field, given by Eqs. (1) and (2), *are computed for all models, at the same triangle centroids, and excluding the self-contribution*. Through the function *hfmm3dpart* (and *lfmm3dpart* for the case ka = 0), the potential and the field are obtained simultaneously. With each head models, the calculations are done for three levels of accuracy:

- 2 digits (iprec = 0)
- 3 digits (iprec = 1)
- 6 digits (iprec = 2)

The frequencies for which the FMM algorithm is tested span over a wide range, which corresponds to *ka* values varying from 0 to 500. Here, *a* is the maximum of the x, y, and z coordinates of the model. Average value of *a* is 107.5754.

3   Results

3.1   Windows Platform (FMM 2017)

3.1.1   Original CAD Models

The relationship between runtimes of FMM calculations on Windows server, averaged over all 16 models, and ka is shown in Figs. 2, 3 and 4, with precisions 0 (two digits), 1 (three digits), and 2 (six digits), respectively. The discrete step for values of ka is 50, starting from ka = 0 and ending at ka = 500, with a more refined resolution within the low-frequency domain, from 0 to 50, where the step is 2.5. As can be seen in the insets in these figures, where the plot for low-frequency domain is magnified, there is always a sharp jump from the runtime for ka = 0 (Laplace case) to the very next value ka = 2.5. After the abrupt jump, FMM time increases steadily in a linear manner within the small ka domain (low frequencies) before growing exponentially

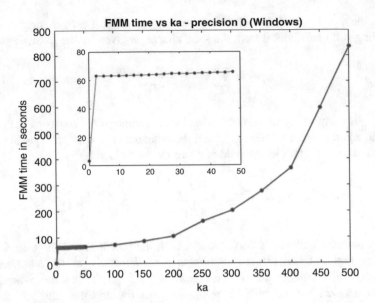

Fig. 2 FMM runtime within MATLAB platform (averaged over all sixteen heads) on Windows vs *ka*. The demanded precision is two digits accuracy. The Laplace case takes on average 3.41 s to complete with precision 0 (two digits)

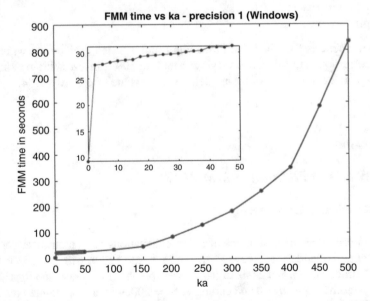

Fig. 3 FMM runtime within MATLAB platform (averaged over all sixteen heads) on Windows vs ka. The demanded precision is three digits accuracy. The Laplace case takes on average 9.23 s to complete with precision 1 (three digits)

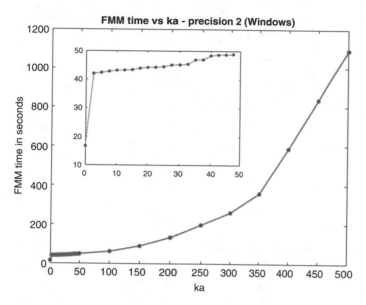

Fig. 4 FMM runtime within MATLAB platform (averaged over all sixteen heads) on Windows vs ka. The demanded precision is six digits accuracy. The Laplace case takes on average 16.70 s to complete with precision 3 (six digits)

at medium and large *ka*. The slope of the time-*ka* dependence in low *ka* with precision 0 (Fig. 2) is $(6.04 \pm 0.19) \times 10^{-2}$, whereas with precision 1, the slope is $(7.86 \pm 0.13) \times 10^{-2}$, and with precision 2, it is $(16.81 \pm 1.03) \times 10^{-2}$. From these numerical estimations, it can be concluded that the higher the demanded precision is, the steeper the time-*ka* slope becomes, and the runtime increases with *ka* in a higher rate.

In Fig. 5, the FMM runtimes for all three precision choices are plotted. *For the Laplace case (ka = 0), it takes on average 3.41 s for the computations to complete with precision 0. If higher level of accuracy is requested, the time taken increases to 9.23 s with precision 1 and 16.70 with precision 2.* This trend, however, is not replicated in the Helmholtz case, particularly at the low-frequency domain. As shown in Figs. 2 and 5, in the small *ka* domain, except for *ka* = 0, FMM is longest with precision 0, the lowest level of accuracy of all. More specifically, with precision 0, FMM runtime increases from an average of 62.88 s at *ka* = 2.5 to 65.33 s at *ka* = 50 (Fig. 1 or 4). Precision 2, the highest accuracy level tested, only takes the second longest amount of time, with 41.92 s for *ka* = 2.5, and rises to 50.77 s at *ka* = 50. Calculations within the low-frequency domain are fastest with precision 1, as it only takes 27.73 s to finish calculating for *ka* = 2.5 and 31.67 s for *ka* = 50. This rather unexpected behavior continues as far as *ka* = 150, where FMM time with precision 2, due to its rapid exponential rise, surpasses the runtime of precision 0. Toward the high end of the frequency range, precision 1 runtime, whose exponential rate is also higher than precision 0 (but not as high as 2), starts approaching before surpassing precision 0's runtime at *ka* = 500. Therefore, at very large values for *ka*, a more intuitively expected trend is observed, where FMM runtime with

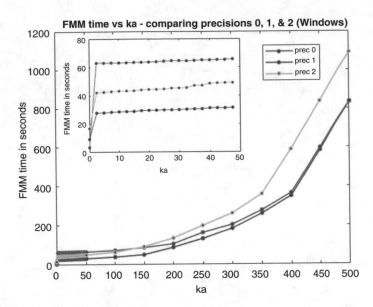

Fig. 5 Comparison among average FMM runtimes within MATLAB platform on Windows for when two, three, and six digits accuracy are demanded

precision 0 is lowest and has the slowest exponential rise as *ka* increases, followed by precision 1, and lastly, precision 2 is most time-consuming and has the quickest exponential rate.

3.1.2   First-Order Mesh Subdivision

In Figs. 6, 7 and 8, FMM runtimes on Windows server with precisions 0, 1, and 2, respectively, averaged over all 16 refined meshes obtained through one iteration of barycentric subdivision done on the original CAD models, are presented. Average mesh size quadruples; it is now 3.464 M facets. *Similar to when the calculations were done on the original head models, for ka = 0 (Laplace case), precision 0 takes the least amount of time, 10.41 s, compared to 41.50 s with precision 1 and 70.86 s with precision 2.* Also similar to the original head models, there are abrupt jumps in runtime from the Laplace case to the Helmholtz calculations, as shown in Figs. 6, 7 and 8. Within low-frequency limit, FMM runtime increases linearly with ka. The higher the requested accuracy is, the steeper the slope is; with precision 0, FMM time increases at the linear rate of $(11.02 \pm 1.90) \times 10^{-2}$ for ka in the low-frequency domain (0–50), while FMM time for calculations done with precision 1 increases at a higher rate, $(21.30 \pm 0.61) \times 10^{-2}$, and precision 2 calculation time increases most rapidly with the rate $(39.21 \pm 6.75) \times 10^{-2}$.

In Fig. 9, the runtimes of FMM applied to the first-order-refined meshes with all three precision choices are plotted. Again, within the low-frequency domain, precision 0 does not yield the fastest runtime. As shown in Fig. 9, at *ka* = 2.5, runtime

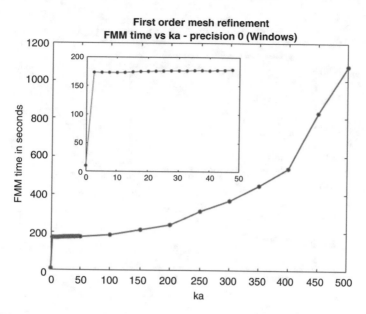

Fig. 6 FMM runtime for the refined models within MATLAB platform (averaged over all sixteen heads) on Windows vs ka. The demanded precision is two digits accuracy. The Laplace case takes on average 10.41 s to complete with precision 0 (two digits)

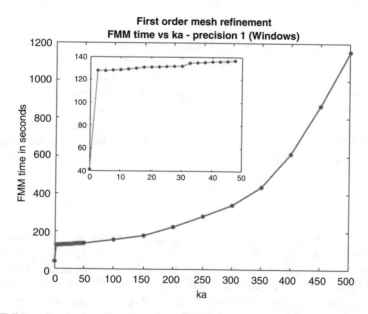

Fig. 7 FMM runtime for the refined models within MATLAB platform (averaged over all sixteen heads) on Windows vs ka. The demanded precision is three digits accuracy. The Laplace case takes on average 41.50 s to complete with precision 1 (three digits)

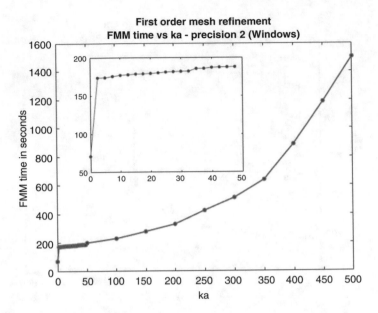

Fig. 8 FMM runtime for the refined models within MATLAB platform (averaged over all sixteen heads) on Windows vs ka. The demanded precision is six digits accuracy. The Laplace case takes on average 70.86 s to complete with precision 2 (six digits)

with precision 0 (172.62 s) is essentially comparable to precision 2 (173.26 s), and both are significantly slower than precision 1 (127.76 s). As *ka* increases, out of the three options, precision 0 has its runtimes increase at the slowest rate. Therefore, by ka = 350, FMM runtimes of precision 0 is surpassed by precision 1, and from then on, its runtimes are quickest, followed by precision 1, and precision 2 takes up the most time.

On the scaling of FMM runtime from the original CAD models that have average of $N_0 = 866,000$ facets to first-order-refined meshes with $N_1 = 3,464,000$ facets, the theoretical factor is as follows:

$$S = \frac{N_1 \log N_1}{N_0 \log N_0} = 4.4 \tag{3}$$

In Fig. 10, throughout the tested range for *ka*, the time ratio for precision 0 is always the smallest out of the three choices for precision. Quite surprisingly, the ratio for precision 2, for most of the time, is smaller than that of precision 1. Also interestingly, all three precision options start at *ka* = 0 with scaling factors relatively close to the theoretical values (perhaps with an exception with precision 0) and then decrease significantly as *ka* increases. Therefore, it appears that the higher the frequency is, the better the scaling in runtime is.

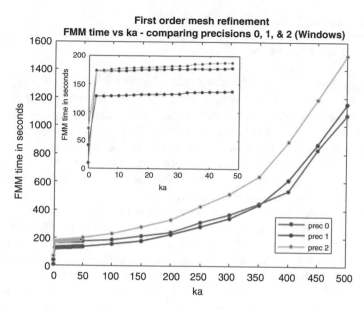

Fig. 9 Comparison among average FMM runtimes for the refined models in MATLAB on Windows for when two, three, and six digits accuracy is demanded

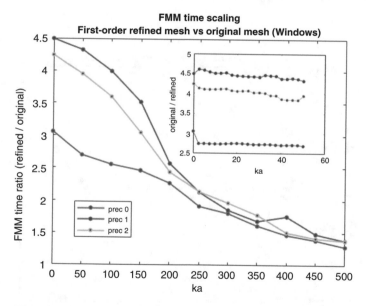

Fig. 10 Ratio between FMM time (Windows) for the refined models and FMM time for the original CAD models is plotted with respect to *ka*

3.2 Linux Platform (FMM 2017)

3.2.1 Original CAD Models

The relationship between runtimes of FMM calculations on Linux server, averaged over all 16 models, and *ka* is shown in Figs. 11, 12 and 13 with precisions 0, 1, and 2, respectively. A summary of runtimes for all three precisions is plotted in Fig. 14. Different from the same calculations done on the Windows platform, the runtime on Linux with precision 0, the lowest level of accuracy tested in this study, takes the least amount of time, while calculations with precision 1 are second, and precision 2, the highest level of accuracy with six digits, consumes the most amount of time. This order is held consistently throughout the entire frequency range from 0 to 500. It is also noticeable that runtimes with precisions 0 and 1, which guarantee accuracy within two and three digits, respectively, are comparable to each other, with calculations with precision 1 take slightly longer than 0. FMM runtime with precision 2, which demands six digits accuracy, takes significantly more time to finish. This rather intuitive behavior, however, is not present in the runtime profiling for Windows, which was discussed in Sect. 3.1. *As a note on how the two servers compare to each other in computing Laplace equation, it takes the Linux server 5.14 s to finish the calculation with precision 0, while for Windows, it is only 3.41 s. The Linux server, however, is notably faster on precision 1, taking 5.33 s to finish as compared*

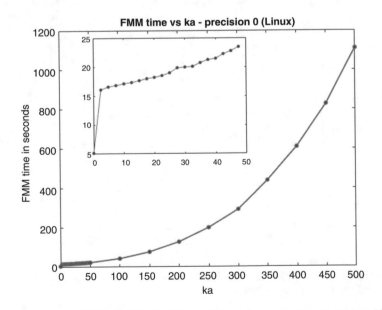

Fig. 11 FMM runtime within MATLAB platform (averaged over all sixteen heads) on Linux vs ka. The demanded precision is two digits accuracy. The Laplace case takes on average 5.14 s to complete with precision 0 (two digits)

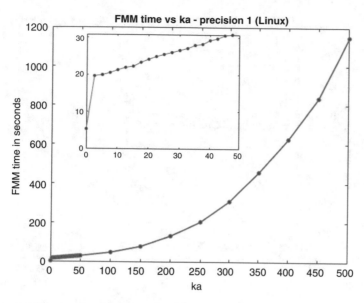

Fig. 12 FMM runtime within MATLAB platform (averaged over all sixteen heads) on Linux vs ka. The demanded precision is three digits accuracy. The Laplace case takes on average 5.33 s to complete with precision 1 (three digits)

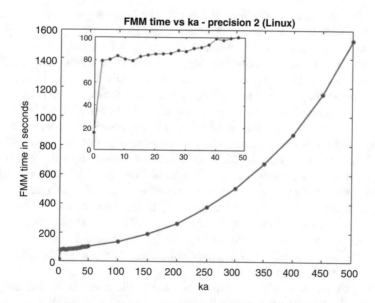

Fig. 13 FMM runtime within MATLAB platform (averaged over all sixteen heads) on Linux vs ka. The demanded precision is six digits accuracy. The Laplace case takes on average 15.31 s to complete with precision 2 (six digits)

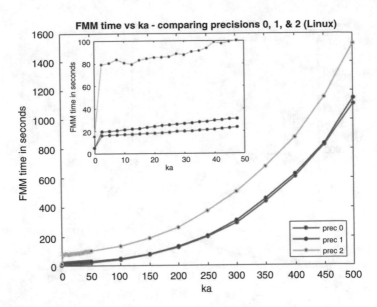

Fig. 14 Comparison among average FMM runtimes in MATLAB on Linux for when two, three, and six digits accuracy are demanded

to 9.23 s by the Windows platform. And finally, the Linux server is slightly better at precision 2, with 15.31 s, whereas Windows takes 16.70 s.

Comparing the calculations done on two platforms, Linux and Windows, we also observe major differences in how the runtime evolves as *ka* varies. First of all, FMM calculations within low-frequency domain are generally faster on Linux than on Windows, especially for low-to-medium accuracies (precisions 0 and 1). In particular, with precision 0, on Linux, it takes on average 19.51 s for *hfmm3dpart* to finish solving the Helmholtz equation on the original CAD models for *ka* within the range 0–50, whereas it takes on average 64.03 s to complete the same task on the Windows server. Similarly, it takes only 25.59 s on Linux to finish the calculations in the low-frequency domain with precision 1. Calculations on Windows, although not drastically slower than Linux as in the case of precision 0, still take 29.71 s to complete. If higher precisions are in demand, in fact, runtimes on Windows will catch up with Linux, and eventually, the speed on Windows will exceed. Evidently, for low-frequency calculations demanding precision 2 (six digits accuracy), it takes only 45.35 s for *hfmm3dpart* to complete computing, while a similar task takes the Linux server 88.88 s to complete.

Scaling of runtime as the frequency (or *ka*) is increased is another important metric. For the Linux server, within the range 0–50 for *ka*, FMM runtime increases linearly with the slope of $(16.06 \pm 0.31) \times 10^{-2}$ with precision 0. This is a significantly faster rate compared to the slope $(6.04 \pm 0.19) \times 10^{-2}$ (already mentioned in Sect. 3.1.1) for the same precision but on Windows. This comparison also holds with precisions 1 and 2, as on Linux the linear rates are $(26.20 \pm 0.07) \times 10^{-2}$ with precision 1 and a whopping $(50.09 \pm 3.81) \times 10^{-2}$

with precision 2. These values are far inferior than the rates $(7.86 \pm 0.13) \times 10^{-2}$ with precision 1 and $(16.81 \pm 1.03) \times 10^{-2}$ with precision 2 on Windows. Therefore, although the Linux server shows an edge over the Windows platform in computing low-frequency Helmholtz equation, the fact that runtimes on Linux increase too quickly with frequency makes it eventually get surpassed by Windows server at medium- and high-frequency domains. At $ka = 500$ (the highest value for ka tested in this study), runtimes on Windows are 834.69 s with precision 0, 833.91 s with precision 1, and 1090 s with precision 2, while on Linux, the numbers are 1108 s, 1148 s, and 1526 s, respectively.

3.2.2 First-Order Mesh Subdivision

In Figs. 15, 16, and 17, FMM runtimes on Linux server with precisions 0, 1, and 2, respectively, averaged over all 16 refined mesh (obtained through one iteration of barycentric subdivision done on the original CAD models), are presented. A summary of runtimes for all three precisions is plotted in Fig. 18. Comparing FMM done for the refined meshes on Linux and Windows, we obtain trends that are mostly similar to what was observed in the calculations done on the original models. *First, comparisons on Laplace calculations yield the same results: Linux with precision 0 takes 16.41 s, considerably slower than Windows, which takes only 10.41 s. For precision 1, the time is 17.7 s on Linux, again significantly better than Windows'*

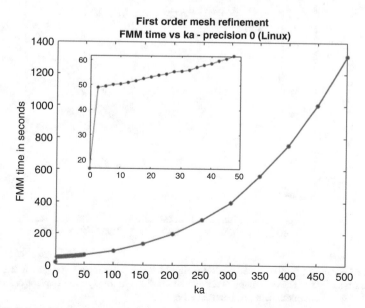

Fig. 15 FMM runtime for the refined models within MATLAB platform (averaged over all sixteen heads) on Linux vs ka. The demanded precision is two digits accuracy. The Laplace case takes on average 16.41 s to complete with precision 0 (two digits)

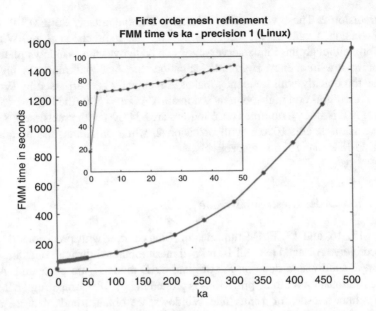

Fig. 16 FMM runtime for the refined models within MATLAB platform (averaged over all sixteen heads) on Linux vs ka. The demanded precision is two digits accuracy. The Laplace case takes on average 17.7 s to complete with precision 1 (three digits)

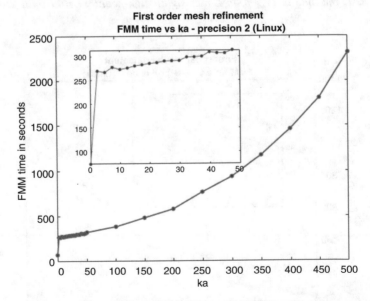

Fig. 17 FMM runtime for the refined models within MATLAB platform (averaged over all sixteen heads) on Linux vs ka. The demanded precision is two digits accuracy. The Laplace case takes on average 73.29 s to complete with precision 2 (six digits)

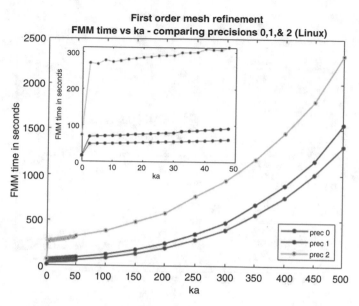

Fig. 18 Comparison among average FMM runtimes for the refined models in MATLAB on Linux for when two, three, and six digits accuracy are demanded

41.50 s, and finally, runtimes for precision 2 of the two platforms are comparable, 73.29 s for Linux and 70.86 s for Windows.

In terms of how runtimes of the three accuracy options compare to each other, as can be seen in Fig. 18, precision 0 takes the least amount of time, tightly followed by precision 1, while precision 2 is a lot more time-consuming. The same behavior was already discussed in Sect. 3.2.1 for calculations done with the original CAD models on Linux. The same conclusion, however, cannot be drawn for calculations done on Windows, as mentioned in Sects. 3.1.1 and 3.1.2.

The rates at which FMM runtimes increase with ka are significantly higher on Linux than on Windows. Within the low-frequency domain, where the FMM time-ka dependence appears to be linear, FMM runtime (with refined meshes) for precision 0 on Linux has the linear rate of $(28.90 \pm 0.45) \times 10^{-2}$, by a large margin higher than the rate $(11.02 \pm 1.9) \times 10^{-2}$ on Windows for the same precision level. Similarly, precision 1's runtimes increase at the rate $(53.91 \pm 0.90) \times 10^{-2}$ in the low-frequency range, while on windows, it is only $(21.30 \pm 0.61) \times 10^{-2}$. And finally, for precision 2, the rate is $(104.5 \pm 4.30) \times 10^{-2}$ on Linux and $(39.21 \pm 6.75) \times 10^{-2}$. Such steep slopes on the runtime-ka dependence in the low-frequency domain of the Linux platform are continued by the rapid exponentiations in the medium- and high-frequency ranges, which result in the Linux server being far inferior to Windows in computing the Helmholtz equations in high-frequency domain.

In Fig. 19, the actual ratio between FMM time (on Linux) of the refined models and the original model is plotted. Unlike the time ratio plot for Windows (Fig. 10), here, we see a more expected trend; the time ratio for precision 0 is lowest, followed

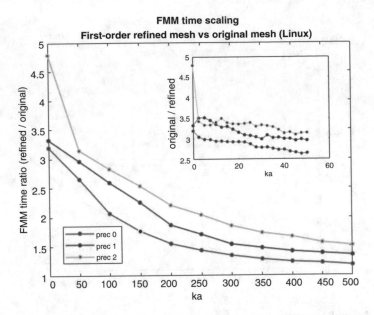

Fig. 19 Ratio between FMM time (Linux) for the refined models and FMM time for the original CAD models is plotted with respect to ka

by precision 1, while precision 2 has the largest ratio, and this behavior is maintained over the entire range $ka = 0–500$. As also shown in Fig. 19, except for only the Laplace case precision 2, all the ratios of the three precisions are below the theoretical scaling factor (see Eq. (3)). As ka increases, a decreasing trend is observed for all three plots. A similar result can be seen in Fig. 10 for Windows.

3.3 Memory Requirements

3.3.1 Original CAD Models

In this section, we discuss the memory consumed by FMM. Due to limitations on tools available, as well as Windows' uncompromising memory recording scheme, only memory information for calculations done on Linux is profiled and analyzed here. However, given the same FMM task, the (approximately) same amount of memory consumption is expected in both platforms. Therefore, valuable insights in memory requirements for performing FMM on Windows can still be drawn. In Figs. 20, 21, and 22, peak physical memory over FMM runtime is plotted with ka for precisions 0, 1, and 2, respectively. In Fig. 23, a summary of memory vs ka is plotted for all three precision choices. As can be seen in the figures, the general shapes of the curves are similar to the runtime plots displayed in previous sections; there is an abrupt jump from memory needed for solving Laplace equation to the Helmholtz

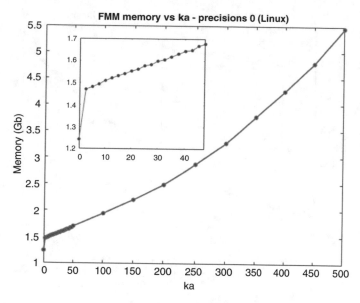

Fig. 20 Average peak memory consumption in MATLAB on Linux is plotted with respect to values of ka. The demanded precision is two digits accuracy

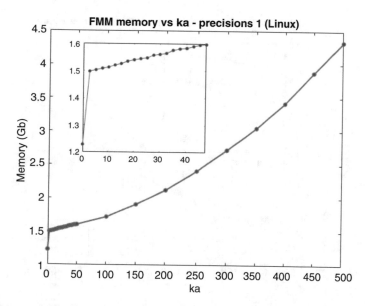

Fig. 21 Average peak memory consumption in MATLAB on Linux is plotted with respect to values of ka. The demanded precision is three digits accuracy

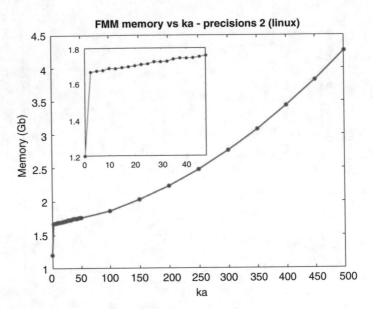

Fig. 22 Average peak memory consumption in MATLAB on Linux is plotted with respect to values of ka. The demanded precision is six digits accuracy

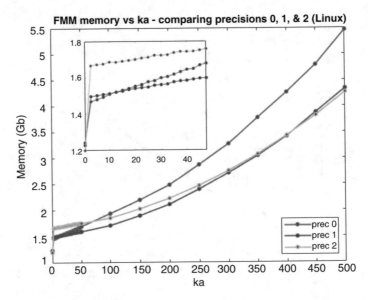

Fig. 23 A comparison among average peak memory consumption in MATLAB on Linux for when two, three, and six digits accuracy are demanded

case. For the Laplace case, calculations with all three levels of precisions require roughly the same amount of memory (approximately 1.2 Gb). Within the low-frequency domain, the memory-*ka* dependence is linear before evolving into an exponential growth in higher frequency ranges.

Perhaps the most astonishing results are that precision 0, the lowest level of accuracy, requires the most amount of memory, especially at high-frequency domain. As shown in Fig. 23, starting with very small values of ka, calculations for precision 0 consume the least amount of memory. However, both its linear rate within the low-frequency range and its exponential rate in the higher-frequency domain exceed that of precisions 1 and 2, resulting in the memory needed for precision 0 to somehow outgrow the supposedly more computationally demanding precision options. The memory plots for precisions 1 and 2, on the other hand, evolve in a more relaxed manner and, over the entire ka range from 0 to 500, tend to stay close to each other.

3.3.2 First-Order Mesh Subdivision

In Figs. 24, 25 and 26, peak physical memory over FMM runtime (performed on refined meshes) is plotted with *ka* for precisions 0, 1, and 2, respectively. In Fig. 27, a summary of memory vs *ka* is plotted for all three precision choices. Similar to the memory recorded for FMM done on the original meshes, here, we again observe that it is precision 0 that consumes the most memory, particularly at high frequencies (Fig. 28).

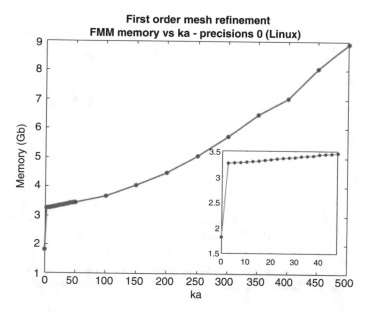

Fig. 24 Average peak memory consumption in MATLAB for the refined models on Linux is plotted with respect to values of ka. The demanded precision is two digits accuracy

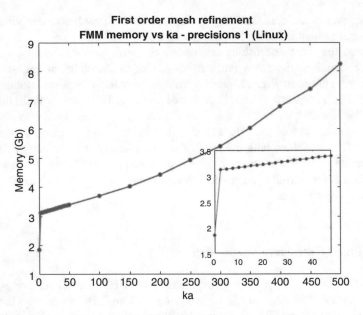

Fig. 25 Average peak memory consumption in MATLAB for the refined models on Linux is plotted with respect to values of ka. The demanded precision is three digits accuracy

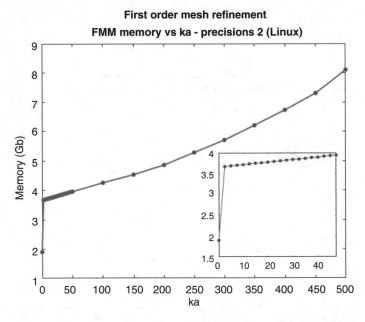

Fig. 26 Average peak memory consumption in MATLAB for the refined models on Linux is plotted with respect to values of ka. The demanded precision is six digits accuracy

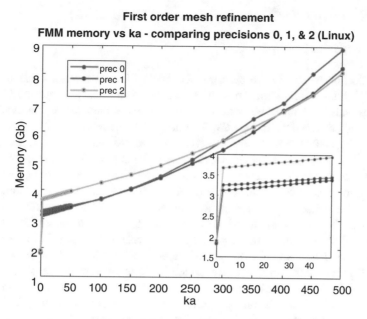

Fig. 27 A comparison among average peak memory consumption in MATLAB for the refined models on Linux for when two, three, and six digits accuracy are demanded

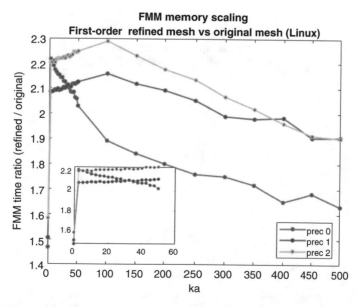

Fig. 28 Ratio between memory requirement (Linux) for the refined models and FMM time for the original CAD models is plotted with respect to ka

3.4   Second-Order Mesh Refinement (FMM 2017)

3.4.1   Windows Platform

In this section, we study the performance of FMM in calculations that use CAD models that have more refined meshes. These models have an average of 13,800,000 triangles and are obtained by performing two levels of barycentric subdivisions on the original head models. In Figs. 29, 30, and 31, FMM runtimes on Windows server with precisions 0, 1, and 2, respectively, averaged over all 16 refined meshes, are presented. In Fig. 32, a summary of runtime vs ka is plotted for all three precision choices. For the studies in this section, due to limited resources in computation, we restrict ourselves with ka values only from 0 to 50. As seen in Figs. 29 and 32, the runtimes for dense meshes are rather unpredictable, as there are no observable patterns for how the runtime of FMM evolves when ka is increased from 0 to 50. This volatility can be seen in the plots for all precisions 0, 1, and 2. In Fig. 33, the ratio between FMM time for the refined models (second level of mesh refinement) and FMM time for the original CAD models is plotted with respect to ka. As seen in Fig. 33, the scaling is best for calculations that require precision 0, followed by precision 2. FMM calculations for precision 1 has the largest scaling factor. This unintuitive scaling result was also seen for first-order mesh refinement (see Fig. 9) and was discussed in Sect. 3.1.2.

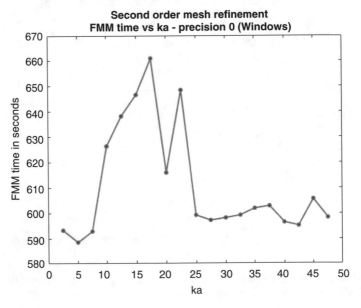

Fig. 29 Average FMM runtime for the doubly refined models in MATLAB on Windows is plotted with respect to values of ka. The demanded precision is two digits accuracy

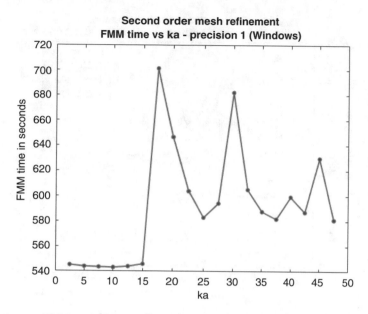

Fig. 30 Average FMM runtime for the doubly refined models in MATLAB on Windows is plotted with respect to values of ka. The demanded precision is three digits accuracy

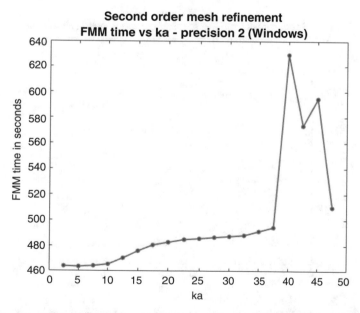

Fig. 31 Average FMM runtime for the doubly refined models in MATLAB on Windows is plotted with respect to values of ka. The demanded precision is six digits accuracy

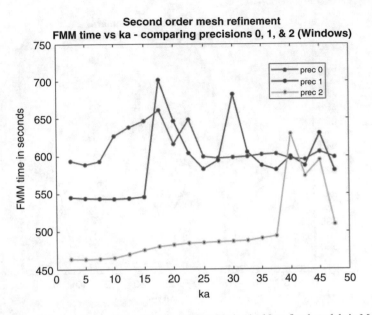

Fig. 32 A comparison among average FMM runtimes for the doubly refined models in MATLAB on Windows for when two, three, and six digits accuracy are demanded

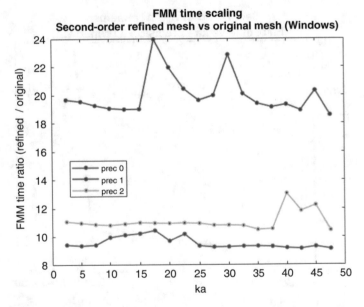

Fig. 33 Ratio between FMM time (Windows) for the doubly refined models and FMM time for the original CAD models is plotted with respect to ka

3.4.2   Linux Platform

In Figs. 34, 35, and 36, FMM runtimes on Linux server with precisions 0, 1, and 2, respectively, averaged over all 16 refined meshes (second order), are presented. In Fig. 37, a summary of runtime vs ka is plotted for all three choices of precision. Unlike the volatile behavior seen in the results for Windows, the FMM runtimes in Linux for meshes of second-order refinement increase linearly as ka increases, as expected. In Fig. 38, the ratio between FMM time for the refined models (second level of mesh refinement) and FMM time for the original CAD models is plotted with respect to ka. Again, we see that the scaling for precision 0 is lowest, while precision 1 has the highest scaling factor.

3.5   Comparisons with the New FMM Package (Summer 2019)

Very recently, a new version of the FMM library was published by the Flatiron Institute [15]. This package was downloaded for testing on June 11, 2019. Shortly after that, the online library was updated; this newer version was downloaded on June 20, 2019. In this section, we compare the performances of these two new versions of the FMM software with the original FMM library by Gimbutas and Greengard [7], last updated on November 8, 2017. Here, we focus on the

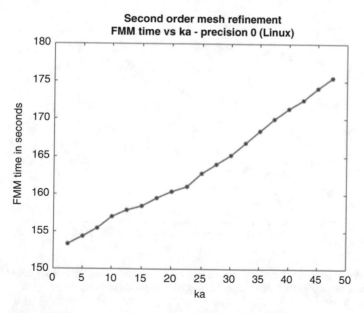

Fig. 34 Average FMM runtime for the doubly refined models in MATLAB on Linux is plotted with respect to values of ka. The demanded precision is two digits accuracy

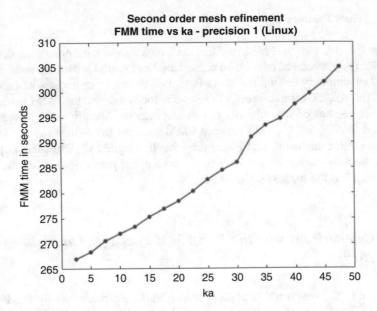

Fig. 35 Average FMM runtime for the doubly refined models in MATLAB on Linux is plotted with respect to values of ka. The demanded precision is three digits accuracy

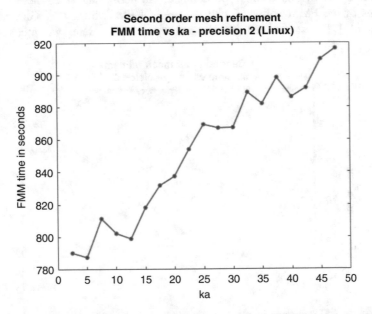

Fig. 36 Average FMM runtime for the doubly refined models in MATLAB on Linux is plotted with respect to values of ka. The demanded precision is six digits accuracy

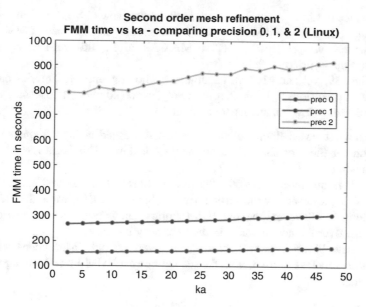

Fig. 37 A comparison among average FMM runtimes for the doubly refined models in MATLAB on Linux for when two, three, and six digits accuracy are demanded

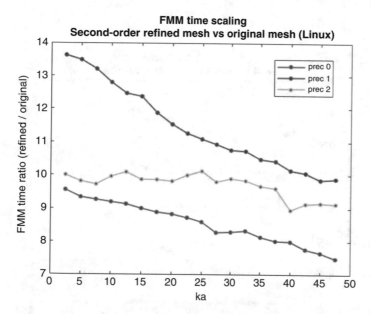

Fig. 38 Ratio between FMM time (Linux) for the doubly refined models and FMM time for the original CAD models is plotted with respect to ka

performance in Linux. *Note the legends in the plots*: "new2" = newly updated MEX file (June 20, 2019), "new1" = MEX file in the original new FMM library (downloaded June 11, 2019), "old" = MEX file in the old FMM library (2017), and zero time = program fails.

In Figs. 39, 40, and 41, FMM runtimes of the new and old libraries on Linux server with precisions 0, 1, and 2, respectively, averaged over all 16 refined meshes, are presented. A few comments are in order:

- The newest FMM library of the three tested crashed at high frequencies. The reason for these crashes is due to memory leaking. This issue has been fixed meanwhile.
- At low frequencies (ka ≤50), the newest library has the best runtime (June 20, 2019), followed by the second newest (June 11, 2019), while the old library is slowest. However, the old FMM library shows significant superiority in runtime over the new libraries as the frequency increases.
- Both new libraries show improvements over the old FMM codes when the Laplace solver is in used, with the second newest FMM library having the best runtime.

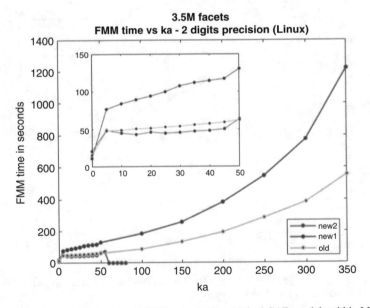

Fig. 39 FMM runtime of the new and old libraries for the original CAD models within MATLAB platform (averaged over all sixteen heads) on Linux vs ka. The demanded precision is two digits accuracy

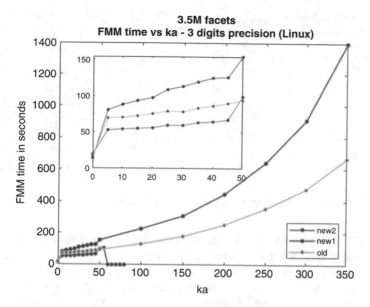

Fig. 40 FMM runtime of the new and old libraries for the original CAD models within MATLAB platform (averaged over all sixteen heads) on Linux vs ka. The demanded precision is three digits accuracy

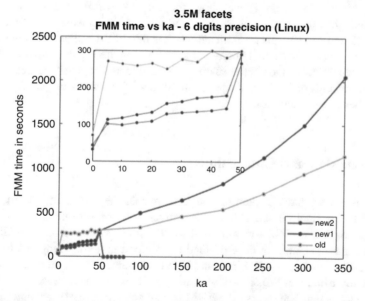

Fig. 41 FMM runtime of the new and old libraries for the original CAD models within MATLAB platform (averaged over all sixteen heads) on Linux vs ka. The demanded precision is six digits accuracy

Table 1 FMM runtimes in
Linux and Windows for different number of sets of
charge distributions are shown

Number of sets, N	Total runtime, T_N		
	Linux1	Linux2	Windows
1	5.37	5.66	2.99
2	6.18	5.64	3.87
4	9.10	7.69	6.43
8	12.82	10.75	11.90
16	19.72	17.94	22.89
32	35.46	35.43	43.70
64	66.76	75.58	90.07
128	X	X	X

3.6 Solving for Multiple Solutions in Parallel

The FMM software updated by the Flatiron Institute also allows one to solve
multiple right-hand sides at a time. In this section, we report the scaling of the
FMM code when it is used to solve the Laplace equation with multiple sets of
charges simultaneously. The original CAD models are used. The performances of
the MEX function compiled for Windows, Linux (June 11, 2019), and the updated
version for Linux (June 20, 2019) are compared against each other. These three
MEX options are, respectively, called Windows, Linux1, and Linux2 in the following table. Runtimes are in seconds, and "X" indicates program failures due to
memory issues. In Table 1, we show the runtimes of FMM run on different
platforms, when different number of sets are included. The scale curves are shown
in Figs. 42 and 43. The most significant result is that FMM run on Windows has the
shortest runtime when a small number of sets are computed. The Linux platform, on
the other hand, has a much better scaling rate, and therefore, the runtimes on Linux
(for both versions of the library) are progressively better than those on Windows as
the number of sets increases.

4 Discussion and Conclusions

In this paper, we have studied the performance of the fast multipole method in
computing the Laplace and Helmholtz source-to-source potentials within human
head topology. The FMM software used for this study was developed by Gimbutas
and Greengard [7], and we have profiled the method in a wide range of frequency
values, mesh density, in both Linux and Windows frameworks, and with all choices
of precision available. We showed that for problems that have reasonably "small"
sizes (up to 3–4 million facets), the FMM runtime and memory consumption evolve
in a predictable manner when run on both Linux and Windows. In particular, the
runtime (and memory usage) varies exponentially as the frequency increases, with a
small linear dependence at small values of frequency. We also observe universally a
sharp, discrete increase in both runtime and memory from when the FMM software
is using the Laplace solver to when the Helmholtz solver is used.

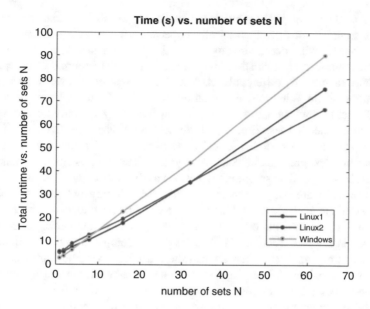

Fig. 42 The plots of runtime vs number of sets of charges are shown for Windows and Linux. Original CAD head models are used

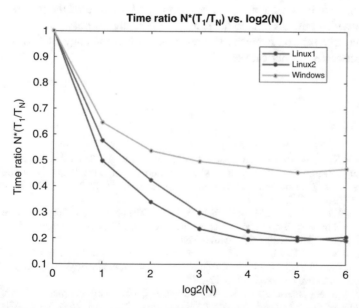

Fig. 43 The log plots of how runtimes on Windows and Linux scale with the number of sets of charges are shown. Original CAD head models are used

Upon studying the scaling efficiency of FMM, we showed that the algorithm deviates slightly from the theoretically expected scaling factors for runtime and memory, along with decreasing trends as the problem size increases. We also observed a number of interesting and unexpected results in terms of comparisons in runtimes required by different level of accuracy. In particular, we showed that for calculations run in Windows, the runtime needed for the three levels of accuracy tested did not follow any particular order at low frequency and only formed a pattern (low accuracy needed less time than high accuracy) when the frequency is sufficiently high. Resources needed for calculations performed in Linux were shown to have much more predictable patterns among different choices of accuracy and problem sizes, as well as smoother evolutions as the frequency changes.

We also compare the performance of this FMM library with a newer package (downloaded June 11, 2019) and its updated version (downloaded June 20, 2019). The results show that although the new library has better performance at low frequency, in Laplace calculations, it scales poorly compared to the old library and therefore is time-wise less efficient than the old FMM codes at high frequencies. Finally, we investigate the scaling rate of the new library with increasing number of right-hand sides (rhs) being solved simultaneously. The overall results show that while the Windows platform has the shorter runtime for small number of rhs, the FMM code compiled for Linux has better scaling rate and therefore has better runtime when the number of rhs increases.

With this study, we have benchmarked the performance of the general-purpose FMM, provided new insights to the behavior of the algorithms in various scenarios, and effectively offered a means to pre-estimate the efficiency of FMM-based or FMM-accelerated numerical methods.

References

1. Makarov, S. N., Noetscher, G. M., Raij, T., & Nummenmaa, A. (2018). A quasi-static boundary element approach with fast multipole acceleration for high-resolution bioelectromagnetic models. *IEEE Transactions on Biomedical Engineering, 65*(12), 2675–2683. https://doi.org/10.1109/TBME.2018.2813261.
2. Barnard, A. C. L., Duck, I. M., & Lynn, M. S. (1967). The application of electromagnetic theory to electrocardiology: I. derivation of the integral equations. *Biophysical Journal, 7*(5), 443–462. https://doi.org/10.1016/S0006-3495(67)86598-6.
3. Kybic, J., Clerc, M., Abboud, T., Faugeras, O., Keriven, R., & Papadopoulo, T. (2005). A common formalism for the integral formulations of the forward EEG problem. *IEEE Transactions on Medical Imaging, 24*(1), 12–28.
4. Makarov, S. N., Noetscher, G. M., & Nazarian, A. (2016). *Low-Frequency Electromagnetic Modeling of Electrical and Biological Systems Using MATLAB*. New York: Wiley. ISBN: 978-1-119-05256-2.
5. Rokhlin, V. (1985). Rapid solution of integral equations of classical potential theory. *Journal of Computational Physics, 60*(2), 187–207. https://doi.org/10.1016/0021-9991(85)90002-6.
6. Greengard, L., & Rokhlin, V. (1987). A fast algorithm for particle simulations. *Journal of Computational Physics, 73*(2), 325–348. https://doi.org/10.1016/0021-9991(87)90140-9.

7. Gimbutas, Z., & Greengard, L. (2015). Simple FMM Libraries for electrostatics, slow viscous flow, and frequency-domain wave propagation. *Communications in Computer Physics, 18*(2), 516–528. https://doi.org/10.4208/cicp.150215.260615sw.
8. Htet, A. T., Noetscher, G. M., Burnham, E. H., Pham, D. N., & Nummenmaa, A. (2019). Makarov SN. Collection of CAD human head models for electromagnetic simulations and their applications. *Biomedical Physics & Engineering Express, 5*(6). https://doi.org/10.1088/2057-1976/ab4c76.
9. Song, J., Lu, C. C., & Chew, W. C. (1997). Multilevel Fast Multipole Algorithm for Electromagnetic Scattering by Large Complex Objects. *IEEE Transactions on Antennas and Propagation, 45*(10), 1488–1493. doi: r S 0018-926X(97)07215-3.
10. Burgschweiger R, Ochmann M, Schäfer I, Nolte B. (2012). Performance-Optimierung und Grenzen eines Multi-Level Fast Multipole Algorithmus für akustische Berechnungen. *38. Jahrestagung für Akustik* (DAGA 2012), März 2012, Darmstadt, Germany.
11. Burgschweiger R, Schäfer I, Ochmann M, & Nolte, B. (2012). Optimization and Limitations of a Multi-Level Adaptive-Order Fast Multipole Algorithm for Acoustical Calculations. *Acoustics* 2012, Mai 2012, Hong Kong.
12. Burgschweiger R, Schäfer I, Ochmann M, & Nolte B (2013). The Combination of a Multi-Level Fast Multipole Algorithm with a Source-Clustering Method for higher expansion orders. *39. Jahrestagung für Akustik* (AIA/DAGA 2013), März 2013, Meran, Italy.
13. Gomez, L., Dannhauer, M., Koponen, L., & Peterchev, A. V. (2018). Conditions for numerically accurate TMS electric field simulation. *bioRxiv*, 505412. https://doi.org/10.1101/505412.
14. Htet, A. T., Saturnino, G. B., Burnham, E. H., Noetscher, G., Nummenmaa, A., & Makarov, S. N. (2019). Comparative performance of the finite element method and the boundary element fast multipole method for problems mimicking transcranial magnetic stimulation (TMS*). Journal of Neural Engineering, 16*, 1–13. https://doi.org/10.1088/1741-2552/aafbb9.
15. Documentation: https://fmm3d.readthedocs.io/en/latest/. Source code: https://github.com/flatironinstitute/FMM3D.

Analytical Solution for the Electric Field Response Generated by a Nonconducting Ellipsoid (Prolate Spheroid) in a Conducting Fluid Subject to an External Electric Field

Andrey B. Yakovlev and Valeriya S. Federyaeva

1 Introduction

In this short study, we retrieve and discuss an analytical solution for the electric field response generated by a nonconducting ellipsoid (prolate spheroid) in a homogeneous conducting fluid subject to an external primary electric field. We assume that the primary field can have any angle of incidence with respect to the longer axis of the ellipsoid. We assume that the ellipsoid has a zero (a nonconducting cell membrane) conductivity.

In the main text, we will utilize the well-known analogy between the electrostatics of dielectrics and DC conduction [1–3]. This analogy means that the basic equations and the corresponding solutions become identical when the ratio(s) of dielectric constants will coincide with the ratio(s) of conductivities. Since the solution of the present problem for dielectric materials does exist [1], its conversion to the conducting case is rather straightforward, but it requires extra steps for computing the induced charge density at the interface.

Here we also note that such an analogy is not the only one: one might consider a relevant fluid dynamics analogy as well. For example, the solution for a potential flow of an ideal incompressible fluid around a sphere with radius R [4] yields the expression for the hydrodynamic potential in the following form (the flow direction is along the x-axis):

A. B. Yakovlev (✉) · V. S. Federyaeva
Department of Mathematics and Mechanics, Saint Petersburg State University, Saint Petersburg, Russian Federation
e-mail: a.b.yakovlev@spbu.ru

© The Author(s) 2021
S. N. Makarov et al. (eds.), *Brain and Human Body Modeling 2020*,
https://doi.org/10.1007/978-3-030-45623-8_22

$$\varphi_v = -v_{0x} r \cos\theta + \frac{B\cos\theta}{r^2} \tag{1}$$

An unknown coefficient B is found from the condition $\frac{\partial \varphi_v}{\partial r} = 0$, which allows us to write the following expression for the potential:

$$\varphi_v = -v_{0x} r \cos\theta - \frac{v_{0x} R^3 \cos\theta}{2r^2} \tag{2}$$

Simultaneously, the tangential velocity at the sphere surface is given by

$$v_\tau = \frac{1}{r}\frac{\partial \varphi_v}{\partial \theta}\bigg|_{r=R} = v_{0x}\sin\theta + \frac{v_{0x} R^3 \sin\theta}{2r^3}\bigg|_{r=R} = \frac{3v_{0x}\sin\theta}{2} \tag{3}$$

This solution is equivalent to the steady electric current solution for a nonconducting sphere in a conducting fluid. In particular, the electric field inside the sphere is given by (cf. [3]):

$$\vec{E}_i = \frac{3}{2}\vec{E}_0 \tag{4}$$

where $\vec{E}_0$ is the primary electric field. On the other hand, for a dielectric sphere with permittivity ε in a dielectric medium with permittivity ε, the corresponding solution for the field inside has the following form [1–3]:

$$\vec{E}_i = \frac{3\varepsilon^{(e)}}{2\varepsilon^{(e)} + \varepsilon^{(i)}}\vec{E}_0 \tag{5}$$

Two solutions (4) and (5) indeed coincide when

$$\varepsilon^{(i)}\big/\varepsilon^{(e)} = 0 \tag{6}$$

2 Materials and Methods

In Ref. [1], the problem is solved for the electric field of an ellipsoid with half axes a, b, c, with permittivity $\varepsilon^{(i)}$ in a dielectric medium with permittivity $\varepsilon^{(e)}$ when a primary or external field $\vec{E}_0$ is applied. This solution will be repeated here; the final result implies that the permittivities should be replaced by conductivities.

We consider an ellipsoid in the form of a prolate spheroid ($a > b = c$). The coordinate system (see Fig. 1) is chosen as follows: the z-axis is directed along a so that the angle between the z-axis and the vector $\vec{E}_0$ is less than 90 degrees. The x-axis is located in a plane defined by the z-axis and vector $\vec{E}_0$. The y-axis is then chosen to construct the right-handed Cartesian system.

In this coordinate system, the depolarization tensor [1] becomes diagonal with the following components:

$$n_{zz} = \frac{abc}{2} \int_0^\infty \frac{ds}{(s+a^2)\sqrt{(s+a^2)(s+b^2)(s+c^2)}},$$

$$n_{yy} = \frac{abc}{2} \int_0^\infty \frac{ds}{(s+b^2)\sqrt{(s+a^2)(s+b^2)(s+c^2)}}, \qquad (7)$$

$$n_{xx} = \frac{abc}{2} \int_0^\infty \frac{ds}{(s+c^2)\sqrt{(s+a^2)(s+b^2)(s+c^2)}}$$

For the prolate spheroid, simplifications are made in the following form [1]:

$$n_{zz} = \frac{1 - e^2}{e^3}(\text{Arthe} - e),$$

$$n_{yy} = n_{xx} = \frac{1}{2}(1 - n_{zz}) \qquad (8)$$

where $e = \sqrt{1 - b^2/a^2}$ is the ellipsoid eccentricity.

Fig. 1 Ellipsoid along with the coordinate system used

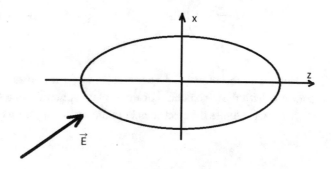

3   Results for the Potential and the Electric Field

Accordingly, the electric potential *everywhere in space* and the electric field *inside* the ellipsoid have the following form [1] (the ratio of dielectric permittivities in the equations given below *must be set to zero* to obtain the result for the nonconducting ellipsoid in the conducting fluid):

$$\varphi^{(i)} = -\left\{ \frac{E_{0x}x}{1 + \left(\dfrac{\varepsilon^{(i)}}{\varepsilon^{(e)}} - 1\right)n_{xx}} + \frac{E_{0y}y}{1 + \left(\dfrac{\varepsilon^{(i)}}{\varepsilon^{(e)}} - 1\right)n_{yy}} + \frac{E_{0z}z}{1 + \left(\dfrac{\varepsilon^{(i)}}{\varepsilon^{(e)}} - 1\right)n_{zz}} \right\}$$

$$\varphi^{(e)} = \frac{-E_{0x}x\left(1 + \dfrac{abc}{2}\left(\dfrac{\varepsilon^{(i)}}{\varepsilon^{(e)}} - 1\right)\displaystyle\int_0^\xi \frac{ds}{(s+c^2)\sqrt{(s+a^2)(s+b^2)(s+c^2)}}\right)}{1 + \left(\dfrac{\varepsilon^{(i)}}{\varepsilon^{(e)}} - 1\right)n_{xx}} +$$

$$+ \frac{-E_{0y}y\left(1 + \dfrac{abc}{2}\left(\dfrac{\varepsilon^{(i)}}{\varepsilon^{(e)}} - 1\right)\displaystyle\int_0^\xi \frac{ds}{(s+b^2)\sqrt{(s+a^2)(s+b^2)(s+c^2)}}\right)}{1 + \left(\dfrac{\varepsilon^{(i)}}{\varepsilon^{(e)}} - 1\right)n_{yy}} +$$

$$+ \frac{-E_{0z}z\left(1 + \dfrac{abc}{2}\left(\dfrac{\varepsilon^{(i)}}{\varepsilon^{(e)}} - 1\right)\displaystyle\int_0^\xi \frac{ds}{(s+a^2)\sqrt{(s+a^2)(s+b^2)(s+c^2)}}\right)}{1 + \left(\dfrac{\varepsilon^{(i)}}{\varepsilon^{(e)}} - 1\right)n_{zz}}$$

$$E_x = \frac{E_{0x}}{1 + \left(\dfrac{\varepsilon^{(i)}}{\varepsilon^{(e)}} - 1\right)n_{xx}}, \quad E_y = \frac{E_{0y}}{1 + \left(\dfrac{\varepsilon^{(i)}}{\varepsilon^{(e)}} - 1\right)n_{yy}}, \quad E_z = \frac{E_{0z}}{1 + \left(\dfrac{\varepsilon^{(i)}}{\varepsilon^{(e)}} - 1\right)n_{zz}}$$

$$(9)$$

Here, ξ is the ellipsoidal coordinate that is a constant for all ellipsoids being confocal with the given one. For the prolate spheroid, one has $\xi = a/\sqrt{a^2 - b^2}$.

Now, we express the electric field in the following form (ϑ is the elevation angle):

$$E_{0z} = E_0 \cos \vartheta,$$
$$E_{0x} = E_0 \sin \vartheta, \qquad (10)$$
$$E_{0y} = 0$$

After substitution of Eq. (10) into Eq. (9) and using Eq. (6), we obtain

$$\varphi^{(i)} = -\left\{ \frac{E_0 x \sin \vartheta}{1 - n_{xx}} + \frac{E_0 z \cos \vartheta}{1 - n_{zz}} \right\}$$

$$\varphi^{(e)} = \frac{-E_0 x \sin \vartheta \left(1 - \dfrac{abc}{2} \displaystyle\int_0^{\xi} \dfrac{ds}{(s+c^2)\sqrt{(s+a^2)(s+b^2)(s+c^2)}} \right)}{1 - n_{xx}} +$$

$$\qquad (11)$$

$$+ \frac{-E_0 z \cos \vartheta \left(1 - \dfrac{abc}{2} \displaystyle\int_0^{\xi} \dfrac{ds}{(s+a^2)\sqrt{(s+a^2)(s+b^2)(s+c^2)}} \right)}{1 - n_{zz}}$$

$$E_z = \frac{E_0 \cos \vartheta}{1 - n_{xx}}, E_x = \frac{E_0 \sin \vartheta}{1 - n_{yy}}, E_y = 0$$

It follows from Eqs. (8) and (11) that the electric field within the ellipsoid is not parallel to the external primary electric field.

4 Results for the Surface Charge Density

For the prolate spheroid, the ellipsoidal coordinates are reduced to the prolate spheroidal coordinates ξ, η, and ψ. They are converted to Cartesian coordinates using the following expressions [5]:

$$x = \frac{d}{2}\sqrt{(\xi^2 - 1)(1 - \eta^2)} \cos \psi$$

$$y = \frac{d}{2}\sqrt{(\xi^2 - 1)(1 - \eta^2)} \sin \psi \qquad (12)$$

$$z = \frac{d}{2}\xi\eta$$

where d is the spacing between two focal points of the ellipsoid, which is equal to $2\sqrt{a^2 - b^2}$.

In order to find the surface charge distribution, one needs to find the normal derivative of the electric potential at the surface. Since the outer normal derivative of the potential is equal to zero, only the inner derivative is needed.

It is convenient to compute the inner derivative using coordinates ξ, η, and ψ. Then,

$$\left.\frac{d\varphi}{dn}\right|_{surf} = -\frac{1}{H_\xi}\left.\frac{\partial\varphi^{(i)}}{\partial\xi}\right|_{surf} \tag{13}$$

where H_ξ is the corresponding Lamé coefficient. We find this coefficient in the following form:

$$
\begin{aligned}
H_\xi &= \sqrt{\left(\frac{\partial x}{\partial\xi}\right)^2 + \left(\frac{\partial y}{\partial\xi}\right)^2 + \left(\frac{\partial z}{\partial\xi}\right)^2} = \\
&= \frac{d}{2}\sqrt{\frac{1-\eta^2}{\xi^2-1}\xi^2 + \eta^2} = \\
&= \frac{d}{2}\sqrt{\frac{\xi^2-\eta^2}{\xi^2-1}}
\end{aligned}
\tag{14}
$$

Here, ξ_0 is the value of ξ on the ellipsoid surface. Following the definition of the prolate spheroid, one obtains $\xi = a/\sqrt{a^2-b^2}$ and $\eta = z/a$. Then,

$$
\begin{aligned}
\left.\frac{-1}{H_\xi}\frac{\partial\varphi^{(i)}}{\partial\xi}\right|_{\xi=\xi_0} &= \left.\left(\frac{E_{0x}}{1-n_{xx}}\frac{1}{H_\xi}\frac{\partial x}{\partial\xi} + \frac{E_{0z}}{1-n_{zz}}\frac{1}{H_\xi}\frac{\partial z}{\partial\xi}\right)\right|_{\xi=\xi_0} = \\
&= \left.\left(\frac{E_{0x}}{1-n_{xx}}\frac{\xi\sqrt{1-\eta^2}}{\sqrt{\xi^2-\eta^2}} + \frac{E_{0z}}{1-n_{zz}}\frac{\eta\sqrt{\xi^2-1}}{\sqrt{\xi^2-\eta^2}}\right)\right|_{\xi=\xi_0} = \\
&= \frac{E_0\sin\vartheta}{1-n_{xx}}\frac{a\sqrt{a^2-z^2}}{\sqrt{a^4+z^2b^2-z^2a^2}} + \\
&+ \frac{E_0\cos\vartheta}{1-n_{zz}}\frac{bz\sqrt{a^2-b^2}}{a\sqrt{a^4+z^2b^2-z^2a^2}}
\end{aligned}
\tag{15}
$$

It follows from here that the surface charge density as a function of z and ϑ (the elevation angle) is obtained in the following form:

$$\sigma = \varepsilon_0\left(\frac{E_0\sin\vartheta}{1-n_{xx}}\frac{a\sqrt{a^2-z^2}}{\sqrt{a^4+z^2b^2-z^2a^2}} + \frac{E_0\cos\vartheta}{1-n_{zz}}\frac{bz\sqrt{a^2-b^2}}{a\sqrt{a^4+z^2b^2-z^2a^2}}\right) \tag{16}$$

which completes the solution. This is the surface charge density residing on the surface of the nonconducting ellipsoid in the conducting fluid.

Consider the primary external field that has only one component parallel to the z-axis, and consider b that tends to zero in Eq. (16). Then, the surface charge density is only different from zero at the tips of the ellipsoid, that is, at $z \to a$ or $z \to -a$. This is a physically meaningful result, which is observed for a thin cylinder in a coaxial external field. Here, the opposite charges are concentrated close to the cylinder tips only.

References

1. Stratton, J. A. (1941). *Electromagnetic theory*. New York/London: McGraw-Hill Book Company.
2. Landau, L. D., Lifshitz, E. M., Pitaevskii, L.P. (1984). *Electrodynamics of continuous media* (Vol. 8, 2nd ed). Butterworth-Heinemann.
3. Makarov, S. N., Noetscher, G. M., & Nazarian, A. (2015). *Low-frequency electromagnetic modeling for electrical and biological systems using MATLAB* (648 p). New York: Wiley. ISBN-10: 1119052564.
4. Landau, L. D., Lifshitz, E. M. (1987). *Fluid mechanics* (Vol. 6, 2nd ed). Butterworth-Heinemann.
5. Chonina S. N. In Russian: Приближение сфероидальных волновых функций конечными рядами. www.computeroptics.smr.ru

Example of Steady-State Electric-Current Modeling of a Complicated Cellular Topology with Boundary Element Fast Multipole Method

Vishwanath Iyer, William A. Wartman, Aapo Nummenmaa, and Sergey N. Makarov

1 Introduction

The recently developed quasistatic formulation of the boundary element fast multipole method or BEM-FMM [1, 2] is based on the integral equation written in terms of induced charge density at the interfaces, which is naturally coupled with the general-purpose fast multipole method [3, 4]. It could potentially be applied to perform quasistatic electromagnetic modeling of complicated 3D surface topologies on the cellular level. Modeling such topologies with the finite element method or FEM is hardly possible in practice, in particular due to the complexity of creation of the required volumetric tetrahedral mesh and the required high mesh resolution near multiple fine model joints.

V. Iyer (✉)
The MathWorks, Inc., Natick, MA, USA
e-mail: vishwanath.iyer@mathworks.com

W. A. Wartman
Electrical and Computer Engineering Department, Worcester Polytechnic Institute, Worcester, MA, USA

A. Nummenmaa
Athinoula A. Martinos Center for Biomedical Imaging, Massachusetts General Hospital, Charlestown, MA, USA

S. N. Makarov
Electrical and Computer Engineering Department, Worcester Polytechnic Institute, Worcester, MA, USA

Athinoula A. Martinos Center for Biomedical Imaging, Massachusetts General Hospital, Charlestown, MA, USA

© The Author(s) 2021
S. N. Makarov et al. (eds.), *Brain and Human Body Modeling 2020*,
https://doi.org/10.1007/978-3-030-45623-8_23

391

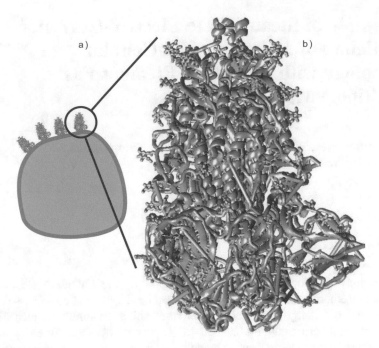

Fig. 1 CAD-based SARS-CoV-2 spike glycoprotein structure [5]. No differentiation between different protein types is made

In this study, we report the performance of the BEM-FMM algorithm for classic (quasi) steady-state electric current modeling around a CAD-based SARS-CoV-2 spike glycoprotein structure [5]. This structure is part of the mechanism by which the coronavirus attaches to a target object, see Fig. 1a. The target object itself is not included into the present study; only the conceptual possibility of quasistatic modeling and the method's numerical performance are reported and discussed.

2 Materials and Methods

Figure 1b shows the protein CAD model [5]. While the overall SARS-CoV-2 virus particle size is reported to be in the range of 80–200 nm [6, 7], the spike glycoprotein structure shown in Fig.1a is much smaller, possibly in the range of 10–20 nm [8]. The electrical conductivity of protein may vary [9]; we assume that the macroscopic conductivity of protein approaches zero while the conductivity of the ambient body fluid is 0.1 S/m. A primary electric field with the amplitude of 1 V/m is applied along the x-axis, as shown in Figs. 2 and 3, respectively.

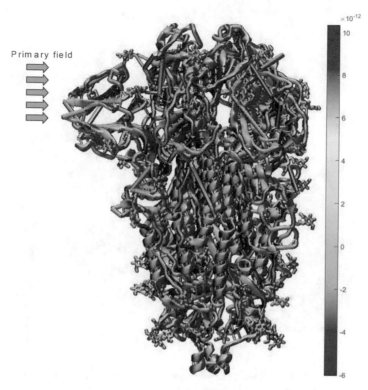

Fig. 2 Primary incident electric field and surface charge density distribution in C/m^2 on the surface of the CAD object

The entire surface CAD model has 2.34 M facets. Nine facets were found to have coincident face centers, although their vertices were somewhat different. These duplicated facets were removed from the mesh prior to simulation.

3   Results

The numerical solution with 100 GMRES iterations executes in approximately 10 min using a general-purpose Intel Xeon E5-2698 v4 CPU (2.10 GHz) Windows server, 256 GB RAM, which runs MATLAB 2019b. The final relative residual of the iterative solution is 4×10^{-7}; the charge conservation law is satisfied to within a 10^{-5} relative error.

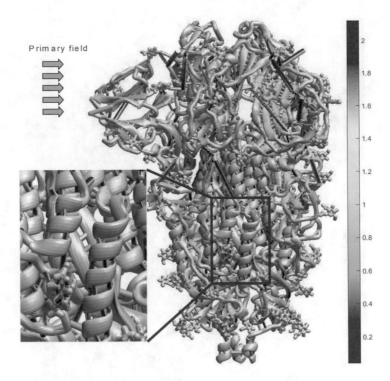

Fig. 3 Total electric field distribution just outside the surface of the CAD object in V/m. Note the inset which shows the zoomed-in field distribution

Figure 2 shows the direction of the primary electric field and the resulting distribution of the surface charge density in C/m^2 on the surface of the CAD object subject to the applied electric field. There seems to be no dedicated domain where the absolute charge concentration would be the largest.

Similarly, Fig. 3 demonstrates the total electric field distribution just outside the surface of the CAD object in V/m. A potentially interesting observation is that the total electric field becomes quite large at certain vertical segments of the protein structure (an inset in Fig. 3), which remain largely perpendicular to the primary field. Regarding the surface fields, the simulations reveal that the total field just outside the surface might exceed the primary field by a factor of up to 80. Further modeling is necessary to justify this result.

The normal field just outside the surface approaches zero (this fact also follows from the current continuity condition) so that the tangential field nearly coincides with the total field.

We further studied the same problem but at different orientations of the primary electric field. Similar convergence results have been observed, but the field distributions are quite different.

4   Discussion and Conclusion

The BEM-FMM approach allows us to model the present rather complicated topology in approximately 10 min using a common Windows server and with a high degree of convergence accuracy. It is straightforward to assign different properties to different constituents of the microscopic structure. It is also possible to clone the structure multiple times and include the effect of a nearby cell object.

Indeed, the performed modeling task remains purely classical and "macroscopic," with no relevant quantum effects included. At the same time, it demonstrates the conceptual possibility of quasistatic modeling via BEM-FMM and demonstrates the method's numerical performance.

References

1. Makarov, S. N., Noetscher, G. M., Raij, T., & Nummenmaa, A. (2018). A quasi-static boundary element approach with fast multipole acceleration for high-resolution bioelectromagnetic models. *IEEE Transactions on Biomedical Engineering, 65*(12), 2675–2683. https://doi.org/10.1109/TBME.2018.2813261.
2. Htet, A. T., Saturnino, G. B., Burnham, E. H., Noetscher, G., Nummenmaa, A., & Makarov, S. N. (2019a). Comparative performance of the finite element method and the boundary element fast multipole method for problems mimicking transcranial magnetic stimulation (TMS). *Journal of Neural Engineering, 16*, 1–13. https://doi.org/10.1088/1741-2552/aafbb9.
3. Greengard, L., & Rokhlin, V. (1987). A fast algorithm for particle simulations. *Journal of Computational Physics, 73*(2), 325–348. https://doi.org/10.1016/0021-9991(87)90140-9.
4. Gimbutas, Z., Greengard, L., Magland, J., & Rachh, M. (2019). Rokhlin V. *fmm3D Documentation.* Release 0.1.0. 2019. Online: https://github.com/flatironinstitute/FMM3D
5. Hernandez, J. (2020). Structure of the SARS-CoV-2 spike glycoprotein (closed state). Fri, 2020-03-13 12:38 PM. *National Institutes of Health 3D Print Exchange.* Online: https://3dprint.nih.gov/discover/3DPX-013123.
6. Chen, N., Zhou, M., Dong, X., Qu, J., Gong, F., Han, Y., Qiu, Y., Wang, J., Liu, Y., Wei, Y., Xia, J., Yu, T., Zhang, X., & Zhang, L. (2020, Feb 15). Epidemiological and clinical characteristics of 99 cases of 2019 novel coronavirus pneumonia in Wuhan, China: a descriptive study. *Lancet, 395*(10223), 507–513. Online: https://www.thelancet.com/action/showPdf?pii=S0140-6736%2820%2930211-7.

7. Kim, J.-M., et al. (2020). Identification of Coronavirus Isolated from a Patient in Korea with COVID-19. *Osong public health and research perspectives, 11*(1), 3–7. Online: https://www. ncbi.nlm.nih.gov/pmc/articles/PMC7045880/pdf/ophrp-11-3.pdf.

8. Walls, A. C., Tortorici, M. A., Snijder, J., Xiong, X., Bosch, B.-J., Rey, F. A., & Veesler, D. (2017, Oct). Tectonic conformational changes of a coronavirus spike glycoprotein promote membrane fusion. *Proceedings of the National Academy of Sciences, 114*(42), 11157–11162. https://doi.org/10.1073/pnas.1708727114. Online: https://www.pnas.org/content/pnas/114/42/11157.full.pdf

9. Zhang, X. Y., Shao, J., Jiang, S. X., Wang, B., & Zheng, Y. (2015 Mar 27). Structure-dependent electrical conductivity of protein: Its differences between alpha-domain and beta-domain structures. *Nanotechnology, 26*(12), 125702. https://doi.org/10.1088/0957-4484/26/12/125702.

Correction to: Brain and Human Body Modeling 2020

Sergey N. Makarov, Gregory M. Noetscher, and Aapo Nummenmaa

Correction to: S. N. Makarov et al. (eds.), *Brain and Human Body Modeling 2020*, https://doi.org/10.1007/978-3-030-45623-8

A chapter was removed from this Open Access volume and three subsequent chapters were moved into sections of the book which was not appropriate. The TOC has now been updated to match the correct order.

The updated online version of the book can be found at:
https://doi.org/10.1007/978-3-030-45623-8

Index

Printed in the United States
By Bookmasters